Kinematics of Human Motion

Vladimir M. Zatsiorsky, PhD

Department of Kinesiology, The Pennsylvania State University

Human Kinetics

Library of Congress Cataloging-in-Publication Data

Zatsiorsky, Vladimir M., 1932-
   Kinematics of human motion / Vladimir M. Zatsiorsky.
     p. cm.
   Includes bibliographical references and index.
   ISBN 0-88011-676-5
   1. Human mechanics. 2. Kinematics. 3. Human locomotion.
  I. Title.
  QP303.Z38   1998
  612.7'6--dc21   97-12025
                                                                 CIP

ISBN: 0-88011-676-5
Copyright © 1998 by Vladimir M. Zatsiorsky

Managing Editor: Lynn M. Hooper-Davenport; **Assistant Editors:** Erin Sprague and Sarah Wiseman; **Editorial Assistant:** Amy Carnes; **Copyeditor:** Donna Tilton; **Proofreader:** Karen Bojda; **Graphic Designer:** Fred Starbird; **Graphic Artists:** Denise Lowry, Julie Overholt, and Tara Welsch; **Cover Designer:** Chuck Nivens; **Printer:** Braun-Brumfield

Printed in the United States of America   10 9 8 7 6 5 4 3 2 1

Human Kinetics
Web site: http://www.humankinetics.com/

United States: Human Kinetics, P.O. Box 5076, Champaign, IL 61825-5076
1-800-747-4457
e-mail: humank@hkusa.com

Canada: Human Kinetics, Box 24040, Windsor, ON N8Y 4Y9
1-800-465-7301 (in Canada only)
e-mail: humank@hkcanada.com

Europe: Human Kinetics, P.O. Box IW14, Leeds LS16 6TR, United Kingdom
(44) 1132 781708
e-mail: humank@hkeurope.com

Australia: Human Kinetics, 57A Price Avenue, Lower Mitcham, South Australia 5062
(08) 277 1555
e-mail: humank@hkaustralia.com

New Zealand: Human Kinetics, P.O. Box 105-231, Auckland 1
(09) 523 3462
e-mail: humank@hknewz.com

To the memory of my parents
Michael T. Zatsiorsky (1903-1945)
Berta L. Bardenstein (1903-1959)

Contents

PREFACE

This textbook is designed for readers seriously interested in human movement, specifically for students in human movement science and such neighboring fields as exercise and sport science, biomechanics, motor control, physiology of movement, biomedical engineering, ergonomics, and physical and occupational therapy. The first volume covers kinematics of human motion and is intended for use in a one-semester course on the biomechanics of human motion. Issues other than kinematics are described in ensuing volumes, each of them corresponding to one-semester biomechanics courses.

At this time a great deal of knowledge on the biomechanics of human motion has been accumulated, but it is rather scattered and not fully organized. Although several good first-level textbooks aimed at undergraduate physical education students, physical therapists, and orthopedists are currently available, there is a substantial gap between the content of these elementary courses and the cutting edge of research. Therefore, most students cannot understand the related scientific literature. The more developed textbooks aimed at graduate students cover either methods of biomechanical research or biomechanics of the musculoskeletal system but not the biomechanics of motion. The objective of this book is to begin filling this void.

I committed myself to writing this book soon after I started teaching a graduate course on biomechanics at Penn State University in 1992. At that time I was well aware that work in such an interdisciplinary field as human movement science is impeded by communication difficulties among people with different backgrounds. In the class, I encountered students from exercise and sport science; biology; psychology; physics; mechanical, electrical, industrial, and biomedical engineering; and even mathematics. Students entering graduate biomechanics courses often have gaps in knowledge that make some ideas difficult to grasp. Biologists and psychologists, for instance, are not familiar with many concepts of classic mechanics. Mechanical engineers are underexposed to biology and are prone to disregard the complexity of living species. Scientists involved in exploring this interdisciplinary field should learn each other's language so they can communicate better. Because human movement science deals with motion, and mechanics is the science of motion, the biomechanics course is a pillar on which the literacy of specialists in the field is based.

This book is intended primarily for graduate students of various backgrounds. Its mathematical level is as simple as is compatible with a sufficiently comprehensive treatment of the subject. It emphasizes understanding the concepts rather than leading the reader through the mathematical intricacies. Still, the text assumes that the student has had basic courses in calculus, as well as at least a one-semester course in matrix theory or linear algebra. However, all of the advanced concepts are explained. It is my hope the text will be intelligible to nonmathematicians. As compared with other books in mechanics, the ratio of text to formulas was greatly increased in favor of the text. Also, throughout the book, many examples are presented. When issues of basic mechanics are being discussed, rigorous proofs are not provided; however, a reference list is given at the end of each chapter. Some students have studied undergraduate courses many years ago and do not recall the fundamentals. To assist them, "Refreshers" are scattered throughout the text outlining the information and facts from undergraduate studies that are referenced in the course. Still, the book is not easy reading. It is directed toward an audience of future specialists who wish to read about human motion from a biomechanical perspective.

The science of biomechanics is based on mechanical models and biological experiments. The text of the book reflects the duality of the subject. Some chapters of the book essentially embrace the regular methods of classic mechanics and can be characterized as "mechanics with biological examples." Other chapters are based mainly on numerous experimental facts collected in biomechanical research. The dual subject necessitates a diverse style. When describing mechanics, the references are limited to such giants as Euler and Coriolis, numerous examples (in the main text) and examples "From the Scientific Literature" (out of the main text) are provided, and the chapters end with sections on biological applications of the relevant mechanics theories. In other chapters the style is more similar to that used in the biological literature.

The following is a brief sketch of the contents of the chapters. Chapters **1** and **2** deal with the kinematic geometry of human motion, i.e., with the description of the human body position and displacement without regard to time, velocity, and acceleration. Chapter **1** considers the kinematic geometry of a single body (e.g., rotation matrices, Euler's angles, helical method, homogeneous transformation matrices). The application of these methods is illustrated by the classic object of biomechanical research, eye movement. Chapter **2** is devoted to the description of joint configuration and kinematic chains. Such chains are commonly used to model human body posture and movement. They are also used in robotics. The second chapter reflects this symbiosis. The chapter is based on the ideas and methods developed within biomechanics itself (e.g., joint configuration convention) as well as on the methods developed in robotics and then borrowed by biomechanists (e.g., Denavit-Hartenberg convention). Chapter **3** addresses differential kinematics of human motion, specifically velocity and acceleration of biokinematic chains. For the sake of teaching it starts from the most simple case of a planar two-link chain. The main theme of the chapter is, however, three-dimensional movement of multilink chains. The first three chapters contain closing sections in which fundamental methods of kinematic geometry and differential kinematics are either elucidated by biological examples (e.g., Section **3.2.1.2** where the refresher is on three-dimensional behavior of the vestibulo-ocular reflex) or complemented by nontraditional methods used in biomechanical research (Section **3.3.1** on the tau hypothesis).

To make complex issues easier for understanding, in the first three chapters joint movement is treated as pure rotation. In Chapters **4** and **5** this assumption is removed and human joint motion is discussed with due attention to its real intricacy. Chapter **4** is intended to familiarize the students with the joint kinematics theory related to all joints, and Chapter **5** is devoted to kinematics of individual joints. Some of the ideas described in Chapter **4**, e.g., the Reuleaux method, were originally conceived in the kinematics of machines and then transplanted into biomechanics.

The references in the text, especially in the first chapters, are limited. Instead, a reasonably comprehensive list of associated references follows each chapter. Authors whose publications have been unintentionally not cited are asked for their forgiveness. Also, I apologize to my colleagues for not mentioning the names in the text. Almost every sentence would have needed at least one reference. Please realize that this is done to preserve the flow of the text and avoid unnecessary commentaries. After all, this is a textbook and not a scientific monograph.

I have done all I can to get rid of slips from earlier drafts, but I have lived long enough to know my imperfections. I assure the readers that their comments calling attention to errors will be greatly appreciated.

ACKNOWLEDGMENTS

It is a pleasure to thank Drs. Boris Prilutsky, Josef Laczkó, Ge Wu, and John Challis for reading the chapters of the manuscript and making valuable comments. I am grateful to the graduate students at the PSU Biomechanical Laboratory, Sherry Werner, William Sloboda, Deborra King, Andrew Hardyk, and Walter Accles, who have helped me to improve the readability of the book. I am particularly indebted to Zong-Ming Li for checking the mathematical accuracy of the entire text. I must especially thank Dr. Richard Nelson for his comments on the early drafts of the book and for continual help and encouragement. I also acknowledge illuminating discussions with Dr. Mark Latash.

My thanks are due to the Pennsylvania State University and its Department of Kinesiology for the opportunity provided to teach graduate courses in biomechanics and to prepare this textbook.

I greatly appreciate the support given to me by my editors and the staff at Human Kinetics.

The book would not have been written without the active support and the enduring tolerance of my wife Rita and our family. Once again—Rita, Betty, Michael, Stacia, Anastasia, and James—thank you.

Vladimir M. Zatsiorsky
State College, Pennsylvania
June 1996

NOTATION AND CONVENTIONS

Sections and subsections of the chapters are numbered and, when referenced, printed in bold. Hence, Section **1.3.2** means: Chapter 1, Section 3, Subsection 2 (this is a section on eye movement).

A vector is denoted by a letter in boldface type, for example, **P**. The same letter in lightface type denotes the magnitude of the vector. Matrices are printed in square brackets, like [R]. Subscripts usually refer to a system of coordinates, for example, $\mathbf{P}_G$ stands for vector **P** represented in a global system of coordinates. Newton's notation is used for the successive derivatives with regard to time, thus, when x is a coordinate, $\dot{x}$ means velocity, and $\ddot{x}$ stands for acceleration. A list of main symbols follows:

X,Y,Z	axes of a global reference system
x,y,z	axes of a local reference system
ℓ	length
$\mathbf{L}_G$	vector giving the origin of the local coordinate system L in the global system G
$\mathbf{P}_G, \mathbf{P}_L$	position vectors of point P in the global and local systems
θ	angle measured in a global reference system
α	angle measured in a local reference system, usually joint angle
$\dot{\theta}, \dot{\alpha}, \omega$	angular velocity
$\ddot{\theta}, \ddot{\alpha}$	angular acceleration
I	moment of inertia
H	angular momentum
[R]	rotation matrix
[T]	transformation matrix
[D]	displacement matrix
[I]	unit matrix
[J] or **J**	Jacobian
v	linear velocity
CNS	central nervous system
DOF	degrees of freedom
ICR/IAR	instantaneous center/axis of rotation

1
CHAPTER

KINEMATIC GEOMETRY OF HUMAN MOTION: BODY POSITION AND DISPLACEMENT

Kinematic geometry is a branch of kinematics that deals with the description of body position and displacement without regard to time. Hence, such notions as velocity and acceleration are not included in discussions of kinematic geometry.

The position of a rigid body is defined by the location of a point on that body and the body's orientation. The human body can be viewed as a system of rigid links connected by joints. Human body parts are not actually rigid structures, but they are customarily treated as such during studies of human motion. The convenience of this approach warrants its use. Based on this postulate, we can define human body *position* by its (a) *location*, or *place;* (b) *orientation*, or *attitude;* and (c) *joint configuration*, or *posture* (Figure 1.1). Chapter 1 is about defining location and orientation. Techniques used to describe body posture are considered later, in Chapter 2.

Section **1.1** deals with defining body location by the coordinate method. It contains a general introduction to the method and discusses Cartesian coordinates compared with oblique coordinates. In addition, the notions of vector components and projections are introduced. Section **1.2** will familiarize students with the methods used to define body orientation, starting with fixation of a local coordinate system within a solid body and then discussing the problems encountered with attaching a reference system to a human body and defining human body rotation. This section also describes the main methods used to characterize attitude: the matrix method, Euler's angles, and the screw method.

The matrix method includes three principal issues, all of which are interrelated.

1. How to define a body position in a given reference system (Where is it?).

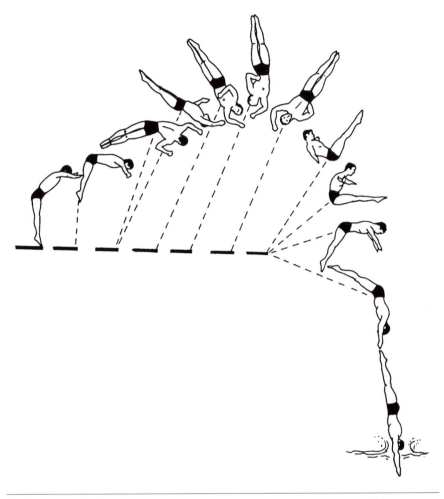

Figure 1.1   When a diver performs a dive, the body changes its position: (1) location in space, (2) orientation (the body is rotating), and (3) joint configuration (the athlete is assuming a piked posture).

2. How to describe a position using an additional reference system. In scientific jargon, this is known as an *alias problem;* that is, the same position is described differently.

3. How to describe the displacement of the body when it assumes different positions. This is called an *alibi problem.*

The advantages and disadvantages of the various angular conventions are discussed in Section **1.2.6**. Finally, eye movement, a vital area of human motion research, is described in Section **1.3**. Note that the area of eye movement research is not only an example of application of the methods of kinematic geometry to the study of human motion, it is a developed area of research with its own facts, theories, and even laws discovered in experiments.

1.1 DEFINING BODY LOCATION

To describe the location of a body in space, a coordinate method is used. Coordinates are a set of numbers that locate a point in a *reference system* or, in other words, in a *reference frame*. Various coordinate systems (e.g., Cartesian, oblique, spherical, cylindrical) have been used in research. Some systems are

••• CLASSICAL MECHANICS REFRESHER •••

Motion

Kinematics is used to describe motion without regard to the force producing the motion.

A *point* has no length, no width, and no thickness. In other words, the dimensions of the point are zero. The *path*, or *locus*, of the point is the line representing successive positions of the point.

Displacement is the difference between the coordinates of a point in its final and initial locations.

Translation is the movement of a body so that any line fixed with the body stays parallel to itself. With translation, all points of the body move along parallel paths and have the same velocity and acceleration at any given instant.

Rotatory, or *angular*, motion is the movement of a body around an axis, called *the axis of rotation*. During rotatory motion, all parts of the body travel in the same direction through the same angle of rotation. With rotation, all parts of the body, except those that lie on the axis of rotation, move in parallel planes along concentric circles centered on the same fixed axis. The angle of rotation is measured on a plane perpendicular to the axis.

General motion is a combination of translation and rotation. At any given instant, the general motion is equivalent to the sum of translation along and rotation about an instantaneous axis.

more appropriate than others in specific situations. For example, spherical co-ordinates are convenient for studying manual exertion. Cartesian reference systems, however, are the preferred choice and have been used much more often than other frames.

1.1.1 The Coordinate Method

The description is performed in three steps:

a. a global reference system of coordinates is defined;

b. any one point P in the body is specified (it is convenient to choose the point of origin of a local system of coordinates, see Section **1.2**); and

c. the location of this point in the global reference system is determined.

Conventionally, a global system is defined as a right-handed orthogonal triad with the origin affixed to some reference point in the vicinity of the performer. Commonly, the global frame is fixed in the ground with the positive X axis horizontal and forward, the positive Y axis vertical and upward, and the positive Z axis horizontal and to the right (Figure 1.2). Another popular choice is to direct the positive X axis to the right, the positive Y axis horizontal and forward, and the positive Z axis vertical and upward. The axes are oriented with respect to the subject, who is facing in the direction of primary interest.

The drawback of the single right-handed system of coordinates is that the same variables have different signs (i.e., positive or negative) for the right and

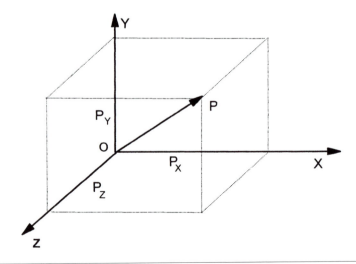

Figure 1.2 Location (place) of the point is defined by the three components, P_x, P_y, and P_z, of the vector **P** in the Cartesian system of coordinates O-XYZ.

left extremity. For example, adduction of the right arm is in a positive direction and the same movement for the left arm is in a negative direction. To avoid this inconvenience, a "Forward, Outward, Upward" (FOU) system is recommended. This system is appropriately designated as either right- or left-handed. The FOU systems are practical for kinematic analysis; however, the use of the left-handed system for solving dynamic equations involving forces and force moments is troublesome.

Location of the point P may be given by its vector **P** or by components of **P**. In a rectangular Cartesian frame, the components of a vector coincide with the projections of this vector onto the three reference axes. The components of the vector, P_X, P_Y, and P_Z, define the location of the point (Figure 1.2).

1.1.2 Cartesian Versus Oblique Coordinates

Cartesian coordinates are orthogonal, meaning the axes of the system are at right angles to each other. When the axes are not orthogonal, the system of coordinates is an *oblique system*. In the study of biomechanics and motor control, oblique reference frames are frequently used when the object of interest is not "how to represent a movement" but "how is the movement represented" (e.g., by a sensory organ or brain). The central nervous system (CNS) operates within coordinates that are intrinsic to the structure of organism. For instance, the semicircular canals of the vestibular organ are not quite at right angles to each other. Hence, the axes of maximal sensitivity of the semicircular canals form a skewed reference system.

In oblique coordinate systems, the location of a point may be represented in two ways, either by components or by projections (Figure 1.3). If the components and projections are not equal to each other, the system is oblique. In a Cartesian coordinate system, the components and projections are identical. The component coordinates are called *contravariant vectors*, and projection coordinates are called *covariant vectors*. Contravariant components of a vector **P**, when summed vectorially according to the *parallelogram rule*, produce vector **P**. Thus, the contravariant representation is also called the *parallelogram representation*. Covariant projections of vector **P**, if summed vectorially, do not produce vector **P**.

If a vector **P** is given in covariant form, that is, by its projections (X_p, Y_p), the contravariant representation of the vector, or its components (X_c, Y_c), can be found by multiplying the covariant vector by a *metric tensor*, which is an n-dimensional matrix. In a planar model, the metric tensor is a 2×2 matrix. These transformations are studied in tensor analysis, a field of mathematics that is not described in this text.

Sensory stimuli are often represented in a covariant form, as projections of the stimulus on coordinate axes. For example, the response of the vestibular

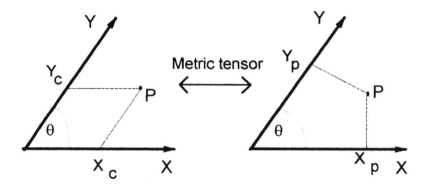

Figure 1.3 Components (Y_c and X_c on the left panel) and orthogonal projections (Y_p and X_p on the right panel) of vector **P** in a skewed reference system. Vector **P** can be easily found as the vector sum of the two components. The vector is not equal, however, to the vector sum of the projections. If the frame were Cartesian, the components and projections would be equal. To convert the coordinates of a point from one form into another, a metric tensor is used. For the system under consideration the metric tensor is a 2×2 matrix

$$\begin{bmatrix} 1 & \cos\theta \\ \cos\theta & 1 \end{bmatrix}$$ The application of matrices for coordinate transformations

will be explained later in the text.

labyrinth, within the inner ear, is proportional to the projection of the axis of rotation of the head on the axis of maximal sensitivity of each semicircular canal. Because the signals from the semicircular canals are proportional to the projections, they cannot be summed vectorially (see Figure 1.3, right panel).

Unlike sensory signals, many motor responses, e.g., forces produced by individual muscles, are summed vectorially in accordance with the parallelogram rule. Thus, the motor responses are believed to be represented in the CNS in a contravariant form and the control of movement includes a series of covariant-to-contravariant transformations.

1.2 DEFINING BODY ORIENTATION

To define body orientation, or attitude, three steps are taken:

1. The global reference system is defined.
2. A local reference system is attached to the body.
3. Orientation of the local reference system relative to the global system

■ ■ ■ *From the Scientific Literature* ■ ■ ■

Projection-to-Component Tranformations in Movement Control

Source: Pellionisz, A.J. (1984). Coordination: A vector-matrix description of transformations of overcomplete CNS coordinates and a tensorial solution using the Moore-Penrose inverse. *Journal of Theoretical Biology, 110,* 353-375.

This model of head movement control (Figure 1.4) includes several transformations of covariant and contravariant vectors. It starts from a three-dimensional sensory vector (in accordance with the number of labyrinth semicircular canals) and ends with a 30-dimensional execution vector, representing the commands sent by the CNS to each of the 30 muscles used in neck movement.

In Figure 1.4, part A shows a sequence of sensory-motor transformations:

1. Because the vestibular canals are nonorthogonal and because the signals from the vestibular labyrinth are proportional to the projections of the head-orientation vectors, the head position is sensed in a covariant form.

2. To orient in a three-dimensional external space, a covariant reception vector is somehow transformed into a contravariant perception vector.

3. The contravariant perception vector is transformed into a covariant intention vector.

4. The covariant intention vector is transformed into a contravariant execution vector.

In Figure 1.4, part B, the orientation of the vestibular semicircular canals is shown with respect to the yaw, roll, and pitch axes in cats. (A, anterior; H, horizontal, P, posterior canal).

Parts C, D, and E are modules schematically displaying transformations implemented by the neural networks. C is a 3×3 sensory contravariant transformation, a metric tensor. D is a 3×30 sensory-motor covariant transformation. E is a 30×30 contravariant transformation, also a metric tensor.

See **TRANSFORMATIONS,** *p. 8*

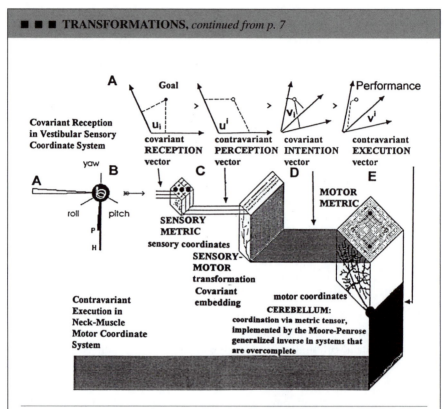

■ ■ ■ TRANSFORMATIONS, *continued from p. 7*

Figure 1.4 Projection versus component (covariant versus contravariant) transformations in movement control. The tensor network model of a vestibular labyrinth reflex participating in head orientation in cats. Note: The reader is not expected to understand, at this time, all the details of this scheme. Focus only on the idea of the covariant-to-contravariant transformations. The Moore-Penrose generalized inverse will be explained later, in Chapter **3**.

From Pellionisz, A.J., Le Goff, B., & Laczkó, J. (1992). Multidimensional geometry intrinsic to head movements around distributed centers of rotation: A neuro-computer paradigm. In A. Berthoz, W. Graf, & P.P. Vidal (Eds.), *The head and neck sensory motor system* (pp. 158-167). New York: Oxford University Press. Copyright © 1992 by Berthoz. Used by permission of Oxford University Press, Inc.

is mathematically defined. Various methods of classic mechanics, briefly described later, are used for this purpose.

When studying human body position, it is convenient to use not one but two local reference systems fixed with the body. The first reference system, called the *moving system,* is oriented similar to the global reference system and moves

••• MATHEMATICS REFRESHER •••

Cross Product of Vectors

A *vector* is a quantity having magnitude and direction. In this textbook vectors are printed in boldface, for example **P** and **Q**. Vectors are treated as geometric objects that are independent of any coordinate system. They may be conveniently represented by arrows, with the length of the arrow proportional to the magnitude. If such an arrow starts at an origin, it terminates at a point. The coordinates of this point (x,y,z) define the vector. More broadly, vectors are defined as unidimensional arrays of numbers $(a_1, a_2, \ldots, a_n)$.

The *cross product* of two vectors **P** and **Q** (Figure 1.5), also called the *vector product*, is defined as the vector **V** that satisfies the following conditions:

1. The line of action for vector **V** is perpendicular to the plane containing vectors **P** and **Q**.

2. The magnitude of vector **V** is the product of the magnitudes of vectors **P** and **Q** and of the sine of the angle formed by **P** and **Q**:

$$V = PQ \times \sin \theta$$

3. The sense of **V** is such that a person located at the tip of **V** will observe as counterclockwise the rotation through angle θ, which

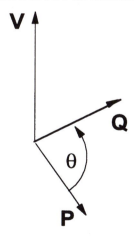

Figure 1.5 Cross product of two vectors **P** and **Q**. The sense of vector **V** is defined by the right-hand thumb rule. The angle θ is formed by vectors **P** and **Q**.

See **MATHEMATICS,** *p. 10*

● ● ● **CROSS PRODUCT,** *continued from p.9*

brings the vector **P** in line with the vector **Q**. This is referred to as the "right-hand thumb rule"; if the fingers of the right hand are curved in the direction from vector **P** to vector **Q**, the extended thumb represents the direction of vector **V** (Figure 1.6).

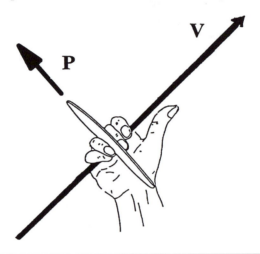

Figure 1.6 The right-hand thumb rule is illustrated. Vector **Q** is not shown; it is directed toward the reader and is perpendicular to the paper. To find the sense of vector **V**, curve your fingers from **P** to **Q**. Your thumb shows the direction of **V**.

The following two properties of vector products are important for applications in biomechanics.

1. Cross products depend on the order of factors; in other words, they are not *commutative*. The sense of the cross product of two vectors is reversed when the order of factors is reversed [**P** × **Q** = – (**Q** × **P**)].

2. When vectors **P** and **Q** are collinear (the angle θ between **P** and **Q** equals 0° or 180°), the magnitude of **V** equals 0 . For this position, called the *singular position,* the cross product is not defined.

with the body while maintaining its orientation in space. The origin of this system is usually in the general center of mass point. The second reference system, called the *somatic system*, is located within the body and changes its orientation in space when body orientation is changed.

1.2.1 Fixation of a Local System With a Rigid Body

A body is called *rigid* or *solid* if the distance between any two points within the body does not change. A rigid-body motion preserves the angle between any two lines of the body. A true rigid body is a mathematical abstraction. As such, rigid bodies do not exist in nature. For each individual case, the researcher must decide whether a body can be modeled as a rigid body or not. For example, in some research projects the trunk is represented as a single rigid body; in others, the trunk is modeled as a system of two or three solid bodies connected by joints or as a single deformable body. The decision of which representation system to use depends on the required accuracy and the movement to be studied and is up to the researcher.

To fix the local reference system within a rigid body, the coordinates of three noncollinear points within the body must be known (Figure 1.7). Two vectors, $\mathbf{r}_1$ and $\mathbf{r}_2$, define a plane. Cross products of these vectors are then calculated to define the local reference frame.

The following routine procedure is employed to fix the reference frame to a rigid body:

1. The cross product of vectors $\mathbf{r}_1$ and $\mathbf{r}_2$ defines vector $\mathbf{r}_3$.
2. The cross product of vectors $\mathbf{r}_3$ and $\mathbf{r}_1$ (or $\mathbf{r}_2$ if you wish) defines vector $\mathbf{r}_4$. At this point, three mutually orthogonal axes (vectors) are known; however, each axis has a different length.
3. Each vector is divided by its own length to determine unit vectors.

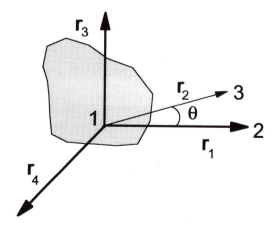

Figure 1.7 To define the local reference system within a rigid body three points, 1, 2, and 3, within the body are given. Vectors $\mathbf{r}_1$ and $\mathbf{r}_2$ are drawn from point 1 to points 2 and 3. Vectors $\mathbf{r}_3$ and $\mathbf{r}_4$ are then determined as cross products.

■ ■ ■ *From the Scientific Literature* ■ ■ ■

Fixating a Local Reference System to a Body Part

In this study by Wei et al. the local systems of coordinates for the femur and the tibia were introduced as follows (Figure 1.8). The origins of the reference frames were anchored to the lateral femoral epicondyle and the

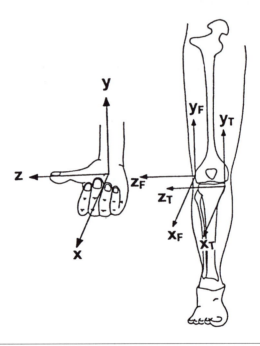

Figure 1.8   The femoral and the tibial reference frames. The positive rotation around the z axis corresponds to knee extension.

Reprinted from Wei, S.-H., McQuade, K.J., & Smidt, G.L. (1993). Three-dimensional joint range of motion measurements from skeletal coordinate data. *Journal of Orthopaedics and Sports Physical Therapy, 18,* 687-691.

medial tibial condyle. The positive x axes are directed from posterior to anterior, the positive y axes are upward, and the positive direction of the z axes is selected according to the right-hand thumb rule, from medial to lateral. The positive direction of rotation is, as usual, counterclockwise. Hence, positive rotation of the tibia with respect to the femur about the x axis is abduction, positive rotation about the y axis is internal rotation, and positive motion about the z axis is extension. According to this convention, knee adduction, external rotation, and flexion occur in the negative direction. The fingers and thumb illustrate the right-hand thumb rule.

••• *ANATOMY REFRESHER* •••

Cardinal Planes and Axes of the Human Body

For a human assuming an *anatomic posture*—standing upright on a horizontal surface with arms hanging straight down at the sides of the body, palms turned forward, and head erect— the following planes and axes are defined (Figure 1.9). Any plane dividing a human body into left and right

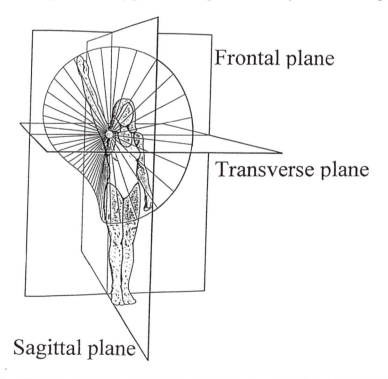

Frontal plane

Transverse plane

Sagittal plane

Figure 1.9 Planes and axes of a human body. The axes intersect at the right shoulder joint. Note that the planes are not cardinal.

See ANATOMY, p. 14

● ● ● **PLANES OF THE BODY,** *continued from p.13*

sections is called a *sagittal plane.* The sagittal plane is positioned in an anteroposterior direction. If the plane divides the body into two symmetrical halves, it is called the *cardinal,* or *principal,* sagittal plane. The term *mid-sagittal* plane is also used to denote this plane.

For bodies in an upright posture, the horizontal plane passing through the body is called the *transverse* plane. The transverse plane that passes through the body's center of mass is the cardinal, or principal, transverse plane.

Finally, the *frontal,* or *coronal,* plane goes from side-to-side and is perpendicular to both the transverse and sagittal planes. The frontal plane divides the body into anterior and posterior sections. When this plane passes through the center of mass, it is called the cardinal, or principal, frontal plane.

The axes on the interception of the planes are called

- *anteroposterior* (interception of the sagittal and transverse planes);
- *longitudinal* (interception of the sagittal and frontal planes); and
- *lateromedial,* or frontal (interception of the frontal and transverse planes).

This system is the most convenient and is systematically used in practice. However, it does not permit the description of joint configurations or limb movements if the body posture is not anatomic, e.g., when the trunk is twisted or bent. Where, for instance, is the frontal plane of a gymnast in the pike position?

The local reference system is now completely defined.

When the local reference system is defined, the description of the body's orientation with regard to the global reference system is reduced to the description of the attitude of the local system (attached to the body) relative to the global one. Thus, the problems of "how to describe a body's attitude" and of "how to describe an attitude of the local frame" are equivalent.

1.2.2 Fixation of a Somatic System With a Human Body

The human body is not rigid. Instead, it is a system of bodies (segments or body parts) whose relative positions vary. The somatic reference system for a human body may be based on (a) anatomic landmarks; (b) mechanical points (e.g., general center of mass) and axes (i.e., principal axes of inertia); or (c) a combination of both (a) and (b). When a somatic system is based on mechanical points, the reference axes change their positions relative to the anatomic landmarks when the body joint configuration changes.

■ ■ ■ *FROM THE SCIENTIFIC LITERATURE* ■ ■ ■

Reference Systems for the Whole Body Based on Anatomic Landmarks

Source: Dapena, J. (1981). Simulation of modified human airborne movements. *Journal of Biomechanics, 14,* 81-89.

To study high jumping, two reference frames attached to the body were defined. One system, RO, was fixed to the upper trunk segment, and the second, RL, was fixed to the lower trunk segment (Figure 1.10). Both

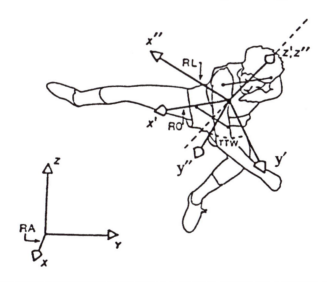

Figure 1.10   Two systems of coordinates are embedded in the upper and lower trunk segments.

Reprinted from *Journal of Biomechanics, 14,* Dapena, J., Simulation of modified human airborne movement, 81-89, 1981, with kind permission from Elsevier Science Ltd, The Boulevard, Langford Lane, Kidlington 0X5 1GB, UK.

See **LANDMARKS,** *p. 16*

■ ■ ■ **LANDMARKS,** *continued from p. 15*

reference frames had one common axis: the longitudinal axis of the trunk ($Z' \equiv Z''$). The other two axes were different and reflected the difference between the orientation of the shoulders and the orientation of the hips. The two trunk segments were free to twist, but not to flex, relative to each other.

••• CLASSICAL MECHANICS REFRESHER •••

Moments of Inertia and Angular Momentum

The *moment of inertia* is a measure of the body's resistance to changes in angular motion. It is analogous to mass (a measure of inertial resistance) in linear motion. The moment of inertia, I, of a particle equals the product of $I = mr^2$ where m is mass of the particle and r is the radius (shortest distance) of the particle from the axis of rotation.

The *principal axes* of a body, or the *principal axes of inertia,* pass through the body's center of mass and are perpendicular (orthogonal) to each other. Compared with other axes that pass through the center of mass, the principal axes are located so that one moment of inertia is the largest and that another is the smallest, among all possible moments of inertia for that body.

The *principal moments of inertia* are the moments about the principal axes.

The *ellipsoid of inertia* is the imaginary surface that defines the body's moment of inertia with respect to any axis passing through its center of mass. Imagine a point, plotted on axis x, at a distance from the center of mass:

$$r = 1 /(I_x)^{0.5}$$

The locus of the points thus obtained forms the ellipsoid of inertia. Projection of the ellipsoid of inertia onto a plane gives the *ellipse of inertia.*

The *angular momentum* of a rigid body equals the product of its moment of inertia and angular velocity.

$$H = I\omega,$$

where H is angular momentum, I is moment of inertia, and ω is the angular velocity.

A human body can be regarded as a system of solid bodies (human body segments). The angular momentum of such a system can be obtained by summing the angular momentum values for each segment.

Furthermore, the following *principle of conservation of angular momentum* is valid: the angular momentum of a system will remain constant regardless of any internal movements and torques, unless external torque is exerted on the system. This is the angular analogue of Newton's first law.

The following anatomically based somatic systems are used most often:

1. *Sacrum.* The sacrum is a large nondeformable bone that can be viewed as a solid. With this approach, the sacrum is regarded as a representative part of the whole body.

2. *Shoulder girdle.* From a purely mechanical point of view, this somatic system is far from ideal. The shoulder girdle consists of many bones that can change their relative positions. However, as employed in some investigations, the internal neural representation of the body's attitude in space is based on the orientation of the shoulder girdle. The internal reference system (from the point of a performer not an external observer) is located in the shoulder girdle, or close to it, rather than fixed with the pelvis.

3. *Trunk.* This is a convenient but not reliable procedure because the trunk is not a rigid body; the parts of the trunk can move relative to each other.

The prevailing local references frames based on mechanical points and axes are (a) the principal somatic system and (b) the "split-body" system. In the principal somatic system, the origin of the system is at the body's center of mass and the reference axes are directed along the principal axes of inertia. An example of application of this system is shown in Figure 1.11. With this local reference system, mathematical equations of body motion (e.g., an airborne movement) can be written in their simplest form. This is the main advantage of the principal reference system. However, it is a cumbersome task to fix the principal somatic system within the body; the location of all the body parts must be known at each instant, and complex calculations must also be performed. In addition, the principal axes cannot be defined when the two princi-

A

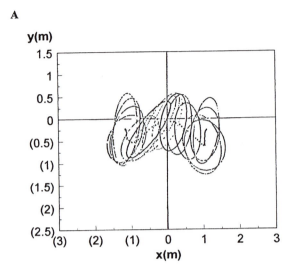

B

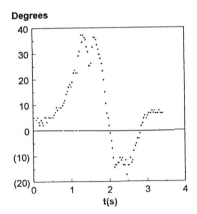

Figure 1.11 Application of a principal somatic system for studying the standing broad jump. The top panel shows the displacement of an ellipse of inertia of the body during the jump. Both the location and orientation of the ellipse are being changed during the flight. Also, the ellipse alters its shape because of the relative relocation of body parts. The bottom panel shows body orientation during the jump and landing. The angle of inclination, λ, is formed by the longitudinal axis of the ellipse of inertia and vertical.

From Mercadante, L.A., Brenzikofer, R., & Cuhna, S.A. (1995). Methodology for tridimensional representation, orientation, and position of the human body. In K. Häkkinen, K.L. Keskinen, P.V. Komi, & A. Mero (Eds.), *XVth Congress of the International Society of Biomechanics, Abstracts* (pp. 620-621). Juväskylä, Finland, July 2-6, 1995. Adapted by permission from the International Society of Biomechanics.

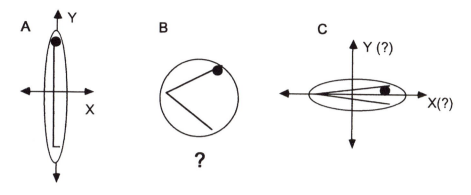

Figure 1.12 The ellipses of inertia are drawn for an athlete in the air in three different postures. By assumption, the athlete does not rotate in the air (the angular momentum equals zero). A, upright posture; B, intermediate posture; C, pike posture.

pal moments of inertia are equal. This may happen when an athlete assumes a tucked, or pike, position in the air (Figure 1.12).

The principal axes of inertia cannot be defined for the "intermediate" posture B (seen in Figure 1.12). With this posture, the initial maximal and former minimal moments of inertia are equal (resulting in a "circle of inertia" rather than the ellipse of inertia). Postures with equal moments of inertia relative to different axes are often assumed by skilled gymnasts and acrobats while performing complex airborne rotations, like a double somersault tucked with a double twist. In gymnastic jargon, this posture is called the "fourth posture" (or "cowboy posture") to differentiate it from the classic layout, tuck, and pike postures.

In the split-body method, the body is divided into upper and lower parts. The center of mass location for each part is then determined. The first axis of the reference system is chosen along the line connecting the selected centers of mass. The origin of the reference system is in the general center of mass—the center of mass of the whole body. The second axis is directed along the lateromedial axis of the pelvis. The third axis is determined as the cross product of the first two. The split-body system is a mixed system based on mechanical and anatomic landmarks: the line connecting the centers of mass of the upper and lower parts of the body and the lateromedial axis of the pelvis. This system does not take into account the possible relative twisting of the pelvic girdle with regard to the shoulder girdle about the longitudinal body axis. The pelvic twist is considered the body twist. Thus, caution should be exercised when the split-body system is used to analyze twisting movements, such as a somersault with one twist. The additional disadvantage of this method is that in some postures the reference axes do not coincide with the principal axes of inertia. To describe body rotation for these postures one must know not only the principal moments of inertia but also the products of inertia. This method is, however, exceptionally suited for studying rotations about the frontal axis, such as a somersault.

When different somatic systems are used, estimations of the attitude of the human body in space may vary. For example, when trunk-based somatic systems are used, the posture shown earlier in Figure 1.12C, compared with the initial upright posture shown in Figure 1.12A, would be regarded as rotated clockwise through 90°. With the split-body system, the rotation is 0°. Consider one more example. Imagine a human who is bending the trunk from an upright posture (Figure 1.13). A legitimate question arises as to whether or not the body, as an entity, is rotating. The answer is different for each somatic system. If the trunk-based somatic system is used, the orientation of the body in the trunk-bent posture (Figure 1.13, bottom row) differs from the initial upright posture by an angle of almost 180°. The orientation of the body at this posture for the split-body system is 90°, and the orientation in the principal system is not precisely defined.

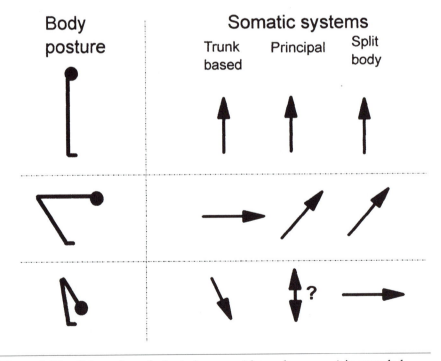

Figure 1.13 Orientation of a body in space (shown by arrows) is regarded differently when various local reference systems are used. See text for explanation.

■ ■ ■ *From the Scientific Literature* ■ ■ ■

An Anatomically Based Split-Body System

Source: Yeadon, M.R. (1990). The simulation of aerial movement. III. The determination of the angular momentum of the human body. *Journal of Biomechanics, 23,* 75-83.

The following system was defined. The first axis is along the line joining the right and left hip centers. The third axis is along the midpoint of the knee centers to the midpoint of the shoulder centers. Finally, the second axis is found from the cross product of the two already-defined unit vectors.

Figure 1.14 depicts a human figure performing a jump in pike position with zero total angular momentum. Axis f_3, drawn along the knee-shoulder line, remains close to zero during the jump. If the axis were defined as the longitudinal axis of the trunk, the somersault angle would increase from 0° to 90°. Hence, the suggested system represents the orientation of the whole body better than a system embedded in one particular body segment. The system is not accurate, however, when an athlete assumes a tucked or asymmetrical position, e.g., bends one leg.

f_3

Figure 1.14 Orientation of the knee-shoulder axis (f_3) during a piked jump.
Reprinted from *Journal of Biomechanics, 23,* Yeadon, M., The simulation of aerial movement, 75-83, 1990, with kind permission from Elsevier Science Ltd, The Boulevard, Langford Lane, Kidlington 0X5 1GB, UK.

1.2.3 Indirect Method of Defining Body Orientation

To avoid the difficulties brought on by changes in body posture, the indirect method of defining the body orientation in space (when the body is in the air) is often used. This method is based on the law of the conservation of the angular momentum, $H = I\omega$, where H is an angular momentum, ω is the angular velocity, and I is the moment of inertia about the same axis. According to the law, H is not altered during flight if air resistance is negligible. Because the product $H = I\omega$ is constant in airborne movements and I can be determined for any given joint configuration, the angular velocity ω can easily be calculated as H/I. With this method, the angular momentum of the body (H) is determined for any phase of the flight when an athlete's posture is not changed. Thus, I and ω can be calculated at any instant in flight.

For a planar movement and when instantaneous values of ω are known, it is possible to determine body orientation as well as the angular distance through which the athlete rotated.

$$\text{Angular distance} = (H/I) \cdot t \qquad (1.1)$$

where t is time.

Despite the fact that angular velocity, angular distance, and orientation of the body can be formally calculated with the indirect method, what exactly rotated remains unclear because the local somatic system is not defined. The indirect method works well in some cases, for example, in describing forward somersaults in the air; however, it provides estimations that go against our intuitive feeling in others. For instance, in a long jump hitch kick, when the athlete rotates his or her arms to prevent body rotation, the method would yield a somersaulting angular velocity ($\omega = H/I$), but the trunk would not change its attitude.

1.2.4 What is "Body Rotation"?

The difficulties encountered with the selection of a somatic system of coordinates are not purely technical ones. They also stem from a more fundamental issue: how to define rotation for a system consisting of several integral parts, such as the human body. Imagine a ski jumper who windmills his arms while airborne to prevent a loss of balance. Is he rotating? Or, is the jumper rotating his arms to prevent rotation of his body? After several arm circles, the jumper assumes his initial body posture and orientation (Figure 1.15). Did he rotate? If yes, what was the angular distance traveled?

Answers to these questions depend on a definition of rotation. If the definition is based on the notion of angular momentum, it does not matter whether all the body parts or only some of them are rotating. If the angular momentum differs from zero and some body parts, say the arms, are rotating, the whole system is said to be rotating, even though the angle of the body would stay at a more or less constant value. Despite the constant body attitude, the jumper is considered to be rotating. However, if body rotation is defined as the rotation of a somatic system attached to the body, the presence of rotation depends on the system selected. If an anatomically based somatic system is used, the jumper has not changed his attitude in space.

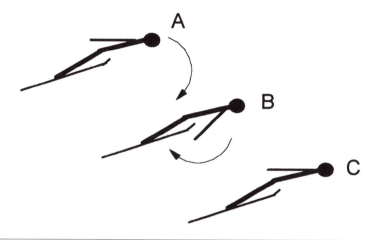

Figure 1.15   For a ski jumper in the air, postures A and C are identical. What is the angular displacement of the jumper's body from posture A to posture B? Should rotation of a segment of the human body be considered body rotation?

1.2.5 Describing Position and Displacement

Representative papers: Berme, Cappozzo, & Meglan, 1990; Kinzel, Hall, & Hilberry, 1972.

Three-dimensional representation of body orientation is considered by many students in biomechanics as a remarkably complicated topic. I do not believe this is true. Even though the mathematics, which stems from the need to interpret body orientation quantitatively, can be confusing, the main ideas are simple. To facilitate understanding of the three-dimensional attitude, the reader may want to manipulate a small object, such as a book.

••• MATHEMATICS REFRESHER •••

Matrices

A *matrix* is a rectangular array of numbers. It is usually written within brackets. Each number in the matrix is an *element of the matrix*. A *column (row) matrix* is a matrix that has only one column (row). A vector can be considered the column (row) matrix. Thus, a vector **A** can be also written as the column matrix [A].

An *order of a matrix*, also called *size* or *dimension* of the matrix, is given by this expression: (the number of rows) by (the number of columns). A matrix with m rows and n columns is called the *(m×n) matrix*. A *square matrix* has the same number of rows and columns.

The *diagonal* of a square matrix consists of the elements a_{ii} (a_{11}, a_{22},...a_{nn}). A *diagonal matrix* is a square matrix in which all the off-diagonal elements are zero. A *unit matrix*, or an *identity matrix* [I], is a diagonal matrix whose diagonal elements equal 1.

A *transpose matrix* $[A]^T$ of the matrix [A] is the matrix with rows and columns interchanged. Transposition of a row matrix gives a column matrix, and vice versa.

The *sum of two matrices* of the same dimension is a matrix with elements that are the sum of the corresponding elements that appear at the same location in each matrix. When the dimensions of two matrices are different, the addition of two matrices is not defined. *Scalar multiplication* of a matrix multiplies each element of the matrix by that scalar. *Dot product* of a (1×n) row matrix and (n×1) column matrix is a real number calculated as follows:

$$[a_1, a_2, ..., a_n] \cdot \begin{bmatrix} b_1 \\ b_2 \\ ... \\ b_n \end{bmatrix} = a_1 b_1 + a_2 b_2 + ... + a_n b_n$$

Dot product is also called *scalar* or *inner* product. The scalar product (of vectors) is a product of the magnitudes of the vectors multiplied by the cosine of the angle between them.

Matrix product of an (m×p) matrix [A] and (p×n) matrix [B], denoted by [AB], is an (m×n) matrix whose element in the *i*th row and *j*th column is the dot product of the *i*th row of matrix [A] and *j*th column of matrix [B]. The number of columns in matrix [A] must match the number of rows of [B]; otherwise matrix multiplication is not defined. Matrix multiplication is not commutative; that is, [AB] is not equal to [BA].

A (left) *inverse matrix* $[A]^{-1}$ is the matrix that satisfies the condition $[A]^{-1}[A] = [I]$, where [A] is a square nonsingular matrix (that is, $|A| \neq 0$) and [I] is the unit matrix.

An *orthogonal matrix* satisfies the condition $[A]^T[A] = [A]^{-1}[A] = [I]$. The left inverse of this matrix is also its right inverse: $[A]^{-1}[A] = [A][A]^{-1} = [I]$. For an orthogonal matrix $[A]^T = [A]^{-1}$. An *orthonormal matrix* is an orthogonal matrix in which all column vectors have unit length; $L_j = \left(\sum_i a_{ij}^2 \right)^{1/2} = 1$, where summation of the squared elements of the *j*th column is done across all the i rows.

The *trace* of a square matrix is the sum of the elements of its diagonal.

When a moving reference system is fixed within a body, its orientation relative to a global system can be determined by three commonly known methods: the matrix method, Euler's method, and the screw method.

1.2.5.1 The Matrix Method

With this method, both the translation and rotation of a local system with regard to the global system can be defined. Thus, the position of the object in the different frames and its displacement can easily be described.

1.2.5.1.1 Relative Orientation of Local and Global Systems (The "Where is it?" Problem)

Location. Defining location is simple (Figure 1.16). Let O-XYZ and o-xyz be a global and a local system of coordinates, respectively. If $\mathbf{L}_G$ is the vector giving the origin of the local system in the global system, the components of this vector, L_X, L_Y, and L_Z, define the location of the local system and can be written as elements of the column matrix $[L_X, L_Y, L_Z]^T$. Translation from point O to point o is defined by the same components, L_X, L_Y, and L_Z.

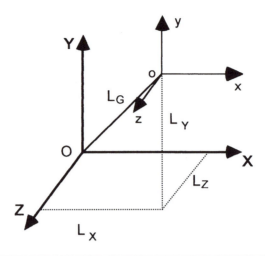

Figure 1.16 Pure translation is defined by the three components of the vector **L**.

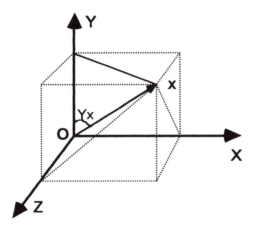

Figure 1.17 Components of unit vector **x** in the global reference system O-XYZ. These components, divided by the length of the vector, give the cosines of the angles that the vector makes with the coordinate axes (the direction angles). For instance, angle Yx is made by the vector **x** and the Y axis. Because the vector under consideration is a unit vector (its length equals one), the direction cosines equal the components of the vector.

Rotation. Each unit vector (**x**, **y**, **z**) of the local system is represented by its components in the global reference system, as depicted in Figure 1.17. Dividing each component by the length of the vector (which is equal to 1) gives the cosine of the angle that the vector makes with each of the coordinate axes of the global system. These angles are called the *direction angles*, and the cosines are termed the *direction cosines* (see Figure 1.17). Direction cosines can be written in a matrix form as elements of a 3×3 matrix. It is customary to use numbers instead of letters to designate reference axes. In this textbook, the numbers and letters are used interchangeably: upper case letters to designate axes of the global system and lower case letters to represent the axes of local systems. With this convention, the *matrix of direction cosines* or, in other words, the *rotation matrix* [R], can be written as follows:

Axes of the local system
(columns)

$$[R] = \begin{bmatrix} \cos_{Xx} & \cos_{Xy} & \cos_{Xz} \\ \cos_{Yx} & \cos_{Yy} & \cos_{Yz} \\ \cos_{Zx} & \cos_{Zy} & \cos_{Zz} \end{bmatrix} \text{ Axes of the global system (rows)}$$

or (1.2)

$$[R] = \begin{bmatrix} \cos_{11} & \cos_{12} & \cos_{13} \\ \cos_{21} & \cos_{22} & \cos_{23} \\ \cos_{31} & \cos_{32} & \cos_{33} \end{bmatrix}$$

In this notation, $\cos_{23}$ means the cosine of the angle formed by the second axis of the global reference frame and the third axis of the local frame. Columns of the rotation matrix are 3×1 unit vectors, representing orientation of the local reference axes in the global frame. The columns correspond to the axes of the local system, and the rows of the rotation matrix match the axes of the global coordinate system.

Translation and rotation. The position of a local system with regard to the global one can be expressed as the sequence of translation and rotation described by the column 3×1 matrix for translation and the 3×3 matrix of direction cosines for rotation. By adding the row (1,0,0,0), which simply means one equals one, a 4×4 matrix can be written. This matrix, called a *position matrix* or *transformation matrix* [T], incorporates both translation and rotation of a local reference system relative to the global one, as follows:

$$[T] = \begin{bmatrix} 1 & 0 & 0 & 0 \\ L_X & \cos_{Xx} & \cos_{Xy} & \cos_{Xz} \\ L_Y & \cos_{Yx} & \cos_{Yy} & \cos_{Yz} \\ L_Z & \cos_{Zx} & \cos_{Zy} & \cos_{Zz} \end{bmatrix}$$

(1.3)

Thus, the position of any coordinate system relative to another may be defined by a 4 × 4 position matrix composed of

- a 3 × 1 column matrix, which gives the location of the origin of one coordinate frame relative to another;
- a 3 × 3 rotation submatrix with nine direction cosines as elements, which gives the attitude of one system relative to another; and
- a 1 × 4 row matrix with elements (1,0,0,0) included for mathematical convenience.

$$[T] = \begin{bmatrix} \cdots & [(1 \times 4)] & \cdots & \cdots \\ \hline \cdots & \vdots & \cdots & \cdots & \cdots \\ [(3 \times 1)] & \vdots & \cdots & [(3 \times 3)] & \cdots \\ \cdots & \vdots & \cdots & \cdots & \cdots \end{bmatrix} \qquad (1.4)$$

One matrix [T] describes any given position of the local reference frame, i.e., the positions of all the points fixed with the frame. If the local frame is attached to a rigid body, the matrix [T] is the same for all the points of the body. However, for each position of the body there is a new matrix [T].

1.2.5.1.2 Defining a Position in Two Reference Frames (Coordinate Transformation, or the Alias Problem)

When a point is given by its radius-vector in a one coordinate system (all the three components of the vector are known), the location of this point with regard to another coordinate system can be found provided that both the translation column matrix and rotation matrix are known.

Pure Translation. If coordinates of a point P in a local reference are (P_x, P_y, P_z) and the location of the local system with regard to the global one is given by a translation column matrix $[L_G]$, the location of the point P in the global frame is determined by the column matrix $[P_G]$ (which is the sum of the column matrices $[P_L]$ and $[L_G]$).

Location of the point P in the global system, $[P_G]$		Location of the local system in the global one, $[L_G]$		Location of the point P in the local system, $[P_L]$		Location of the point P in the global system, $[P_G]$
$\begin{bmatrix} P_X \\ P_Y \\ P_Z \end{bmatrix}$	$=$	$\begin{bmatrix} L_X \\ L_Y \\ L_Z \end{bmatrix}$	$+$	$\begin{bmatrix} P_x \\ P_y \\ P_z \end{bmatrix}$	$=$	$\begin{bmatrix} L_X + P_x \\ L_Y + P_y \\ L_Z + P_z \end{bmatrix}$ (1.5)

Location of the point P in the global system, $[P_G]$		Location of the local system in the global one, $[L_G]$				Location of the point P in the local system, $[P_L]$		Location of the point P in the global system, $[P_G]$	

$$
\begin{bmatrix} 1 \\ P_X \\ P_Y \\ P_Z \end{bmatrix} = \begin{bmatrix} 1 & 0 & 0 & 0 \\ L_X & 1 & 0 & 0 \\ L_Y & 0 & 1 & 0 \\ L_Z & 0 & 0 & 1 \end{bmatrix} \cdot \begin{bmatrix} 1 \\ P_x \\ P_y \\ P_z \end{bmatrix} = \begin{bmatrix} 1 \\ L_X + P_x \\ L_Y + P_y \\ L_Z + P_z \end{bmatrix} \quad (1.6)
$$

It is more convenient, however, to represent the translation through multiplication rather than a summation of the matrices. To do that a 4×4 matrix $[L_G]$ is used instead of the 3×1 column matrix, and matrices $[P_L]$ and $[P_G]$ are written as 4×1 column matrices with element 1 in the first row. Thus, the translation is described by the product of

The advantage of this approach is the consistency of describing both translation and rotation through one mathematical operation, the matrix product (see the subsection on translation and rotation below). Systems permitting such a unified description are called *homogeneous*. Note that in this textbook the column vector convention is adopted. Some authors use the row vector convention. With the row vector convention, transformations are performed by

---

••• MATHEMATICS REFRESHER •••

Determinants

A *determinant* is a certain scalar (a number) associated with a square matrix. It is indicated by vertical lines. The determinant of a matrix **M** is written: $\det[M] = |M|$.

The third order determinant of a 3×3 matrix is

$$
\det[A] = \begin{bmatrix} a_{11} & a_{12} & a_{13} \\ a_{21} & a_{22} & a_{23} \\ a_{31} & a_{32} & a_{33} \end{bmatrix}
$$

$$
= a_{11}a_{22}a_{33} - a_{11}a_{32}a_{23} + a_{21}a_{32}a_{13} - a_{21}a_{12}a_{33} + a_{31}a_{12}a_{23} - a_{31}a_{22}a_{13}
$$

The determinant of the matrix is zero when some row or column of the matrix is a linear function of other rows or columns. Such a matrix is called a *singular*.

postmultiplying a position vector by a transformation matrix. The vectors and matrices used in the two conventions are transpositions of each other.

Rotation. Consider first a pure rotation around one reference axis, say, axis Z. In this case, rotation takes place in the XY plane. The vector **P** in the local system is given by its components (Figure 1.18). The problem is to find the components of the same vector in the global system.

Direction cosines of the axis x from the local system, with regard to the X and Y axes of the global system, are cos α and cos $(90° - \alpha)$; in turn, the direction cosines of the axis y are cos $(90° + \alpha)$ and cos α. Multiplying the

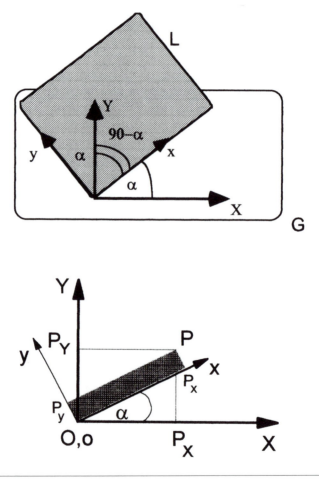

Figure 1.18   The upper panel shows rotation of the local system, L, relative to the global frame, G, around the third axis (not shown). In the bottom panel, the coordinates of point P (P_x, P_y) are given in the local system. Its coordinates in the global system, P_X and P_Y, have to be found.

components of the vector **P** from the local system, which are supposed to be known, by their direction cosines, the respective components of the vector **P** along the X and Y axes in the global system can be calculated. The components of vector **P** along these axes are

$$P_X = P_x \cos \alpha + P_y \cos (90 + \alpha)$$
$$P_Y = P_x \cos (90 - \alpha) + P_y \cos \alpha \qquad (1.7)$$

Using matrix notation this equation appears as

$$[P_G] = \begin{bmatrix} P_X \\ P_Y \end{bmatrix} = \begin{bmatrix} \cos \alpha & \cos (90 + \alpha) \\ \cos (90 - \alpha) & \cos \alpha \end{bmatrix} \cdot \begin{bmatrix} P_x \\ P_y \end{bmatrix} = [R] \cdot [P_L] \qquad (1.8)$$

where $[P_G]$ and $[P_L]$ are column matrices of the components of the vector **P** in the global and local reference systems, respectively, and $[R]$ is the rotation matrix. Elements in the columns of matrix $[R]$ are the direction cosines of the local reference axes with respect to the axes of the global system of coordinates. Because $\cos (90° + \alpha) = -\sin \alpha$ and $\cos (90° - \alpha) = \sin \alpha$, the rotation matrix $[R]$ can be also written as

$$[R] = \begin{bmatrix} \cos \alpha & -\sin \alpha \\ \sin \alpha & \cos \alpha \end{bmatrix} \qquad (1.9)$$

For example, rotating the local reference system around the Z axis by an angle α brings the point P (1,0) to

$$\begin{bmatrix} \cos \alpha & -\sin \alpha \\ \sin \alpha & \cos \alpha \end{bmatrix} \begin{bmatrix} 1 \\ 0 \end{bmatrix} = \begin{bmatrix} \cos \alpha \\ \sin \alpha \end{bmatrix} \qquad (1.10)$$

In a three-dimensional setup, a vector given by its components in a local reference system can also be expressed by its components in a global one if the direction cosines are known:

$$\begin{bmatrix} P_X \\ P_Y \\ P_Z \end{bmatrix} = \begin{bmatrix} \cos_{Xx} & \cos_{Xy} & \cos_{Xz} \\ \cos_{Yx} & \cos_{Yy} & \cos_{Yz} \\ \cos_{Zx} & \cos_{Zy} & \cos_{Zz} \end{bmatrix} \cdot \begin{bmatrix} P_x \\ P_y \\ P_z \end{bmatrix} = [P_G] = [R][P_L] \qquad (1.11)$$

To transform the coordinate of a vector given in a global system to the local one, the transpose matrix of the matrix $[R]$ should be used:

$$[P_L] = [R]^T [P_G] \qquad (1.12)$$

From equations 1.11 and 1.12 it follows that

$$[R]^T[R] = [I] \qquad (1.13)$$

Thus, $[R]^T = [R]^{-1}$ and matrix $[R]$ is an orthogonal matrix. It is also an orthonormal matrix because all the column vectors have unit length. The determinant of the matrix $[R]$ equals +1 (in a right-hand coordinate system). Orthogonal matrices whose determinant equals +1 are called *proper orthogonal*. Proper orthogonal 3 × 3 matrices represent solid-body rotation.

Translation and rotation. When one reference system is related to another by translation and rotation and the position of a point P is given by its vector components in the local reference system, the location of this point in the global system can be expressed if the translation and rotation matrices are known. The position of point P in the global system of coordinates is the vector sum of the vector $\mathbf{L}_G$, giving the origin of the local system in the global system, and the vector $\mathbf{P}_L$ (Figure 1.19).

Before these two vectors can be added, they should be expressed in the global system.

$$\mathbf{P}_G = \mathbf{L}_G + [R]\,\mathbf{P}_L \qquad (1.14)$$

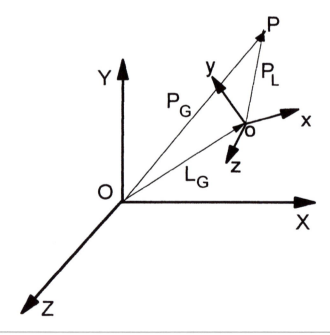

Figure 1.19   Position of point P in the global frame is given by vector $\mathbf{P}_G$, which is the sum of vectors $\mathbf{P}_L$ (for position of the point in the local system) and $\mathbf{L}_G$ for position of the local frame with respect to the global one.

or

$$
\begin{bmatrix} P_X \\ P_Y \\ P_Z \end{bmatrix} = \begin{bmatrix} L_X \\ L_Y \\ L_Z \end{bmatrix} + \begin{bmatrix} \cos_{Xx} & \cos_{Xy} & \cos_{Xz} \\ \cos_{Yx} & \cos_{Yy} & \cos_{Yz} \\ \cos_{Zx} & \cos_{Zy} & \cos_{Zz} \end{bmatrix} \cdot \begin{bmatrix} P_x \\ P_y \\ P_z \end{bmatrix}
\tag{1.15}
$$

Equations 1.14 and 1.15 provide the coordinate transformation from the local frame, L, to the global frame, G. The transformation is given in terms of the vector $\mathbf{L_G}$, which represents the location of the origin of the local frame relative to the global one, and the matrix [R], representing the orientation of the local frame. Using a 4 3 4 transformation matrix [T], translation and rotation can be simultaneously represented by one matrix multiplication.

$$
\begin{bmatrix} 1 \\ P_X \\ P_Y \\ P_Z \end{bmatrix} = \begin{bmatrix} 1 & 0 & 0 & 0 \\ L_X & \cos_{Xx} & \cos_{Xy} & \cos_{Xz} \\ L_Y & \cos_{Yx} & \cos_{Yy} & \cos_{Yz} \\ L_Z & \cos_{Zx} & \cos_{Zy} & \cos_{Zz} \end{bmatrix} \cdot \begin{bmatrix} 1 \\ P_x \\ P_y \\ P_z \end{bmatrix}
\tag{1.16}
$$

or

$$
[P_G] = \begin{bmatrix} 1 & \vdots & 0 & 0 & 0 \\ \cdots & \cdots & \cdots & \cdots & \cdots \\ [L] & \vdots & & [R] & \end{bmatrix} [P_L] = [T][P_L]
\tag{1.17}
$$

The transformation matrix is a homogeneous matrix: both translation and rotation are described by one common mathematical operation. In general, if n is a dimension of the space (n = 2 for planar movement and n = 3 for three-dimensional movement), the (n + 1) × (n + 1) matrix [T], describing translation and rotation, is called the *homogeneous transform*. The position of any point given by its components in one reference system can be easily defined in another system if the homogeneous transform [T] is known.

The 4 × 4 transformation matrix is not orthogonal. Its inverse, which gives the position of the global frame with regard to the local, is

$$
[T]^{-1} = \begin{bmatrix} 1 & \vdots & 0 & 0 & 0 \\ \cdots & \cdots & \cdots & \cdots & \cdots \\ -[R]^T[L] & \vdots & & [R]^T & \end{bmatrix}
\tag{1.18}
$$

■ ■ ■ *From the Scientific Literature* ■ ■ ■

Describing Orientation of the Scapula

Source: Veeger, H.E.J., van der Helm, F.C.T., & Rozendal, R.H. (1993). Orientation of the scapula in a simulated wheelchair push. *Clinical Biomechanics, 8,* 81-90.

Kinematic data were expressed in a global reference system with a fixed thorax orientation. The origin of the global system was at the suprasternal notch, with the axes X, Y, and Z pointing outward, upward, and backward, respectively. Because orientation of the scapula in the anatomic posture is not defined, its orientation was calculated from a virtual reference position in which the spine of the scapula is along the X axis of the global system and the scapular plane is parallel to the plane XY. Note that the virtual position is physically not attainable. The x, y, and z axes of the local system were directed medially, downward, and backward accordingly. Hence, the orientation matrix was

$$[R_{G1}] = \begin{bmatrix} -1 & 0 & 0 \\ 0 & -1 & 0 \\ 0 & 0 & 1 \end{bmatrix}$$

The rotation of the scapula was defined around the axes of the global system and the matrix $[R_{12}]$ was found from the equations $[R_{G2}] = [R_{12}][R_{G1}]$ and $[R_{12}] = [R_{G2}][R_{G1}]^T$ (compare with equation 1.29). The matrix $[R_{12}]$ was then split into three elementary rotation matrices containing three Euler's, or Cardan's, angles. The following sequence of rotations was selected: (1) protraction-retraction around the vertical axis Y, angle β; (2) lateromedial rotation around the axis z', angle γ; and (3) tipping around the axis x", angle α. The complete decomposition was

$$[R_{12}] = [R_{12}]_\beta \, [R_{12}]_\gamma \, [R_{12}]_\alpha$$

or

$$[R_{12}] = \begin{bmatrix} \cos \beta & 0 & \sin \beta \\ 0 & 1 & 0 \\ -\sin \beta & 0 & \cos \beta \end{bmatrix} \begin{bmatrix} \cos & -\sin \gamma & 0 \\ \sin \gamma & \cos \gamma & 0 \\ 0 & 0 & 1 \end{bmatrix} \begin{bmatrix} 1 & 0 & 0 \\ 0 & \cos \alpha & -\sin \alpha \\ 0 & \sin \alpha & \cos \alpha \end{bmatrix}$$

1.2.5.1.3 Defining a Displacement (The Alibi Problem)

Figure 1.20 depicts a local frame, L, that changes its position in space. Its initial position in the global frame is described by the 4×4 matrix $[T_{G1}]$. Its final position, position 2, with regard to the global frame is defined by the matrix $[T_{G2}]$. With regard to the initial position 1, position 2 is defined by the matrix $[T_{12}]$. The displacement matrix $[D_{12}]$, relating global coordinates at positions 1 and 2, can also be defined. Note that there is a new matrix, $[T_{G}]$, for each position of the local frame; however, at a given position the matrix $[T_{G}]$ is the same for all points of the body to which the local frame, L, is attached. Oftentimes, matrix $[T_{G1}]$ is called a *position matrix*, and matrix $[T_{12}]$ is called a *transformation matrix*. Matrix $[T_{G1}]$ defines an initial position of the local frame and matrix $[T_{12}]$ defines transformation from the initial to the final position. If only rotation occurs, matrices $[R_{G1}]$ and $[R_{12}]$ are called the *orientation matrix* and *rotation matrix*, respectively. Matrix $[R_{G1}]$ defines orientation of the local frame with regard to the reference attitude $[0,0,0]$. Matrix $[R_{12}]$ defines rotation from attitude 1 to attitude 2.

Of the four mentioned transformation matrices, $[T_{G1}]$, $[T_{G2}]$, $[T_{12}]$, and $[D_{12}]$, two are considered given and two are unknown. Three problems are typical:

1. What is the final position of the local system with regard to the global system, $[T_{G2}]$, when both the initial position, $[T_{G1}]$, and the relative position, $[T_{12}]$, are known?

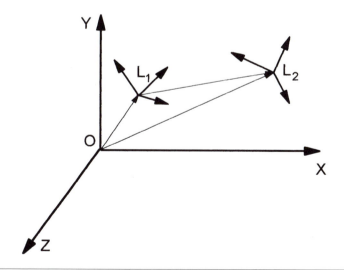

Figure 1.20 A local frame, L, moves from position L_1 to position L_2. The initial position in the global system is given by the matrix $[T_{G1}]$; the final position by the matrix $[T_{G2}]$. The final position of L_2 relative to L_1 is described by the matrix $[T_{12}]$. The displacement can also be described by the matrix $[D_{12}]$.

2. Both the initial, $[T_{G1}]$, and the final, $[T_{G2}]$, positions are given. What is the relative position of L_2 with regard to L_1, in other words, what is $[T_{12}]$?

3. Again, $[T_{G1}]$ and $[T_{G2}]$ are given. What is the displacement $[D_{12}]$ expressed in the global frame?

The solutions to the problems are as follows:

1. Any position of the local system L relative to the global one can be viewed as a displacement of L from an initial position coinciding with the global reference frame G to its present position. Thus, the final position can be considered the composition of two displacements: $G \rightarrow L_1 \rightarrow L_2$, where a horizontal arrow stands for a displacement, or, in matrix form:

$$[T_{G2}] = [T_{G1}][T_{12}] \qquad (1.19)$$

In general, the composition of several displacements is given by multiplication of the corresponding transformation matrices:

$$[T] = [T_{G1}] = [T_{12}], \dots, [T_{(n-1)n}] \qquad (1.20)$$

where n is the number of displacements. To apply equations 1.19 and 1.20, the coordinates (x,y,z) should be measured in the local coordinate frames. That is, the homogeneous transform $[T_{(k-1)k}]$ is applied to the data points measured in the L_k local reference frame. It converts the data into the L_{k-1} reference system. Note how simple it is to represent consecutive transformations with the homogeneous transforms.

When coordinate transformation is a pure rotation, instead of equation 1.19, a similar equation can be written:

$$[R_{G2}] = [R_{G1}][R_{12}] \qquad (1.21)$$

Rotation $[R_{12}]$ is performed around the axes of the local reference system L_1. As a result, the local system L_2 coincides with the local system L_1. The matrix $[R_{G1}]$ is an orientation matrix, it defines orientation of the local system L_1 in the global system G. Hence, postmultiplying of an orientation matrix, for example, $[R_{G1}]$, by a rotation matrix, $[R_{12}]$, indicates rotations around the axes of the local coordinate system. When the order of rotations is changed,

$$[R_{G2}]' = [R_{12}][R_{G1}], \qquad (1.22)$$

the rotation described by the matrix $[R_{G1}]$ brings reference system L_1 in coincidence with, or occupying the same space as, the global system G. This rotation is evidently performed around the axes of the global system of coordinates. After the rotation, the axes of the systems L_1 and G coincide, and the subsequent rotation, $[R_{12}]$, is also performed around the axes of the global system. Hence, premultiplying an orientation matrix by a rotation matrix represents rotations around the axes of the global system. This is one of the general properties of rotation matrices: (1) when the rotation matrix is placed before the orientation matrix it indicates rotations around the axes of the global reference frame and (2) when the rotation matrix is placed after the orientation matrix it indicates rotations around the axes of the local frame. In general, [R] and [R]' are not equal, they define different rotations. When a rotation matrix [R] is split into elementary rotation matrices, it can be interpreted differently when reading from left to right (successive rotations around the local axes that change their orientation because of the earlier rotations or from right to left (rotations around global nonmoving axes). In this textbook, the consecutive rotations are interpreted as local rotations and should be read from left to right.

2. When the global coordinates of L are available at positions 1 and 2, the matrix of the relative position $[T_{12}]$ can be found. From equation 1.19 it follows

$$[T_{12}] = [T_{G1}]^{-1} [T_{G2}] \tag{1.23}$$

This matrix transforms local coordinates in the frame L_2 to coordinates in the local frame L_1. Applying equation 1.17 we can write

$$[P_L]_1 = [T_{12}] [P_L]_2 \tag{1.24}$$

where $[P_L]_1$ and $[P_L]_2$ are the coordinates of the point P in the local reference systems L_1 and L_2, respectively. Equation 1.24 is essentially the same as equation 1.17. The only difference is that in equation 1.17 the transformation from a local to the global frame is considered, and in equation 1.24 the same is described for two positions of the local system.

3. To find a displacement matrix $[D_{12}]$, which transforms global coordinates at position 1 to global coordinates at position 2, the following equations are analyzed:

$$[P_G]_1 = [T_{G1}] [P_L]_1 \tag{1.25}$$

$$\left[P_G\right]_2 = \left[T_{G2}\right]\left[P_L\right]_2 \tag{1.26}$$

where $[P_G]_1$ and $[P_G]_2$ are the global coordinates of point P at positions 1 and 2 and $[P_L]_1$ and $[P_L]_2$ are the local coordinates of the point. Because $[P_L]_1$ and $[P_L]_2$ are identical in all the local frames, they can be written as $[P_L]$. Thus, equation 1.25 can be rearranged as

$$\left[P_L\right] = \left[T_{G1}\right]^{-1}\left[P_G\right]_1 \tag{1.27}$$

Substitution of equation 1.27 into equation 1.26 provides the following:

$$\left[P_G\right]_2 = \left[T_{G2}\right]\left[T_{G1}\right]^{-1}\left[P_G\right]_1 \tag{1.28}$$

The displacement matrix, $[D_{12}]$, can be found by

$$\left[D_{12}\right] = \left[T_{G2}\right]\left[T_{G1}\right]^{-1} \tag{1.29}$$

Equations 1.23 and 1.29 differ only in the order of multiplication: $[T_{12}] = [T_{G1}]^{-1}[T_{G2}]$ and $[D_{12}] = [T_{G2}][T_{G1}]^{-1}$. The matrix $[T_{12}]$, given by equation 1.23, transforms the local coordinates defined in frame 2 to coordinates of frame 1. The displacement matrix $[D_{12}]$, from equation 1.29, transforms global coordinates given in position 1 into global coordinates in position 2. Thus, depending on the sequence of multiplication, the coordinates can be expressed either in the global or the local frame, and vice versa. Sometimes, the transformation matrix $[T_{12}]$ is called the *local transformation matrix*, and the displacement matrix $[D_{12}]$ is called the *global transformation matrix*. Because in the scientific literature the symbols for the matrices may vary, the reader should be aware whether the local or the global matrices are being used.

In summary, the transformation matrices can represent

- the position, i.e., the place and attitude, of a local frame L relative to the global frame G;
- the coordinate transformation from L coordinates to G coordinates; and
- the displacement of the body from one position to another, as well as the composition of several displacements.

In human movement biomechanics, the global reference systems are customarily fixed with the external environment and the local systems with body parts. When a succession of body part orientations is defined with regard to the global system it is often called *segment*, or *bone, rotation.* When the same movement is described with regard to the local system fixed within the adjacent body segment, it is called *joint rotation.*

■ ■ ■ *FROM THE SCIENTIFIC LITERATURE* ■ ■ ■

Using the Matrix Method for Describing Arm Movement

Source: Engin, A.E. (1980). On the biomechanics of the shoulder complex. *Journal of Biomechanics, 13,* 575-590.

The relative motion of the arm with respect to the torso was studied (Figure 1.21). Note that the movement is a combination of rotation and translation. For the purpose of kinematic analysis, both the trunk and the arm were assumed to be rigid bodies. The fixed system of coordinates,

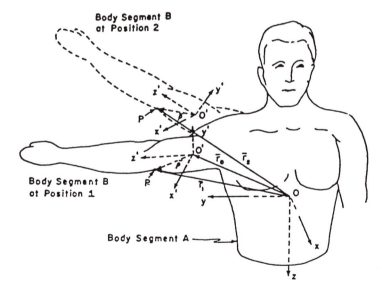

Figure 1.21   Motion of the arm (segment B) with respect to the trunk (segment A) in three dimensional space. Displacement of the arm from position 1 to position 2 is illustrated.

Reprinted from *Journal of Biomechanics, 13,* Engin, A.E., On the biomechanics of the shoulder complex, 575-590, 1980, with kind permission from Elsevier Science Ltd, The Boulevard, Langford Lane, Kidlington 0X5 1GB, UK.

See **MATRIX,** *p. 40*

■ ■ ■ **MATRIX,** *continued from p. 39*

O-XYZ, is attached to the torso, segment A. The local system of coordinates, o'-x'y'z' is attached to the arm, segment B. Position of any arbitrary point P in body segment B is designated by the vectors **p** and **r** (**r**$_1$ at position 1 and **r**$_2$ at position 2). Because point P is fixed in a moving body segment, vector **p** in the o'-x'y'z' system remains constant; whereas vector **r** in the O-XYZ system changes when the arm moves with respect to the torso.

The initial position of point P in the fixed frame is given by the equation

$$\mathbf{r}_1 = \mathbf{r}_O + [R_1] \cdot \mathbf{p}$$

where [R$_1$] is the 3×3 rotation matrix representing orientation of o'-x'y'z' in position 1 with respect to the fixed reference frame, and **r**$_O$ represents translation of o'-x'y'z' with respect to the fixed frame (in the main text, instead of **r**$_O$ the symbol **L**$_G$ is used). This is essentially equation 1.14. For position 2, the equation is

$$\mathbf{r}_2 = \mathbf{r}_O + [R_2] \cdot \mathbf{p}$$

The coordinates of P in position 2 can be expressed through the coordinates in position 1

$$\mathbf{r}_2 = \mathbf{r}_{12} + [R_{12}] \cdot \mathbf{p}$$

where the subscript 12 refers to the translation and rotation from position 1 to position 2. In a homogeneous form, the last transformation can be written as

$$\mathbf{r}_2 = [D_{12}] \cdot \mathbf{r}_1$$

where [D$_{12}$] is the 4×4 displacement matrix (compare with equations 1.28 and 1.29).

1.2.5.1.4 Projection Angles

Instead of the direction angles, *projection angles* can be used. In some applications, it is easier to measure the projection angles than the direction ones. The projection angles are formed by (a) the projections of a vector on the orthogonal planes of the global reference frame and (b) the axes of this frame (Figure 1.22).

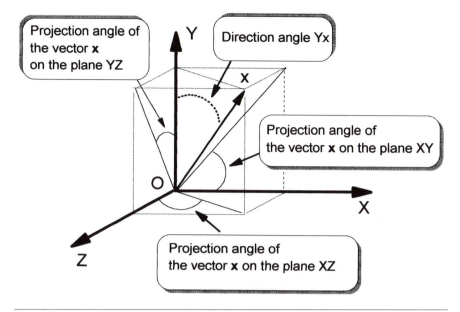

Figure 1.22 Three projection angles of vector **x** are shown (thin solid curves). The direction angle Yx is drawn also (thick broken curve). The tangent of the angle formed by the projection of the vector **x** on the YX plane equals the ratio of the two direction cosines, cos (Yx) and cos (Xx).

Tangents of the projection angles can be easily represented through the direction cosines. For example, for the axis x of the local reference system

$$\tan (YZ)_x = Z/Y = \cos (Zx)/\cos (Yx)$$
$$\tan (YX)_x = Y/X = \cos (Yx)/\cos (Xx) \qquad (1.30)$$
$$\tan (XZ)_x = X/Z = \cos (Xx)/\cos (Zx)$$

where $\tan (YZ)_x$, $\tan (YX)_x$, and $\tan (XZ)_x$ are the tangents of the projection angles of the axis x on the planes YZ, YX, and XZ, and cos (Xx), cos (Yx), and cos (Zx) are direction cosines of the angles formed by the axis x of the local frame and the axes X, Y, and Z. If the direction cosines are known, the corresponding projection angles can be easily determined using equation 1.30. The projection angles are not independent. From equation 1.30 it follows

$$\tan (YZ)_x \cdot \tan (YX)_x \cdot \tan (XZ)_x = \frac{Z}{Y} \cdot \frac{Y}{X} \cdot \frac{X}{Z} = 1 \qquad (1.31)$$

Thus, only two projection angles of the three can be chosen arbitrarily. Equation 1.31 imposes a constraint on the projection angles. In total, there are nine projection angles. Only three of them are independent. When projection angles are given (measured), the direction cosines can be easily calculated. After that, the rotation matrix is used in the usual manner. This procedure is conventionally used in experimental research.

The matrix method is the basic mathematical tool used primarily for computing body position and/or movement. However, this method is not convenient for an immediate interpretation of the relative movement between two bodies. Nine direction cosines are obviously redundant, because the position of a body in space can be described with a smaller number of angles. Three independent angles (corresponding to the three rotational degrees of freedom) need to be determined. These angles are called Euler's angles.

1.2.5.2 Euler's method

Finite rotations in three-dimensional space are noncommutative; in other words, they must be performed in a specific order. This concept may easily be demonstrated with a book or other small object. Place a book on a desk and rotate it 90° two times relative to different axes of rotation, say frontal and anteroposterior. Repeat this simple experiment changing the order of rotations: rotate first around the anteroposterior axis and afterward around the frontal one. The book will no doubt be oriented in space differently after each set of rotations. Because finite rotations are not commutative ($a \cdot b \neq b \cdot a$) they cannot be considered vectors.

The change of orientation can be described as a sequence of three successive rotations from an initial position, at which two reference frames G and L coincide. Euler's angles are defined as these three successive angles of rotation about preset axes.

Let us examine a gymnast performing the dismount, a somersault with a twist, from a horizontal bar (Figure 1.23). The gymnast's body

a. rotates around the axis parallel to the bar. Note that this axis cannot be called "frontal" because the frontal axis is defined relative to the body, not to the environment. The axis of rotation coincides with the frontal axis of the gymnast's body only occasionally (e.g., at the moment of release) and does not change orientation in the global reference frame. So, the first rotation takes place relative to an axis defined in the global reference system. This rotation is called *precession*.

b. inclines and declines relative to the sagittal plane. This motion is called a *nutation*, or *tilt*. The axis of the tilt, however, is not fixed with regard to both the global reference frame and the local reference frame (with the athlete's body); thus, it is sometimes called the "floating" axis. The floating axis is always orthogonal to both the first and the third axes.

Figure 1.23   Two dismounts from the high bar: one with one twist (left figure) and one with two twists (right figure). The numbers correspond to sequential body positions. When performing the somersault with a twist, a gymnast rotates around three different axes: (a) somersault axis, (b) tilt axis, and (c) twist axis. In general, the corresponding rotations are coined: precession, nutation, and spin or, in other words, twist.

  c. rotates around its longitudinal axis. The axis changes orientation in space but is fixed with the body. The rotation is called a *twist* or *spin.*

A specific definition of Euler's angles, in which rotations take place in a described sequence (precession [in the sagittal plane], nutation [motion away or toward the sagittal plane], and twist, or spin [along the axis fixed in the body]), is commonly used in biomechanics. However, it is possible to change the ordered sequence of Euler's angles and to define, for example, as the first angle the rotation of the body about the anteroposterior axis. This designation may prove useful when studying gymnastic exercises on a pommel horse. In this case, the tilt and twist should also be redefined. Also, the so-called *nautical* angles, *yaw, pitch,* and *roll,* are often used. The yaw is defined as rotation about the vertical Y axis of the global, fixed frame. The pitch is the second rotation about the axis z of the moving frame, and the roll is the rotation about the x axis of the local frame.

In total, there are 12 sequences of rotations (Euler's conventions). However, the general succession is the same: The first rotation is defined relative to an axis oriented in the global reference frame, the third (not the second) is defined with regard to an axis fixed within the rotating body, and the second is performed relative to the floating axis. The second and the third rotations are about local axes transformed by previous rotations. These local axes are denoted below by single prime (') and double prime (") , according to the number of preceding rotations defining their orientation. For example, a sequence Xy'x" means that the second rotation is performed around the local y axis, which was rotated previously around the global X axis. The third rotation is around the local x axis, which was rotated previously around the global X axis and then around the y axis.

The terminal coordinate axis in the sequence of rotations may be either identical (e.g., Xy'x", Zx'z") or different (e.g., Xy'z", Yx'z"). Some authors use the term "Euler's angles" to mean the six sequences with identical terminal coordinates and the term "Cardan's angles," or "Bryant's angles," for the six sequences with different terminal axes. In this textbook, all the 12 sequences, when they are considered as one set, are denoted as "Euler's angles." When the order of rotations is important, the term "Cardan's angles" is also used.

Euler's or Cardan's angles can easily be visualized if one imagines a body mounted in a system of gimbals nested within each other (Figure 1.24). In the starting position of the Euler's suspension, all three axes of rotation are located in one plane, for example, the axes Yx'y" are in the XY plane. Cardan's suspension comprises three different axes, for example, Y, x', and z". The Cardan's angular convention is also referred to as the *three-axes system* or the *gyroscopic system*. Euler's convention is referred to as the *two-axes system*. Euler's or Cardan's angles can be considered the angles of rotation of the gimbals maintaining the body. In various conventions, the hierarchy of the nesting gimbals is different. For instance, a horizontal axis may be nested within a vertical axis (Figure 1.24A), or the vertical axis is nested within the horizontal (Figure 1.24B). The various conventions give rise to different sets of coordinates.

When the same body attitude is measured in various Euler's or Cardan's conventions, the angular values are different. Consider the example in Figure 1.25. At the top of the figure, an object twists around the Z axis, which is perpendicular to the plane of this paper. In the bottom panel, the object, starting from the same position, undergoes three rotations about the Y, x', and y" axes. In both cases, the final orientation is the same. This phenomenon is seen in human movement. For example, starting from the dependent posture, the

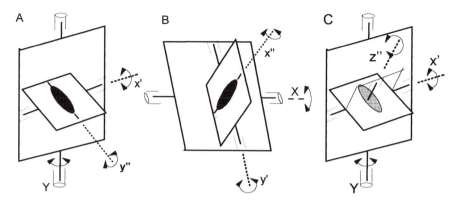

Figure 1.24 Representation of Euler's and Cardan's angles as the angles of a gimbal suspension. In part A, Euler's suspension, Yx'y" convention is shown. The outer gimbal rotates about the vertical Y axis of the global reference system; the internal gimbal rotates about the local y" axis fixed with the gimbal; the intermediate gimbal rotates about the x' axis, which is not fixed firmly with either the global or with the local system. In part B, Euler's suspension, Xy'x" convention is shown. The conventions differ in the order in which the three coordinate angles are measured. In part C, Cardan's suspension, Yx'z" convention is shown.

same attitude of the arm can be reached either by flexing 180°, or by abducting 180° followed by externally rotating (supinating) 180°. The arm flexion (rotation about the frontal axis) brings the arm into the same position as rotation around first the anteroposterior axis and then around the arm's longitudinal axis. Thus, depending on the convention adopted (the sequence of rotations), the same attitude of the arm can be represented either as $(\pi, 0, 0)$ or as $(0, \pi, \pi)$. In general, the values calculated for the second and third rotations are affected by the preceding rotations. For example, when the same body orientation is described with the triads Xy'x" and Xz'x", the values of rotation around the x" axes are different. Hence, to define a body orientation, all the three Euler's angles should be known.

The expression "sequence of rotations" can refer to either hierarchy of the nested gimbals—in the foregoing paragraphs exactly this meaning was adopted—or to the temporal order of the gimbal rotations. The gimbal rotations can be performed starting from the outer gimbal, the upstream rotation, or from the inner gimbal, the downstream rotation. If the hierarchy, or sequence, of the nested gimbals is established, the final orientation does not depend on whether the rotations were performed in the upstream or downstream order. Hence, for a given Euler's or Cardan's angular convention, the finite angular displacements are independent of sequence. Thus, if a three-axial goniometer is constructed rightly, i.e., is a gyroscopic mechanism, the results of

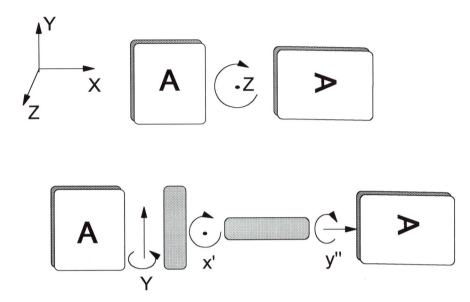

Figure 1.25 Illustration of the fact that the same body orientation can be represented differently in various angular conventions. The final body attitude is reached by rotating –90° about the Z axis or by triple rotating (90°, –90°, 90°) about Y, x', and y'' axes. The body's final coordinates are (–90°, 0°, 0°) in any Cardan's convention with the Z axis as the first axis of rotation, and they are equal to (90°, –90°, 90°) in the Yx'y'' (Euler's) convention. Note that torsion of the body relative to the Z axis can be represented as the sequence of rotations about two other axes, X and Y.

the measurement do not depend on the rotation sequence. When reading the literature, the student should pay strict attention to the real meaning of the expression "sequence of rotation" in a particular paper. He or she should note whether the selection of an angular convention or the rotations around the preselected axes are being discussed.

To view the Euler's angles as successive angles of rotation is convenient but not mandatory. They may be defined geometrically without invoking the image of consecutive rotations from some initial position. For example, the attitude of two frames can be defined relative to the *nodes axis*, which is an intersection of two planes, one from the global system and the second from the local (Figure 1.26). This approach is used in the joint rotation convention, which will be described in detail in Chapter 2. In this convention, one of the axes of rotation, the *floating axis*, coincides with the line of nodes.

Defining the relative angular position between two bodies with Euler's angles has several advantages. The angles are easily understood and can be directly measured. For example, the Euler's angles between two segments of a human body can be measured with goniometers (Figure 1.28).

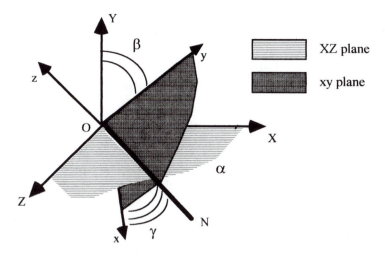

Figure 1.26   Euler's angles. Two planes, XZ and xy, intersect in a nodes axis N. An attitude of the frame xyz relative to the frame XYZ can be defined with three angles: α is the angle between X and N, β is the angle between Y and y, and γ is the angle between N and x. These angles are independently sequenced.

With Euler's angles, however, only an attitude is defined. Thus, the Euler's angles do not form a homogeneous system, and translation and rotation must be handled separately. Yet, the Euler's angles can be readily expressed as elements of the 3×3 rotation matrix, $[R] = [R_1][R_2][R_3]$, where $[R_1]$, $[R_2]$, and $[R_3]$ are the matrices of sequential rotations, and, then, an augmented 4×4 transformation matrix $[T]$ may be constructed and used in the usual way. For a Zy'x" rotation sequence,

$$[R] = [R_z][R_y][R_x] = \begin{bmatrix} \cos\alpha & -\sin\alpha & 0 \\ \sin\alpha & \cos\alpha & 0 \\ 0 & 0 & 1 \end{bmatrix} \begin{bmatrix} \cos\beta & 0 & \sin\beta \\ 0 & 1 & 0 \\ -\sin\beta & 0 & \cos\beta \end{bmatrix} \begin{bmatrix} 1 & 0 & 0 \\ 0 & \cos\gamma & -\sin\gamma \\ 0 & \sin\gamma & \cos\gamma \end{bmatrix}$$

$$= \begin{bmatrix} \cos\alpha\cos\beta & \cos\alpha\sin\beta\sin\gamma - \sin\alpha\cos\gamma & \cos\alpha\sin\beta\cos\gamma + \sin\alpha\sin\gamma \\ \sin\alpha\cos\beta & \sin\alpha\sin\beta\sin\gamma + \cos\alpha\cos\gamma & \sin\alpha\sin\beta\cos\gamma - \cos\alpha\sin\gamma \\ -\sin\beta & \cos\beta\sin\gamma & \cos\beta\cos\gamma \end{bmatrix}$$

$$(1.32)$$

Elements in the combined matrix (equation 1.32) represent direction cosines between the axes of two reference frames. They are expressed as functions of the Euler's angles. This operation is called *decomposition* of the Euler's angles; the angles are decomposed into their projections onto the axes of the

■ ■ ■ *From the Scientific Literature* ■ ■ ■

Using Euler's Angles to Describe Knee Joint Movement

Source: An, K.-N. & Chao, E.Y.S. (1991). Kinematic analysis. In K.-N. An, R.A. Berger, & W.P. Cooney III (Eds.), *Biomechanics of the Wrist Joint* (pp. 21-36). New York: Springer-Verlag.

Euler's angles are often used for describing joint kinematics (see example in Figure 1.27). Flexion and extension take place around a mediolaterally directed axis attached to the femoral condyle. Axial rotation occurs around the longitudinal axis of the tibia. The third axis defines abduction and adduction and can be determined as the cross product of the two axes fixed to the proximal and distal segments. This axis changes its orientation in space when the knee is flexed, but it is always orthogonal to both the flexion-extension and axial rotation axes. When the joint is not in a neutral position, e.g., flexed and abducted, the two body segment–fixed axes are not orthogonal to each other.

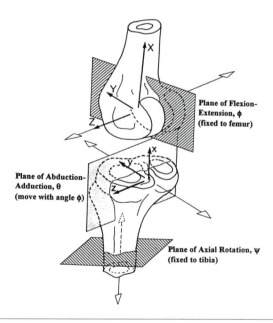

Figure 1.27 Description of the knee joint motion by the Euler's angle system.

Reprinted, by permission, from An, K.-N. & Chao, E. Y. S. (1991). Kinematic Analysis. In K.-N. An, R. A. Berger, & W.P. Cooney III (Eds.), *Biomechanics of the Wrist Joint* (pp 21-36). New York: Springer-Verlag.

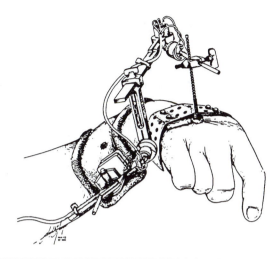

Figure 1.28 Triaxial electrogoniometer for wrist joint motion measurement. The goniometer measures the Euler's angles. The hierarchy of the angles— defined by the goniometer design—is: flexion-extension→ abduction-adduction→ axial rotation. However, because the goniometer assembly is located outside of the joint, the goniometer readings may deviate from the real joint angles (see "Converting Data From a Technical to a Somatic System" on page 84).
From An, K.-N. & Chao, E.Y.S. (1991). Kinematic analysis. In K.-N. An., R.A. Berger, & W.P. Cooney III (Eds.), *Biomechanics of the Wrist Joint* (pp. 21-36). New York: Springer-Verlag. Reprinted by permission.

global frame. An inverse operation is also possible. When a matrix [R], rather than the Euler's angles, is given, the elements of the matrix can be interpreted in terms of the Euler's angles if one assumes a certain order of rotations. From equations 1.2 and 1.32, it follows that the Euler's angles for the Zy'x" sequence can be obtained by the following equation:

$$A(Z) = \tan^{-1}\left(\frac{\cos_{21}}{\cos_{11}}\right)$$

$$A(y') = \tan^{-1}\left(-\frac{\cos_{31}}{\sqrt{\cos_{11}^2 + \cos_{11}^2}}\right) \qquad (1.33)$$

$$A(x") = \tan^{-1}\left(\frac{\cos_{32}}{\cos_{33}}\right)$$

■ ■ ■ *FROM THE SCIENTIFIC LITERATURE* ■ ■ ■

Avoiding Singular Positions by Proper Selection of Reference Systems and Angular Convention

Source: Zatsiorsky, V.M. & Aleshinsky, S.Y. (1975). Simulation of human locomotion in space. In P.V. Komi (Ed.), *Biomechanics V-B* (pp. 387-398). Baltimore: University Park Press.

The following systems of coordinates were introduced (Figure 1.29):

a. a global system, O-XYZ;

b. moving systems at the body links; the systems maintain constant orientation in space with the axes parallel to the axes of the global system, O_k-$X_k Y_k Z_k$, where k is a link number; and

c. somatic systems attached to the body links, o_k-$x_k y_k z_k$.

To avoid singular positions when studying walking, the following angles were used for defining the orientation of link k:

- θ_k, an angle between the positive direction of the X_k axis and the z_k axis;

- ψ_k, an angle formed by the negative direction of the Z_k axis and projection of the z_k axis onto plane $Y_k O_k Z_k$; and

- ϕ_k, an angle formed by the line of intersection of planes $X_k O_k Y_k$ and $x_k o_k y_k$ (the line of nodes) and the x_k axis.

The orientation angles are found as follows:

$$\theta_k = \arccos(X_k, z_k)$$

$$\psi_k = \begin{cases} \pi, & \text{when } \cos(Y_k, z_k) = 0 \\ \pi - \arccos(\cos(Y_k, z_k)/\sin\theta_k), & \text{when } \cos(Y_k, z_k) < 0 \\ \pi + \arccos(\cos(Y_k, z_k)/\sin\theta_k), & \text{when } \cos(Y_k, z_k) > 0 \end{cases}$$

$$\phi_k = \begin{cases} \pi, & \text{when } \cos(X_k, y_k) = 0 \\ \pi - \arccos(\cos(X_k, x_k)/\sin\theta_k), & \text{when } \cos(X_k, y_k) < 0 \\ \pi + \arccos(\cos(X_k, x_k)/\sin\theta_k), & \text{when } \cos(X_k, y_k) > 0 \end{cases}$$

This convention works well when studying walking. It would yield a singular position for the gymnast performing a handstand, a stunt at which the body is oriented head down with the arms extended.

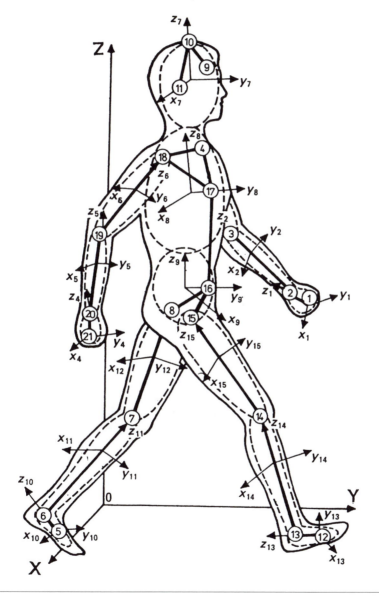

Figure 1.29   Reference systems used to study gait.

where A(Z), A(y'), and A(x") are the angles of rotation around the axes Z, y', and x".

For particular angular positions, when the nutation angle is zero and the first and third axes are parallel, the Euler's angles cannot be defined. These angular positions are called *singular*, or *gimbal-lock* positions. If a body is in a singular position, the values of the first and third angles cannot be determined. Only their sum (or difference) is measurable. For example, when a human standing at military attention turns 90°, it is impossible to discern the axis of rotation: was it in the global system, the Y axis, or in the local system, the y axis? The singular positions are different for each set of Euler's angles and frequently can be avoided by proper selection of a reference frame or angular convention. For some movements, however, neither Euler's convention is suitable.

An inescapable disadvantage of any Euler's convention is the dependence of results from a chosen hierarchy of the nested gimbals. The same body orientation, when represented in various Euler's coordinate frames, is characterized by different values of angular coordinates. A change in the sequence of rotations generally leads to an altered result. In experimental settings, when there is no reason to treat any axis differently from any other, the selection of a reference system that does not depend on arbitrary decision of the researcher seems less biased.

1.2.5.3 Screw (Helical) Method

Representative papers: Kinzel et al. 1972; Ramakrishnan & Kadaba, 1991.

The screw method permits a description of body attitude without referring to arbitrarily chosen axes of rotation.

• • • CLASSICAL MECHANICS REFRESHER • • •

Euler's and Chasles' Theorems

Euler's Theorem: Any motion of a rigid body with one point fixed is a single rotation around an axis through that fixed point. In other words, the theorem states that several rotations around different axes passing through a point are equivalent to a single rotation.

Chasles' Theorem: Any rigid-body motion can be obtained as the rotation around an axis, known as the *screw axis,* and a translation parallel to the screw. The rotation and translation may occur in any sequence.

Consider first a planar movement. For any planar displacement, except pure translation, there exists a point, called the *pole* or *center of rotation*, that does not move. In the case of pure translation, the coordinates of the pole approach infinity. Coordinates of the pole are the same in both reference frames, G and L (Figure 1.30). The instantaneous center of rotation (ICR) is defined as a point on the plane maintaining a constant distance from every point of the body, or as the point with zero velocity during infinitesimally small motion. In the literature, the ICR is also called the *instantaneous center of zero velocity*. The displacement is defined by the position of the pole and by the rotation around the pole. A sequence of body positions may then be equated with a sequence of finite rotations about properly chosen poles. The locus of points that define the center of rotation during the body motion is called *centrode*. The centrode can be viewed as a trajectory of the instant center of rotation.

Consider now a three-dimensional motion. At any given instant, there exists a line, rather than a point, that maintains its position in space. This line is called the *screw axis* (an old term) or the *helical axis* (a relatively new term). Any point on the screw axis is constrained to move along the line that is the axis of rotation. Generally, the helical axis does not lie within the body. The screw (helical) method is based on the proposition that any general motion can be represented as the sum, or sequence, of translation and rotation (Chasles' theorem). At any given instant, the translation and rotation occur along and around a helical axis (Figure 1.31). The helical axis may change its position in space. If the position of the helical axis, i.e., its location and direction, is known, the position of the body can be described with regard to the axis. Helical motion is described by

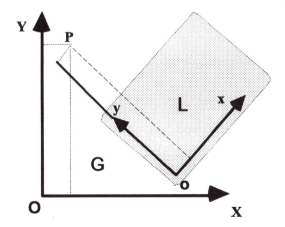

Figure 1.30 Coordinates of the pole P are equal in both the references, $P_X = P_x$ and $P_Y = P_y$.

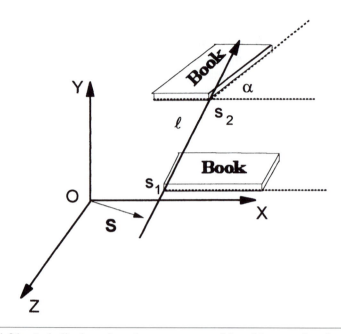

Figure 1.31 In helical motion, the movement of the object can be visualized as the translation along the screw axis from point s_1 to point s_2, and the rotation about this axis at angle α. The position of a point on the helical axis can be defined by a vector **S**. The direction (unit) vector of the screw axis should also be given. Thus, six independent parameters are needed to define a helical motion.

1. the position of the helical axis relative to the global frame, i.e.,
   a. the position vector, and
   b. the direction (unit) vector, and
2. the position of the body relative to the helical axis, i.e.,
   a. the position of the body along the axis (ℓ), and
   b. the rotation of the body about the axis (α).

The following six parameters are needed to define a body position:

1. and 2. Two coordinates of the piercing point of the screw axis with any one of the three coordinate planes, e.g., the XY plane.
3. and 4. Two direction cosines of the screw axis.
5. and 6. The translation along and the rotation about the screw axis.

The translation of the body along and its rotation about the helical axis can be jointly described by a *pitch of the screw*, which is defined as the following

■ ■ ■ *FROM THE SCIENTIFIC LITERATURE* ■ ■ ■

Using the Helical Method for Describing Arm Movement

Source: Engin, A.E. (1980). On the biomechanics of the shoulder complex. *Journal of Biomechanics, 13,* 575-590.

Arm movement with respect to the trunk can be described with the screw method. Because of the large displacement of the glenohumeral joint, arm translation should be taken into account in addition to arm rotation. Hence, rotation matrices or Euler's angles alone cannot adequately specify arm motion. Arm motion can be viewed as translation of the glenohumeral joint along the screw axis and arm rotation at this joint.

Figures 1.21 and 1.32 illustrate two ways of describing the same arm movement, through transformation matrices and the screw method, respectively.

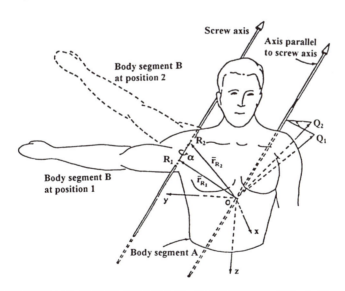

Figure 1.32 Representation of the arm motion from position 1 to position 2 in terms of rotation about and translation along the screw axis. A pure rotation of a point Q about an axis parallel to the screw axis is also displayed. See Figure 1.21 for comparison.

Adapted from *Journal of Biomechanics, 13,* Engin, A.E., On the biomechanics at the shoulder complex, 575-590, 1980, with kind permission from Elsevier Science Ltd, The Boulevard, Langford Lane, Kidlington 0X5 1GB, UK.

ratio: Pitch $= 2\pi\ell/\alpha$. The pitch of the screw is the magnitude of translation per one revolution. The so-called *reduced pitch of the screw* is also used. It is defined as the ratio ℓ/α and represents the magnitude of translation per one radian of the rotation. The pitch is used to normalize the helical translation relative to the helical rotation. When the pitch is zero ($\ell = 0$) the motion is a pure rotation; when the pitch is infinite ($\alpha = 0$), the motion is a translation along the axis. Any helical motion can be achieved in two steps, translation and rotation, in either order. For a helical motion, the matrix multiplication is commutative. A 4×4 transformation matrix whose elements are a function of the helical parameters is called the *helical matrix*. The helical matrix is not described here. It may be found in many advanced kinematics texts.

When the helical axis method is used to define rotation, $S = 0$ and only three parameters are needed. Rotation in three dimensions is completely determined by the direction of the rotation axis (two parameters) and the angle of rotation. Various conventions for the description of orientation and rotation are commonly used. In particular, any rotation can be represented as a vector. The direction of this vector is along the axis of rotation, whereas its length is proportional to the magnitude of rotation. The three-dimensional unit vector along the rotation vector is called the *direction vector*. The orientation of the direction vector is given by its direction cosines with regard to a global frame. If u_X, u_Y, and u_Z represent the direction cosines of a direction vector U, C refers to the cosine of angle α, and S refers to the sine of angle α, then the rotation matrix becomes

$$[R] = \begin{bmatrix} u_X \cdot u_X - u_X \cdot u_X \cdot C + C & u_X \cdot u_Y - u_X \cdot u_X \cdot C - u_Z \cdot S \\ u_X \cdot u_Y - u_X \cdot u_Y \cdot C + u_Z \cdot S & u_Y \cdot u_Y - u_Y \cdot u_X \cdot C + C \\ u_X \cdot u_Z - u_X \cdot u_Z \cdot C - u_Y \cdot S & u_Y \cdot u_Z - u_Y \cdot u_Z \cdot C + u_X \cdot S \end{bmatrix}$$

$$\times \begin{bmatrix} u_X \cdot u_Y - u_X \cdot u_Z \cdot C + u_Y \cdot S \\ u_Y \cdot u_Z - u_Y \cdot u_Z \cdot C - u_X \cdot S \\ u_Z \cdot u_Z - u_Z \cdot u_Z \cdot C + C \end{bmatrix} \qquad (1.34)$$

To find the rotation vector, the direction unit vector should be multiplied by the amount of rotation. Several measures of the magnitude of rotation are used for this. In the most simple case, the direction cosines of the helical axis are multiplied by the rotation angle α, measured in radians:

$$a = u_X\, \alpha; \qquad b = u_Y\, \alpha; \qquad c = u_Z\, \alpha \qquad (1.35)$$

where u_X, u_Y, and u_Z are the direction cosines of the direction vector U. This representation, however, is not convenient for computing. Therefore, instead of the angle of rotation α, the sine or tangent values of $\alpha/2$ are used. The rotation vector is calculated as $U \cdot \sin(\alpha/2)$ or $U \cdot \tan(\alpha/2)$. The direction of a rotation vector is determined according to the right-hand thumb rule: when the right thumb points in the direction of **U**, the fingers curl in the direction the body has turned. The rotation vectors are easily manipulated mathematically.

Final (total) rotation resulting from sequential elementary rotations can be readily calculated from several conventions based on the idea of helical displacement. These representations, which are not covered in detail in this textbook, are *Euler's parameters, Rodriguez's parameters,* and *quaternions.* Short descriptions of these techniques follow.

A. *Euler's parameters* are a set of four variables calculated as follows:

$$a = u_X \cdot \sin(\alpha/2) \qquad b = u_Y \cdot \sin(\alpha/2)$$
$$c = u_Z \cdot \sin(\alpha/2) \qquad d = \cos(\alpha/2) \tag{1.36}$$

The parameters are considered a 4×1 column vector, $[E] = [a, b, c, d]^T$. The Euler's parameters are subjected to the following constraint equation: $a^2 + b^2 + c^2 + d^2 = 1$. Hence, the parameters are not independent. Only three of them can be chosen arbitrarily. The first three Euler's parameters, a, b, and c, are the Cartesian components of the rotation vector $\mathbf{U} \cdot \sin(\alpha/2)$.

B. *Rodriguez's parameters* are a set of three independent variables:

$$A = u_X \cdot \tan(\alpha/2) \qquad B = u_Y \cdot \tan(\alpha/2) \qquad C = u_Z \cdot \tan(\alpha/2) \tag{1.37}$$

The Rodriguez's parameters are the Cartesian components of the rotation vector $\mathbf{U} \cdot \tan(\alpha/2)$.

C. *Quaternions* are nontraditional mathematical objects. Formally, they are written as a sum of a scalar and a vector with the Euler's parameters as coefficients, $q = ai + bj + ck + d$, where i, j, and k are the so-called quaternion units. Quaternions can also be seen and understood as complex numbers depending on four units, i, j, k, ℓ. Mathematical operations with quaternions are defined by special rules. The quaternion resulting from two sequential rotations is the product $q_t = q_1 q_2$ where q_t is the quaternion expressing the total rotation. In biomechanics, quaternions are used mainly to describe sequential eye movements.

From Euler's theorem it follows that any rotation in space can be represented as the rotation about a certain axis. If this axis is chosen as an axis of the reference frame, the body rotates in a plane perpendicular to the reference axis. Hence, the rotation is planar and its mathematical description is simple. For the rotation about the Z axis, the 3×3 rotation matrix is

$$[R] = \begin{bmatrix} \cos_{Xx} & \cos_{Xy} & 0 \\ \cos_{Yx} & \cos_{Yy} & 0 \\ 0 & 0 & 1 \end{bmatrix} = \begin{bmatrix} \cos\alpha & -\sin\alpha & 0 \\ \sin\alpha & \cos\alpha & 0 \\ 0 & 0 & 1 \end{bmatrix} \tag{1.38}$$

where α is the angle of rotation (compare with equations 1.10 and 1.11). The rotation matrix given in equation 1.38 is called a *canonical form* of the rotation matrix. Because the rotation angle and the direction vector remain unchanged under different choices of coordinate axes they are called the *invariants* of rotation. When a rotation axis is not parallel to the reference axis, the rotation matrix [R] contains nine elements expressed in terms of trigonometric functions. Although [R] completely describes the body orientation, this representation is not convenient for immediate perception. When the rotation matrix is in the canonical form, it is much simpler to grasp the body position. Hence, it is very convenient to direct the reference axis along the axis of rotation. Because any rotation matrix [R] represents a rotation about a finite axis U at angle α, the matrix [R] can be written as [R (U,α)], specifying both the direction and magnitude of rotation. In what follows, the notion of eigenvectors and eigenvalues are used.

Consider briefly the following problems:

1. The rotation matrix [R] is given. Find the axis and the angle of rotation.

2. The axis and the angle of rotation are known. Obtain the matrix [R] with reference to a specific system of coordinates.

Answers

1. A rotation 3×3 matrix [R] is a proper orthogonal matrix. Hence, the length of the eigenvector associated with eigenvalue +1 does not change after rotation. The eigenvector of the rotation matrix [R] associated with

••• *MATHEMATICS REFRESHER* •••

Matrix Algebra: Eigenvectors and Eigenvalues

Let an $n \times n$ matrix [T] be a transformation matrix. Also, an $n \times 1$ non-zero vector $\mathbf{V}$ and a scalar λ exist that satisfy the equation

$$[T]\mathbf{V} = \lambda\mathbf{V}$$

The vector $\mathbf{V}$ and the scalar λ are called an *eigenvector* and an *eigenvalue,* respectively, of [T]. The expression $\lambda\mathbf{V}$ can be interpreted as stretching along the direction of vector $\mathbf{V}$. Thus, the expression $[T]\mathbf{V} = \lambda\mathbf{V}$ means that when the transformation [T] is applied, vector $\mathbf{V}$ does not change its direction. Vectors not changing their direction by a coordinate transformation are called invariant vectors.

Proper orthogonal matrices have one eigenvalue, +1.

its +1 eigenvalue is the direction vector of the body rotation. To find the direction vector, the eigenvector of the rotation matrix should be determined. The direction vector of the rotation matrix [R] is given by the following expression:

$$\mathbf{U} = \begin{bmatrix} \cos_{Yz} - \cos_{Zy} \\ \cos_{Zx} - \cos_{Xz} \\ \cos_{Xy} - \cos_{Yx} \end{bmatrix} \tag{1.39}$$

The reader can find a sample proof in textbooks of theoretical kinematics, for example, Bottema and Ross (1979). For the rotation matrix shown in equation 1.38, the eigenvector is $[0,0,-2\sin\alpha]$. In this example, the direction vector lies along the axis Z and has length $-2\sin\alpha$. The rotation angle also can be readily obtained. From the canonical form of [R] given in equation 1.38, it is evident that $\mathrm{Tr}[R] = 1 + 2\cos\alpha$, where $\mathrm{Tr}[R]$ is the trace of the rotation matrix. It follows that

$$\cos\alpha = \frac{1}{2}\left(\mathrm{Tr}[R] - 1\right) \tag{1.40}$$

Because the trace of matrix [R] is invariant under rotation, the rotation angle can easily be computed without transforming matrix [R] into its canonical form.

2. The relationship between the finite rotation variables and the elements of the rotation matrix can be readily found. The form of relationship depends on the helical parameters used (Euler's parameters, Rodriguez's parameters, or quaternions). For Euler's parameters, the relationship assumes the following form:

$$[R] = \begin{bmatrix} 2(a^2 + d^2) - 1 & 2(ab - cd) & 2(ac + bd) \\ 2(ab + cd) & 2(b^2 + d^2) - 1 & 2(bc - ad) \\ 2(ac - bd) & 2(bc + ac) & 2(c^2 + d^2) - 1 \end{bmatrix} \tag{1.41}$$

The elements of matrix [R] in equation 1.41 are quadratic, whereas they are trigonometric when expressed in terms of direction cosines and Euler's angles. The quadratic elements make the calculations simpler.

The finite rotation vectors are used to define either the body orientation or the body displacement. The difference is in the initial position. When the movement starts at the reference attitude $[0,0,0]$, the rotation vector defines the body orientation. If movement starts at any other point, the rotation vector defines the body displacement.

The helical method allows certain advantages. For example, translation along the helical axis represents a real translation in three-dimensional space.

1.2.6 Advantages and Disadvantages of the Various Angular Conventions

All the described angular conventions permit definition of the angular position of a body in space, but they possess certain features that make some more convenient than others for use under specific circumstances. Following are the primary requirements of defining body attitude from a practical standpoint:

- Direct measurability.
- Immediate physical meaning. The body coordinates should be easily understood by the users.
- Graphical representation. Changes in the attitude should be easily displayed by graphs. The fulfillment of the last two requirements is rather subjective.
- Singularity, or the lack of singular positions. In some conventions, the

Table 1.1   Comparison of the Different Angular Coordinate Systems

Method	Direct measurability	Immediate physical meaning	Graphs	Singularity	Periodicity	Describing sequential rotations
Matrix	No	Yes/No[1]	No	No	No	Yes
Euler's	In some cases only	Yes	Yes	Yes	Yes	No
Screw	No	Yes/No[2]	Yes	Yes	Yes	No[3]

[1]The meaning of the individual matrix elements may be obvious, but it is often difficult to visualize the body attitude when all matrix elements are applied.

[2]The representation of body movement as a combination of translation and rotation is easily understood. However, when linear and angular acceleration of the body are determined, the helical method is not very convenient for immediate interpretation of these parameters. It is difficult to imagine what angular acceleration about an instantaneous helical axis means when the axis constantly changes its orientation in space.

[3]Final rotation resulting from several consecutive elementary rotations can be found if quaternions or Rodriguez's parameters are used.

value of one or more coordinates cannot be defined for certain angular positions.

- Periodicity, or the lack of periodicity problem. Particular angular positions approaching 180° (e.g., 179° and 181°) may be represented as if they are distant from each other, such as 179° and −179°.
- Convenience of describing sequential rotations.

The various coordinate systems are compared in Table 1.1.

A very important feature of a coordinate system is its susceptibility to measurement errors. Which system is the most robust is still a matter of research and discussion in biomechanics literature.

1.2.7 Determining Body Position from Experimental Recordings

Representative papers: Challis, 1995; Spoor & Veldpaus, 1980; Veldpaus et al., 1988; Woltring, 1991; Wu, 1995.

In experimental research, the position of a body, i.e., its location and orientation, is determined from spatial coordinates of markers. To determine a body orientation, no less than three markers must be fixed with the segment, $n \geq 3$, and the markers must not lie on a straight line. The coordinates of the markers are measured with inevitable errors, which are customarily called *noise*. The noise may be caused by skin marker displacement with regard to the bony skeleton, imprecise optical measurements, or other factors. When a body position is calculated from coordinates that have been affected by noise, the error extends. The error can be minimized, however, if proper procedures are employed. In experimental research, two main approaches have been commonly practiced:

1. The noise is ignored; only three markers per segment are used.
2. The error due to noise is minimized; more than three markers are used.

In the first case, the researcher assumes the markers do not change their positions with regard to the body segment to which they are attached. Hence, their local coordinates are constant and can be written as a 3×3 matrix, $[P_L]$. To determine the rotation matrix (determining a translation given by vector $\mathbf{L}_G$ is an elementary operation), the global coordinates of the three markers are recorded. Then, for each marker $P^{(i)}$ ($i = 1, 2, 3$) a difference, $\mathbf{P}_M^{(i)} = \mathbf{P}_G^{(i)} - \mathbf{L}_G$, is calculated, where $\mathbf{P}_G^{(i)}$ is a position vector of the point $P^{(i)}$ in the global reference frame G. Elements of vector $\mathbf{P}_M^{(i)}$ are coordinates of a given marker in the moving reference system, i.e., in the local coordinate system that moves in translation together with the body segment but maintains a constant orientation in space, the reference axes being parallel to the axes of the global system. The coordinates of all the markers in the moving system are represented by the 3×3 matrix $[P_M]$. The moving system and somatic system differ only in attitude and, hence, equation 1.11 can be applied. Because the equation holds for any point on the body, it holds for the markers and can be rewritten as

$$[P_M] = [R][P_L] \qquad (1.42)$$

The rotation matrix [R] can be found from

$$[P_M] = [R][P_L] \qquad (1.43)$$

In the second case, the accuracy is improved by increasing the number of markers fixed to a body segment and by estimating body position parameters with optimization techniques. When more than three markers are used, matrix $[P_L]$ is not a square matrix and equation 1.43 cannot be directly applied. An optimal solution for the rotation matrix [R] can be found either by using the least-squares approach or by employing the so-called Moore-Penrose pseudoinverse matrices. In particular, determination of the rotation matrix [R] and the translation vector $\mathbf{L}_G$ can be achieved by minimizing

$$\frac{1}{n} \sum_{i=1}^{n} \left([R] P_L^{(i)} + \mathbf{L}_G - \mathbf{P}_G^{(i)} \right)^T \left([R] \mathbf{P}_L^{(i)} + \mathbf{L}_G - \mathbf{P}_G^{(i)} \right) \rightarrow \min \qquad (1.44)$$

where $\mathbf{P}_L^{(i)}$ and $\mathbf{P}_G^{(i)}$ are the coordinates of point $P^{(i)}$ in the local and global frames, and n is the number of nonlinear points measured in both reference frames (n > 3). Various techniques for solving equation 1.44 have been suggested in the literature. Similar approaches have been developed for the Euler's angles and helical axes methods.

When Moore-Penrose matrices are used for finding an optimal solution (these matrices are explained in Chapter **3**, Section **3.1.1.3**), the rotation matrix [R] is determined from the following equation:

$$[R] = [P_G][P_L]^+ = [P_G][P_L]^T \left([P_L][P_L]^T \right)^{-1} \qquad (1.45)$$

where $[P_G]$ and $[P_L]$ are the $3 \times n$ matrices of the markers' coordinates in the global and local frame, and $[P_L]^+ = [P_L]^T ([P_L][P_L]^T)^{-1}$ is a pseudoinverse of $[P_L]$. Equation 1.45 is used instead of equation 1.42 when the number of markers exceeds three.

1.3 THREE-DIMENSIONAL REPRESENTATION OF HUMAN MOVEMENT: EYE MOVEMENT

Representative publication: Carpenter, 1977.

Human movements are essentially three dimensional. In the foregoing paragraphs, it has been shown that three-dimensional position and displacement can be described in various ways. When studying inanimate objects, the researcher prefers the method that is most convenient for him or her, that is, for the observer. However, when the object of interest is motor control, it is important to uncover the representation used by CNS, that is, by the performer. Systems of coordinates used by scientists and those used by nature may be different. For example, the CNS may use only planar movements, although, mechanically, three-dimensional movements are permissible. The difference between the "engineering" (the observer's) and "human scientist" (the performer's) approach should be taken into account.

Movement of the eyes is a good example of the kinematic problems the biomechanist must tackle. Study of eye movement is an advanced field of research concerning human motion. Only a short synopsis of the issue is presented here. When considering eye movement, we highlight first how the movement is described, and then how the eye really moves.

1.3.1 Eye Orientation

An eyeball is capable of rotatory movements around three orthogonal axes and, when considered together with the eye socket, is kinematically analogous to a spherical ball-and-socket joint. Under ordinary conditions of normal vision, eye translation is negligibly small and is not considered here. For the following discussion we assume that (a) the eye rotates about an unmovable center, and (b) the center of rotation coincides with the geometric center of the eye globe. These assumptions are very close to reality.

To define an eye position we need two systems of coordinates: the global system, fixed with the head, and the local system, fixed with the eyeball. The global, or head-centered, reference system is usually defined in the following way. The line joining the centers of rotation of the two eyes, called a *base line*, is taken as the first reference axis (X). For an erect subject looking at the horizon, the base line is horizontal. In this posture, the second axis (Y) is also horizontal and perpendicular to the base line, and the third axis (Z) is vertical and perpendicular to the first two. The frontal vertical plane through the center of rotation is called *Listing's plane* (Figure 1.33).

The local, or eye-centered, system of coordinates (x,y,z) is defined in a similar manner. When the head is held erect and the eye is looking straight ahead,

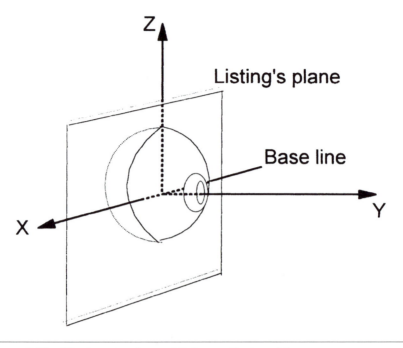

Figure 1.33 Head-centered, or orbital, coordinates of the eye.

the axes of the local system x,y,z coincide with the axes of the global system X,Y,Z. The horizontal plane through the eye center of rotation is called the *retinal horizon*. This is the xy plane of the local system, which moves with the eye. The vertical plane through the center of rotation, which is the yz plane of the eye-centered system, is the *retinal vertical*. Normally, both eyes are directed to fixate on the same external point. The line drawn from the point of fixation to the eye's center of rotation is called the *line of fixation,* or the *line of sight.* The axis (y) of the local system is directed along this line. The line of fixation may be turned in "up-down" and "left-right" directions. The plane containing both the object of regard, at which the axis of fixation is directed, and the base line is called a *plane of fixation.* Note that the plane of fixation is aligned with the base line and the retinal horizon is fixed with the eye and moves with it. Hence, when the head does not move, the plane of fixation may be tilted, and the retinal horizon may be tilted and twisted. It is important to understand that the plane of fixation does not belong to either the global system of coordinates, X,Y,Z, or to the local system, x,y,z. This is the plane X,y of an "intermediate" reference frame X,y,z. When the head is erect and stationary, the plane of fixation X,y can revolve around axis X but not around axis y. The angle between the retinal horizon and the plane of fixation is called the *angle of actual torsion,* or *cyclorotation.*

Geometrically, the eye actual torsion is similar to the *twist* of human body segments. (For human body segments other than the eye, the term "twist" denotes rotation about the longitudinal axis of the segment and the term "torsion" is the rotation about an axis fixed in the space. The axis of torsion coincides with the longitudinal axis of the segment in a reference, or starting, position.) The angle between the retinal horizon and the real horizon of external space, or between the retinal vertical and the vertical (the line of gravity), is called the *angle of torsion*, or *false torsion*. An eye position is customarily described by the gaze direction and the torsion. The gaze direction characterizes where the eye is pointing, the attitude of the line of sight. To specify the gaze direction only two parameters are needed. The cyclotorsion determines the eye rotation (twist) along the line of sight without modifying its orientation.

The eye positions are then classified as

1. the primary position, at which the axes of the global and local system coincide. The angular coordinates of the eye in the primary position are (0,0,0).

2. the secondary positions, at which the eye can arrive after rotation around either the horizontal or vertical axis. After such a rotation, for example, upward vertical rotation, the horizontal axis x is in the same position, but the axis of fixation (y) and the axis that was oriented vertically in the primary position (z) is tilted in the vertical plane. The amount of the false torsion at secondary positions is zero: an eye-fixed retinal vertical and the earth-related vertical are parallel.

3. the tertiary positions, or all other positions. At the tertiary positions, the false torsion is nonzero: an eye-fixed vertical direction is not the same as the earth-vertical direction. The false torsion is seen even when the eye does not rotate about the line of sight (this phenomenon will be treated in more detail later).

In modern literature, the distinction between the primary position and the reference position is made. The primary position is defined, according to Listing's law (described in Section **1.3.2**), as a unique position from which any secondary or tertiary positions can be reached through rotation around an axis in Listing's plane. The reference position is the position at which the eye angular coordinates are (0,0,0). In this book, we follow a traditional approach and simplify the situation by assuming that the primary position and the reference position coincide.

Rotations in three-dimensional space are not commutative. Therefore, the eye attitude after consecutive rotations about different axes depends on the sequence of rotations. If the eye rotates first around the vertical axis Z and then around another axis, the eye position will not be the same as in the case when the order of rotations is reversed. To specify a tertiary position, the sequence of rotations should be defined beforehand. Two conventions, introduced in the middle of the 19th century, are most broadly used: Fick's convention and Helmholtz's convention. With Fick's convention, the horizontal position (*longitude*) is specified first, and the vertical component (*latitude*) second. In the Helmholtz's convention, the *elevation*, i.e., the vertical rotation of the line of sight and the plane of regard, is initially specified. Then the eye is turned sideways in the plane of regard (*azimuth*). In both conventions, the second rotation is considered to occur relative to the axis fixed with the eye (x or z) rather than relative to the head-centered axes, X or Z. With this approach, the eye is treated as if it is moved in nested gimbals (Figure 1.34), and both the Fick's and Helmholtz's systems can be regarded as examples of the Euler's-Cardan's angles discussed previously.

The main difference between the systems is the hierarchy of the nesting of the axes of rotation. The Fick's gimbal has a horizontal axis nested within a vertical axis. The vertical axis is fixed relative to the head, which is considered unmovable, and does not tilt. During horizontal movements, the horizontal axis rotates with the gimbal's pointing direction. Any subsequent vertical movement occurs around the horizontal axis. As a result of these two rotations, the retinal horizon remains horizontal and parallel to the plane of fixation. Both the actual torsion and the false torsion equal zero. The advantage of Fick's

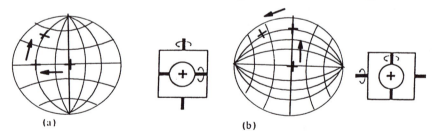

(a)   (b)

Figure 1.34 Fick's and Helmholtz's conventions of specifying tertiary eye positions. The position is specified by the two rotations: horizontal and vertical. A stereographic projection, similar to that in geographical maps, is used. Crosses mark the cornea and represent the eye orientation about the line of sight (cyclorotation). (a) In Fick's convention, the longitude is specified first and the latitude second; the equivalent gimbal is shown on the right. (b) In Helmholtz's convention, first the elevation is specified and then the azimuth; the equivalent gimbal is on the right. In both cases, the second rotation is taken with respect to a local reference axis, which moves with the eye. Note the different final orientation of the cross. The horizontal line of the cross coincides with the vertical meridians existing in that position.
Reprinted from Carpenter, R.H.S. (1988). *Movement of the eyes*. 2d ed. London: Pion Limited.

system is that it maintains lines perpendicular or parallel to the horizon at all gaze directions.

In the Helmholtz's gimbal, the vertical axis is nested within the horizontal. This originates a system of coordinates at right angles to the Fick's system. After the first (vertical) movement, the retinal horizon is still horizontal, but the local axis z is tilted. The second rotation, around the (inclined) axis z, is performed in the plane of fixation. The retinal horizon is still in parallel with the plane of fixation, hence, no actual torsion occurs. This rotation, however, shifts the retinal horizon out of the horizontal. As a result, the retinal vertical is inclined with regard to the external vertical (the line of gravity). A false torsion is observed.

The term "false torsion" has been introduced in the literature to differentiate the eye torsional angle taken with regard to the global reference frame from the eye rotation about the line of fixation. During described maneuvers, the rotations took place around vertical and horizontal axes of the nested gimbals but not around the line of fixation. The order of rotations was, however, different. Notwithstanding the zero actual torsion, the eye arrived at different torsional positions with regard to the environment. We may conclude that the occurrence of the eye false torsion depends on the route by which the eye arrived at the destination.

When an eye position is given, different system of coordinates yield different numerical values. In particular, for the same eye position, the angle of torsion is not the same when measured in Fick's and Helmholtz's systems. There is nothing strange about it. Differences always happen when the position of the same object is defined in various systems of coordinates. For instance, when the position of a point on a plane is described in Cartesian and polar coordinates, the numerical values are different (see Figure 1.25).

The classic Fick's and Helmholtz's coordinates have been widely used in clinical practice and research since the 19th century. However, these systems are not convenient to study the eye movement in depth. Contemporary methods of registering eye movement lend themselves more naturally to using rotation vectors. The rotation vector is parallel to the axis of rotation, and its length is a function of the angular displacement (Figure 1.35).

The rotation vectors provide computational efficiency, they do not require arbitrary preselection of the coordinate axes and the order of rotation, and they permit easy graphical representation.

1.3.2 Motions Actually Made by the Human Eye (Donders's Law and Listing's Law)

The eye is mechanically free to assume any angular position within the range of the motility within the orbit. The eye muscles can turn the eye around any axis. However, under the conditions of normal vision, the eye does not per-

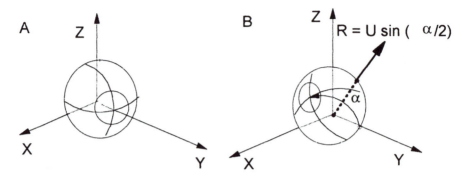

Figure 1.35 Representation of the eye position with the rotation vector. A. The eye is in the reference position. B. Eye position with the gaze to the upper right. The rotation vector is along an imaginary axis, about which the eye could be rotated straight from the reference position. The angle of rotation is α. The length of the rotation vector is proportional to $\sin(\alpha/2)$.

form all the movements it is capable of executing. Two famous laws govern the eye movement.

Donders's law (1847) states that the angle of torsion is constant for any given orientation of the line of sight. In other words, the angle of torsion is a function of the angles determining the gaze direction, for example, the angle of elevation and the angle of azimuth (in Helmholtz's convention). The angle of torsion does not depend on the way the eye travels to the position of interest. In ordinary people, the angle of torsion cannot be changed voluntarily without special long-term training. Donders's law is equivalent to the statement that the eye is restricted to a two-dimensional movement rather than to a three-dimensional one. The eye does not make use of all three possible rotations. Only two parameters are needed to describe an eye position. Hence, the eye position can be conveniently represented by a two-dimensional graph. The law is not a consequence of mechanical constraints imposed on the eye and extraocular muscle mechanics.

Both mechanically and anatomically, the eye is permitted to rotate about three axes. The rotations are coupled, however, in such a way that the gaze direction completely determines the magnitude of cyclotorsion. Donders's law is a result of a specific pattern of muscle activity. This is the law of motor control rather than the law of mechanics. The law is an example of a synergy used by the CNS to reduce the number of available patterns of muscle activity (the so-called *Bernstein's problem,* which will be enunciated exactly later). Because of Donders's law, at a given head position and gaze direction, an external object is always projected on the same elements of the retina. Listing's law further specifies the eye movement. Donders's law can be considered a corollary of Listing's law: if Listing's law holds, Donders's law holds too.

Listing's law of ocular movements (1855) may be stated as follows: (a) the eye assumes only those positions that may be reached by a single rotation from the primary position and (b) the axis of this rotation lies in the plane perpendicular to the gaze direction in the primary position. The defined plane is called Listing's plane (Figure 1.36). The law does not mean that the eye really turns around a fixed axis. On the contrary, the travel can be done in any way. However, of all the possible eye positions, only those that satisfy Listing's law are assumed. To visualize Listing's law, imagine a wall clock. The plane of the clock face acts as Listing's plane. A hand of the clock changes orientation. Now, further imagine rotation of a body (the eyeball) around the longitudinal axis of the hand. Evidently, only two parameters instead of three are needed to

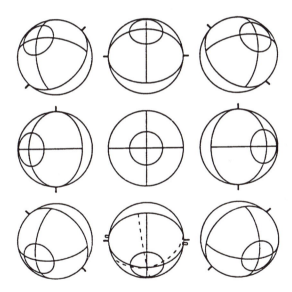

Figure 1.36 Listing's law for the eye. The primary position is in the center. Listing's plane is the plane of the paper. The eye orientations drawn in solid lines conform with Listing's law. They can be attained by a direct rotation from the primary position around an axis lying in Listing's plane. The orientation drawn in dashed lines at the bottom center violates Listing's law. The corresponding axis of rotation, drawn as a dashed line, tilts out of Listing's plane. Reprinted from *Vision Research, 30,* Tweed, D., & Villis, T., Geometric relations of eye position and velocity vectors during saccades, 111-127, 1990, with kind permission from Elsevier Science Ltd, The Boulevard, Langford Lane, Kidlington 0X5 1GB, UK.

describe the "eye" position: the orientation of the rotation axis (the hand of the clock) and the angular displacement around the axis.

When the validity of Listing's law is tested experimentally, it is convenient to use rotation vectors. They are defined with regard to the reference orientation (primary position) and, thus, are the position vectors. If the law holds, all the vectors of rotation lie in a flat plane. An example is shown in Figure 1.37.

Similar to Donders's law, Listing's law is upheld by the CNS. It does not hold under all conditions; for example, it fails during sleep. The law is an example of a muscle synergy. There are six extraocular muscles that can rotate the eye around any axis. Both mechanically and anatomically the desired gaze direction can be achieved by various patterns of eye movement and muscle activity. However, only those that satisfy Listing's law are used. Other combinations of muscle innervation are prohibited. Hence, the dimensionality of the six-muscle system is reduced to two. In the current literature, both Donders's and Listing's laws are being actively discussed. Two issues have been addressed: first, what is the purpose of these laws, and second, how are they implemented? In both cases, kinematic analysis provides a factual basis for the discussion.

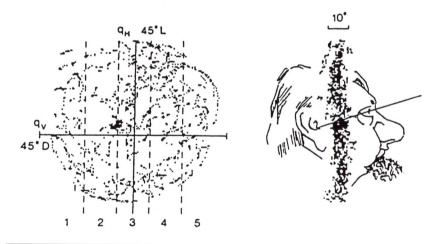

Figure 1.37 The projections of the rotation vectors on the frontal (left) and sagittal plane. The eye position was registered with an electromagnetic device (eye-coil magnetic field technique). The subjects were required to change the gaze direction following the targets arrayed homogeneously throughout the visual space. The rotation vectors were determined for each eye angular position. In accordance with Listing's law, the rotation vectors lie in a flat plane. The plane is, however, rather thick. Hence, the eye follows Listing's law only approximately.

Adapted from *Vision Research, 30,* Tweed, D., & Villis, T., Geometric relations of eye position and velocity vectors during saccades, 111-127, 1990, with kind permission from Elsevier Science Ltd, The Boulevard, Langford Lane, Kidlington 0X5 1GB, UK.

1.3.3 Rotation Surfaces: The Laws Obeyed by the Pointing Head and Arm Movements

Representative papers: Hore, Watts, & Vilis, 1992; Miller, Theeuwen, & Gielen, 1992; Tweed & Vilis, 1992.

Kinematically, the eye is free to rotate about three orthogonal axes. Under ordinary conditions, however, the eye rotates only around the axes that lie in one flat plane. Hence, the eye does not make use of its potential mobility. Many of the eye's orientations, which are mechanically attainable, are never assumed in real life. A certain analogy to this fact is seen in the gaze-directing head movements and the pointing movements performed with a straight arm. Both the head and the arm do not usually assume all of the kinematically possible orientations. The following issues have been addressed in experimental research:

- Donders's law for pointing and gaze-directing movements. When the arm or head adopts any single *pointing* direction, does it assume a unique three-dimensional orientation? The answer is "yes." Whenever the arm or nose points in a particular direction, the three-dimensional attitude of the arm or head is approximately constant. Thus, Donders's law holds for the pointing head and arm movements.

- The generalized Listing's law. Do the axes of arm or head rotation during pointing movements lie in a single surface? Is this surface a flat plane? If the answer to both questions is "yes," then the pointing arm or head movements obey Listing's law. If the answer to the first question is "yes," but the answer to the second question is "no," then this is an example of the generalized Listing's law. It has been reported that the rotation vectors of the pointing arm or head movements are confined to a specific surface. Hence, the generalized Listing's law is valid for this class of movements. It also seems that the classic Listing's law holds for the head: during gaze-shifting head movements, the rotation vectors lie in a single plane. An important difference exists, however, between the Listing's law for the eye and for head or arm movement. We cannot move the eyes voluntarily in another way, at least not without special training. We can easily move the head and the arm differently. Nonetheless, we prefer to move them in accordance with the (generalized) Listing's law.

For the pointing arm movements, however, Listing's law holds only over a narrow range of joint motion, about 30°. For a broader range of motion, the surface containing the rotation vectors is not a flat plane; it is curved. This surface is called the *rotation surface*. Hence, Listing's plane is just one example of the rotation surface. For pointing arm movements, the rotation surface is similar to that produced by rotations around axes of a Fick gimbal—the vertical axis is fixed and the horizontal axis moves with the gimbal. In biomechanics, rotation surfaces are a popular tool of kinematic analysis of assemblages of angular movements in space.

1.4 SUMMARY

This chapter deals with the different methods used for describing location and orientation of a body in space. Two approaches have been used in research to describe a human body's orientation in space. In the first, a local system of coordinates (a *somatic system*) is affixed with the body and, then, orientation of the somatic system with regard to the global system is determined. The second ("indirect") approach is based on calculation of the quotient "angular momentum/moment of inertia." In the framework of the indirect approach, the rotation of a body part is considered to be rotation of the whole body. Such an approach may lead to estimations that are psychologically difficult to accept.

When a local frame is defined, its orientation relative to the global frame is described by classic methods of kinematics. The following methods are considered: matrix method, Euler's angles, and the helical method. These methods permit (a) the interpretation of a body position in a given reference system; (b) the definition of the body position, given in one frame, in an alternate reference system; and (c) the description of a body's displacement from one position to another. The relative position of two frames can be viewed as a combination of translation and rotation. If both the translation and rotation are described by one common mathematical operation, the corresponding matrix is called the *homogeneous transform*. Pros and cons of various angular representations used in biomechanical research are discussed.

In Section **1.3**, three-dimensional representation of the eye movement is considered. Helmholtz's and Fick's conventions, used to define eye movement, are highlighted and the main empirical laws governing eye movement, Donders's and Listing's laws, are discussed. Rotation vectors are described and recommended as a convenient tool of analysis. Parallels existing between the eye movement and the pointing head or arm movement are briefly described.

1.5 QUESTIONS FOR REVIEW

1. Name the three main characteristics defining a human body position.
2. What is the difference between the principal somatic system of coordinates and the split-body system?

3. Describe the indirect method of defining human body rotation.

4. Define a homogeneous transformation matrix.

5. A global reference system, G, is fixed within a trunk, which is assumed to be a rigid body. A local reference system, L, is fixed within an upper arm. The attitude of the upper arm with regard to the trunk is described by a 3 × 3 orientation matrix [T]. Solve the following problems (write equations in matrix form):

    a. A position of the tip of the finger in the local system is described by a vector $\mathbf{P}_L$. Find a position of the finger with regard to the trunk.

    b. A position of the tip of the finger in the global system is described by a vector $\mathbf{P}_G$. Find a position of the finger with regard to the upper arm.

    c. After arm rotations, the upper arm assumes a new orientation. The initial position of the upper arm as well as the joint angle changes, e.g., flexion of 45° and internal rotation of 30°, are known. Write an equation to find the final orientation of the upper arm with regard to the trunk.

    d. Both the initial and the final position of the arm with regard to the trunk are registered. Write an equation to find the joint angular displacement, i.e., the final position of the arm with regard to its initial position in terms of the joint angles.

    e. As in case d, both the initial and the final position of the arm with regard to the trunk are registered. Write an equation to find an upper arm displacement in the global coordinate system, in terms of the angles defined in the global reference.

6. The movement of the oar during rowing is usually described with three angles: (1) the oar swing angle in the horizontal plane projection, (2) the angle measured in the vertical plane projection, and (3) the angle of the oar rotation about its longitudinal axes. Can these angles be called Euler's angles? Are they independent? What should be specified explicitly to define the second angle unambiguously?

7. What is the difference between Euler's and Cardan's angles?

8. In the literature, some authors have described Euler's angles as sequence-dependent. Contrarily, other authors have mentioned that the angles are generally independent and, thus, sequence-independent. What is the reason for this controversy? What does the expression "sequence-dependent" really mean?

9. Compare direction and projection angles.

10. Are the joint angles used for clinical description of human body posture—flexion-extension, abduction-adduction, and internal-external rotation—Euler's angles? If they are not, why?

11. Describe the parameters used in the helical method for defining a body position.

BIBLIOGRAPHY

This list mainly includes representative publications in which three-dimensional methods of kinematics have been used for studying human movements.

An, K.-N., & Chao, E.Y.S. (1984). Kinematic analysis of human movement. *Annals of Biomedical Engineering, 12*, 585-597.

An, K.-N., & Chao, E.Y.S. (1991). Kinematic analysis. In K.-N. An, R.A. Berger, & W.P. Cooney III (Eds.), *Biomechanics of the wrist joint* (pp 21-36). New York: Springer-Verlag.

Andersen, R.A., Snyder, L.H., Li, C.-S., & Stricanne, B. (1993). Coordinate transformations in the representation of spatial transformation. *Current Opinion in Neurobiology, 3,* 171-176.

Berme, N., Cappozzo, A., & Meglan, J. (1990). Kinematics. In N. Berme & A. Cappozzo (Eds.), *Biomechanics of human movement: Applications in rehabilitation, sports and ergonomics* (pp. 89-102). Worthington, Ohio: Bertec Corporation.

Bottema, O., & Roth, R. (1979). *Theoretical kinematics.* New York: Dover Publications, Inc.

Carpenter, R.H.S. (1977). *Movement of the eyes.* London: Pion Ltd.

Challis, J.H. (1995). A procedure for determining rigid body transformation parameters. *Journal of Biomechanics, 28,* 733-737.

Chao, E.Y.S., Rim, K., Smidt, G.L., & Johnston, R.C. (1970). The application of 4×4 matrix method to the correction of the measurements of hip joint rotations. *Journal of Biomechanics, 3,* 459-471.

Dapena, J. (1981). Simulation of modified human airborne movements. *Journal of Biomechanics, 14,* 81-89.

Drerup, B., & Hierholzer, E. (1987). Automatic localization of anatomical landmarks on the back surface and construction of a body-fixed coordinate system. *Journal of Biomechanics, 20,* 961-970.

Engin, A.E. (1980). On the biomechanics of the shoulder complex. *Journal of Biomechanics, 13,* 575-590.

Gervais, P., & Marino, G.W. (1983). A procedure for determining angular positional data relative to the principal axes of the human body. *Journal of Biomechanics, 16,* 109-113.

Haslwanter, T. (1995). Mathematics of three-dimensional eye rotation. *Vision Research, 35,* 1727-1739.

Hore, J., Watts, S., & Vilis, T. (1992). Constraints on arm position when pointing in three-dimensions: Donders' law and the Fick gimbal strategy. *Journal of Neurophysiology, 68* (2), 374-383.

Horn, B.K.P. (1987). Closed-form solution of absolute orientation using unit quaternions. *Journal of the Optical Society of America, 4,* 629-642.

Kinzel, G.L., Hall, A.S., & Hillberty, B.M. (1972). Measurement of the total motion between two body segments—1. Analytical development. *Journal of Biomechanics, 5,* 93-105.

Mercadante, L.A., Brenzikofer, R., & Cuhna, S.A. (1995). Methodology for tridimensional representation, orientation, and position of the human body. In: K. Häkkinen, K.L. Keskinen, P.V. Komi, & A. Mero (Eds.). *XVth Congress of the International Society of Biomechanics, Abstracts* (pp. 620-621). Juväskylä, Finland.

Miller, L.E., Theeuwen, M., & Gielen, C.C. (1992). The control of arm pointing movements in three dimensions. *Experimental Brain Research, 90,* 415-426.

Misslisch, H. (1994). Rotational kinematics of the human vestibuloocular reflex. III. Listing's law. *Journal of Neurophysiology, 72,* 2490.

Opstal, J. van (1993). Representation of eye position in three dimensions. In A. Berthoz (Ed.), *Multisensory control of movement* (pp. 27-41). Oxford, United Kingdom: Oxford Science Publication.

Panjabi, M.M., Krag, M.H., & Goel, V.K. (1981). A technique for measurement and description of three-dimensional six degree-of-freedom motion of a body joint with an application to the human spine. *Journal of Biomechanics, 14,* 447-460.

Paul, J.P. (1990). Definition of the neuromuscular skeletal system for locomotion analysis. In *Models: Connection with experimental apparatus and relevant DSP techniques for functional movement analysis* (CAMARC, Computer-Aided Movement in a Rehabilitation Context, deliverable F), (pp. 35-42). Ancona, Italy: University di Ancona.

Pellionisz, A. (1984). Coordination: A vector-matrix description of transformations of overcomplete CNS coordinates and a tensorial solution using the Moore-Penrose inverse. *Journal of Theoretical Biology, 110,* 353-375.

Pellionisz, A.J., Le Goff, B., & Laczkó, J. (1992). Multidimensional geometry intrinsic to head movements around distributed centers of rotation: A neurocomputer paradigm. In A. Berthoz, W. Graf, & P.P. Vidal (Eds.). *The head and neck sensory motor system* (pp. 158-167). New York: Oxford University Press.

Ramakrishnan, H.K. & Kadaba, M.P. (1991). On the estimation of joint kinematics during gait. *Journal of Biomechanics, 24,* 969-977.

Reynolds, H.M., & Hubbard, R.P. (1980). Anatomical frames of references and biomechanics. *Human Factors, 22,* 171-176.

Robinson, D.A. (1982). The use of matrices in analyzing the three-dimensional behavior of the vestibulo-ocular reflex. *Biological Cybernetics, 46,* 53-66.

Saltzman, E. (1979). Levels of sensorimotor representation. *Journal of Mathematical Psychology, 20,* 91-163.

Shapiro, R. (1978). Direct linear transformation method for three-dimensional cinematography. *The Research Quarterly, 49,* 197-205.

Siegler, S., Chen, J., & Schneck, C.D. (1988). The three-dimensional kinematics and flexibility of the human ankle and subtalar joints—Part 1: Kinematics. *Journal of Biomechanical Engineering, 110,* 364-373.

Siegler, S., Chen, J., Selikar, R., & Schneck, C.D. (1990). A system for investigating the kinematics of the human ankle and subtalar joint based on Rodriguez' formula and the screw axes parameters. In N. Berme & A. Cappozzo (Eds.), *Biomechanics of human movement: Applications in rehabilitation, sports and ergonomics* (pp. 278-283). Worthington: Bertec Corporation.

Soechting, J.F., Buneo, C.A., Herrmann, U., & Flanders, M. (1995). Moving effortlessly in three dimensions: Does Donder's law apply to arm movement? *Journal of Neuroscience, 15,* 6271-6280.

Soechting, J.F., & Flanders, M. (1992). Moving in three-dimensional space: Frames of reference, vectors and coordinate systems. *Annual Review of Neuroscience, 15,* 167-191.

Spoor, C.W. (1984). Explanation, verification, and application of helical-axis error propagation formulas. *Human Movement Science, 3,* 95-117.

Spoor, C.W., & Veldpaus, F.E. (1980). Rigid body motion calculated from spatial coordinates of markers. *Journal of Biomechanics, 13*(4), 391-393.

Tupling, S.J., & Pierrynowski, M.R. (1987). Use of Cardan angles to locate rigid bodies in three-dimensional space. *Computer Methods & Programs in Biomedicine, 25,* 527-532.

Tweed, D., & Vilis, T. (1987). Implications of rotational kinematics for the oculomotor system in three dimensions. *Journal of Neurophysiology, 58,* 832-849.

Tweed, D. & Vilis, T. (1990). Geometric relations of eye position and velocity vectors during saccades. *Vision Research, 30,* 111-127.

Tweed, D., & Vilis, T. (1992). Listing's law for gaze-directing head movements. In A. Berthoz, W. Graf, & P.P. Vidal (Eds.), *The head-neck sensory and motor systems* (pp. 387-391). New York: Oxford University Press.

Veeger, H.E.J., van der Helm, F.C.T., & Rozendal, R.H. (1993). Orientation of the scapula in a simulated wheelchair push. *Clinical Biomechanics, 8,* 81-90.

Veldpaus, F.E., Woltring, H.J., & Dortmans, L.J.M.G. (1988). A least-squares algorithm for the equiform transformation from spatial co-ordinates. *Journal of Biomechanics, 21,* 45-54.

Wei, S.-H., McQuade, K.J., & Smidt, G.L. (1993). Three-dimensional joint range of motion measurements from skeletal coordinate data. *Journal of Orthopaedics and Sports Physical Therapy, 18,* 687-691.

Woltring, H.J. (1991). Representation and calculation of 3-D movement. *Human Movement Science, 10,* 603-616.

Woltring, H.J., & Huiskes, R. (1985). A statistically motivated approach to instantaneous helical axis estimation from noisy, sampled landmark coordinates. In D.A. Winter, R.M. Norman, R.P. Wells, K.C. Hayes, & A.E. Patla (Eds.), *Biomechanics IX-B* (pp. 274-279). Champaign, IL: Human Kinetics.

Woltring, H.J., Huiskes, R., Lange, A. de, & Veldpaus, F.E. (1985). Finite centroid and helical axis estimation from noisy landmark measurements in the study of human joint kinematics. *Journal of Biomechanics, 18,* 379-389.

Wu, G. (1995). Kinematics theory. In R.L. Craik & C.A. Oatis (Eds.), *Gait analysis* (pp. 159-182). St. Louis: Mosby.

Yeadon, M.R. (1990a). The simulation of aerial movement. I. The determination of orientation angles from film data. *Journal of Biomechanics, 23,* 59-66.

Yeadon, M.R. (1990b). The simulation of aerial movement. III. The determination of the angular momentum of the human body. *Journal of Biomechanics, 23,* 75-83.

Youm, Y., & Yoon, Y.S. (1979). Analytical development in investigation of wrist kinematics. *Journal of Biomechanics, 12,* 613-621.

Zatsiorsky, V.M. & Aleshinsky, S.Y. (1975). Simulation of human locomotion in space. In P.V. Komi (Ed.), *Biomechanics V-B* (pp. 387-398). Baltimore: University Park Press.

2
CHAPTER

KINEMATIC GEOMETRY OF HUMAN MOTION: BODY POSTURE

This chapter is about methods used to define a body posture (joint configuration). The position of a body segment can be defined relative to

- a global system of coordinates (e.g., segment coordinates, absolute coordinates, World coordinates),
- a somatic system attached to the whole body (somatic coordinates), or
- an adjacent segment (joint coordinates).

Mixed systems are also used. In such a system the orientation of one body segment is determined through the global reference system, and the orientation of the second segment is realized through the joint coordinates. For example, the thigh position measured relative to the vertical can be plotted versus the shank position taken relative to the thigh. A system of coordinates is often referred to as a *space*. Hence, the expression "joint space" is a synonym for "joint coordinates."

Section **2.1** addresses the techniques used for describing relative orientation of adjacent body segments. It starts by explaining the difference between the technical and somatic reference systems; then the various techniques used for describing joint orientation are discussed, particularly the clinical reference system, globographic presentation, segment coordinate system, and—the most popular approach—joint rotation convention. Kinematic chains are discussed in Section **2.2**. In this section, the notion of degrees of freedom and the mobility of kinematic chains are discussed and various modifications of the Gruebler's formula are used to determine the total number of degrees of freedom of kinematic chains. This section also concentrates on open kinematic chains and end-effector mobility. Kinematic models and mobility of the human body are also described, as are constraints on human movement and the position of kinematic chains. In the discussion of kinematic chain position, two simple chains are considered: a two-link planar and three-link planar. Multilink chains are also addressed and transformation analysis of kinematic chains is highlighted. In this Section, I will also consider the Denavit-Hartenberg convention broadly used for describing kinematic chains. Section **2.3** is devoted to several applications of kinematic geometry in biomechanics. The discussion focuses on the internal representation of the immediate extrapersonal space and body posture.

2.1 JOINT CONFIGURATION

Representative papers: Kinzel, Hall, & Hilberry, 1972; Woltring, 1991.

Generally, the relative motion of one human body segment with regard to an adjacent one is a combination of translation and rotation. However, if the object of the study is gross human motion, such as walking or a gymnastic exercise, the translation can be disregarded due to its small magnitude and joint motion is analyzed as pure rotation. In Chapters **2** and **3**, the joint motions are considered pure rotations around fixed axes. The following simplifying assumptions are repeatedly made:

- If the joint exhibits rotation around more than one axis, the axes of rotation intersect at one point.
- The axes of rotation coincide with the joint reference system of coordinates.
- The axes of rotation coincide with the anatomic axes (for instance, the flexion-extension axis coincides with the frontal axis passing through the joint center).

More complex forms of joint motion, as well as specific systems used for particular joints, will be described in Chapters **4** and **5**.

2.1.1 Technical and Somatic Systems

Representative papers: Cappozzo & Gazzani, 1990; Söderkvist & Wedin, 1993.

Methods used to describe joint configuration or body posture differ primarily in the manner in which the local reference frame is fixed within a body segment. All local systems can be classified as either technical or somatic.

Technical reference systems are fixed with technical devices, such as goniometers or skin markers. For instance, if goniometers or other transducers are used, they are located externally on the body surface rather than fixed directly at the origin of the segment's local reference frame. Technical reference systems are related to somatic systems by translation and rotation. Experimental data registered in a technical system of coordinates are subsequently expressed in the corresponding somatic frame. The 4 × 4 transformation matrix method, described in Section **1.2.5.1**, can be used for this purpose.

The main problem associated with technical reference systems is accuracy. The errors made when recording human body movement are divided into *errors of measurement,* for example skin marker position in space, and *errors of transformation,* which occur when data are transformed from a technical reference frame (skin markers) to a somatic reference frame (body segment position). To minimize the errors of measurement, various smoothing procedures, which are not described in this textbook, are used. The errors of measurement depend on, among other factors, the configuration of the markers. For instance, if all of the skin markers are located along a straight line, accuracy decreases. It has been shown that accuracy depends on the *condition number* of the configuration. To define the condition number, assume that we have n skin markers on a rigid body. Determine the projected distance of each landmark from the geometric center of the configuration along axes x, y, and z. The sum of the squared distances taken along axis x is σ_x^2. The axes are selected in such a way that $\sigma_z^2 \geq \sigma_x^2 \geq \sigma_y^2$. For an absolute error, i.e., for the error in determining body position in an external frame, the condition number is determined by

$$k_A = \left(\sigma_x^2 + \sigma_y^2\right)^{-1/2} \tag{2.1}$$

The condition number k_A depends on the size of the body segment because it is impossible to place the markers far apart on a small body segment. Therefore, it seems reasonable to normalize the condition number with regard to the body segment size. To do that, let us first define a 3 × n matrix, [A], with the distances of the skin markers to the center of configuration as the elements. Then, the relative condition number is defined as

$$k_R = \frac{\|A\|}{\left(\sigma_x^2 + \sigma_y^2\right)^{1/2}} \tag{2.2}$$

where ‖A‖ is the *Frobenius norm* of matrix [A]. Matrix [A] is found by

$$\|A\| = \left(\sum_{i,k} a_{i,k}^2\right)^{1/2} \tag{2.3}$$

To calculate the Frobenius norm, the sum is computed over all the elements of the matrix [A]. A practical recommendation is to place the markers as far apart as possible.

The errors of transformation depend on two factors: first, accuracy of the relative location of the skin markers and bony landmarks, and second, displacement of the skin markers during human motion. Location of the technical reference frame with regard to the body, especially the skeletal bones, changes during human movement. This is due to movement of the skin and underlying tissues relative to the bony landmarks. There are two sources for such displacement: angular joint motion and inertial forces caused by high accelerations and impacts, like those caused by heelstrikes in walking. In experimental settings, the displacement of skin markers and transducers should be reduced as much as possible or the errors caused by this displacement should be corrected mathematically.

Somatic systems are classified into three types based on (a) anatomic landmarks; (b) mechanical points and axes; and (c) a combination of (a) and (b) used to define the reference frame. Results may differ substantially when alternative systems are employed (Figure 2.1). This is because the reference frames are slightly rotated and translated relative to each other. Also, a movement pattern that is considered as pure rotation around a reference axis in one frame may be expressed as a complex general motion in another local system.

2.1.2 The Clinical Reference System

Representative publication: *American Academy of Orthopaedic Surgeons,* **1965.**

This system is most often used by practitioners and is commonly taught in anatomy and kinesiology classes. The clinical system is defined as follows:

1. Sagittal, frontal, and transverse planes are defined when the body is in the anatomic position (see "Cardinal Planes and Axes of the Human Body" on page 13 and Figure 1.9).

2. The components of segmental motion with regard to the anatomic position are defined as

   a. motion of the segment in a sagittal plan (flexion-extension),

   b. motion of the segment away from or toward the midsagittal plane (abduction-adduction), and

   c. rotation about the long axis of the segment (internal-external rotation or supination-pronation; the terminology differs depending on the joint).

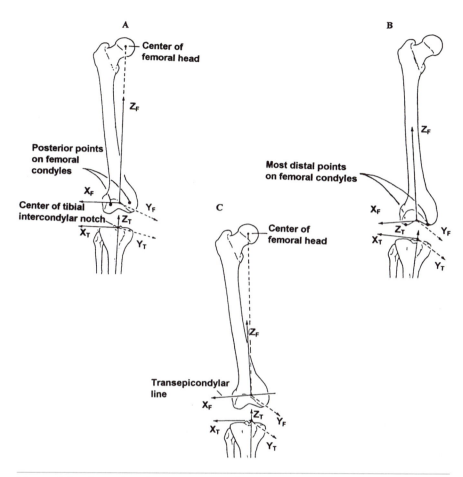

Figure 2.1 Somatic reference systems for a knee joint. A. A mechanically based system. B. This system is based on the anatomic axes of the femur and the tibia. C. This system is based on the transepicondylar line (anatomic) and the tibial mechanical axis. Experimentally measured differences in abduction-adduction at maximal swing phase flexion during walking equal 4° between A and C and 7° between B and C.

A. is revised from Grood, E.S. & Suntay, W.J. (1983). A joint coordinate system for the clinical description of three-dimensional motion: application to the knee. *Journal of Biomechanical Engineering, 105,* 136-144. Adapted by permission from ASME International.

B. is adapted by permission from Lafortune, M.A. (1984*). The use of intra-cortical pins to measure the motion of the knee joint during walking.* Unpublished doctoral dissertation, Pennsylvania State University. Adapted by permission from the author.

C. is adapted from *Journal of Biomechanics, 23,* Pennock, G.R. & Clark, K.J., An anatomy-based system for the description of the kinematic displacements in the human knee, 1209-1218, 1990, with kind permission from Elsevier Science Ltd, The Boulevard, Langford Lane, Kidlington 0X5 1GB, UK.

■ ■ ■ *FROM THE SCIENTIFIC LITERATURE* ■ ■ ■

Converting Data From a Technical to a Somatic System

Source: Chao, E.Y.S. (1980). Justification of triaxial goniometer for the measurement of joint rotation. *Journal of Biomechanics, 13,* 989-1006.

Converting data from a technical to a somatic system may be a laborious problem. Consider, as an example, triaxial goniometers. Ideally, the axes of the goniometer should coincide with the joint rotation axes (Figure 2.2). If the goniometer assembly could be placed at the center of the joint, it would measure the joint angles without distortion. Placing the goniometer in the center of the joint is technically impossible, however. The goniometer assembly is fixed to the side of the joint. The axes of the goniometer, especially the axis of internal-external rotation, do not coincide

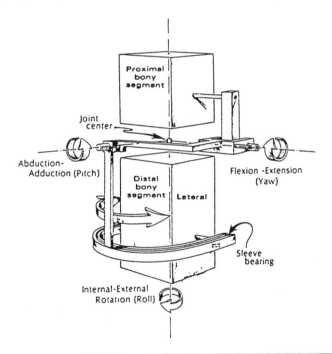

Figure 2.2 A schematic diagram of a skeletal joint and its three-dimensional rotation as measured by an ideal triaxial goniometer.

Reprinted from *Journal of Biomechanics, 13,* Chao, E.Y.S., Justification of triaxial goniometer for the measurement of joint rotation, 989-1006, 1980, with kind permission from Elsevier Science Ltd, The Boulevard, Langford Lane, Kidlington 0X5 1GB, UK.

with the joint axes of rotation. This arrangement creates error and makes the results of the individual angular measurements interdependent.

For example, if the body segment is undergoing internal rotation, the goniometer measurements of the flexion angle and the abduction angle are not equal to the actual joint angles. This difference is called *cross-talk,* and it can be large (Figure 2.3).

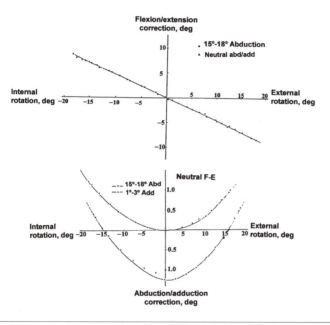

Figure 2.3 The influence ("cross talk") of the internal-external rotation on the recorded values of the flexion-extension angle (upper panel) and the abduction-adduction angle (bottom panel). The actual flexion angle α (not marked on the figure) can be obtained from the goniometer readings as $\alpha = \alpha_{gon} \pm 0.48\psi_{gon}$, where α_{gon} is the flexion angle and ψ_{gon} is the axial rotation measured by the goniometer. The correction pattern for the abduction angle is nonlinear and depends on the initial angle of abduction. The actual

abduction angle β can be calculated as $\beta = \beta_{gon} + c\left(1 - e^{k|\psi_{gon}|}\right)$, where β_{gon}

is abduction measured at the goniometer, $|\psi_{gon}|$ is the absolute value of axial rotation measured at the goniometer, and c and k are constants depending on the initial abduction-adduction angle at the joint.

Reprinted from *Journal of Biomechanics, 13,* Chao, E.Y.S., Justification of triaxial goniometer for the measurement of joint rotation, 989-1006, 1980, with kind permission from Elsevier Science Ltd, The Boulevard, Langford Lane, Kidlington 0X5 1GB, UK.

See **DATA,** *p. 86*

■ ■ ■ **DATA,** *continued from p. 85*

To arrive at these results, the author performed a commendable job. The goniometer linkage and the skeletal joint system were modeled as a seven-bar, nine degree of freedom (DOF), spatial mechanism (Figure 2.4). Denavit-Hartenberg transformation matrices were employed to solve for the unknown joint motion based on the recorded goniometer readings. Both the notion of DOF and the Denavit-Hartenberg convention will be explained later in this chapter (see Section **2.2.1** and Section **2.2.5.3**).

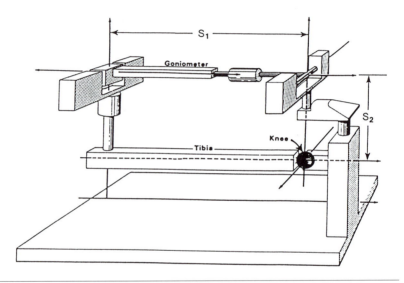

Figure 2.4 The nine-DOF spatial model used to simulate the triaxial goniometer as applied for the measurement of human knee joint motion. The model has eight DOF in rotation and one in translation.

Reprinted from *Journal of Biomechanics, 13,* Chao, E.Y.S., Justification of triaxial goniometer for the measurement of joint rotation, 989-1006, 1980, with kind permission from Elsevier Science Ltd, The Boulevard, Langford Lane, Kidlington OX5 1GB, UK.

These motions are defined with regard to the cardinal anatomic planes. A combination of flexion-extension and abduction-adduction resulting in a circular movement is called *circumduction*.

Included and anatomic joint angles are also defined (Figure 2.5). The *included joint angle* is located between the longitudinal axis of the two segments defining a joint. The *anatomic joint angle* is that angle through which the joint would have to be moved to take it from the anatomic position to the position of interest. From a purely geometric standpoint, the included angles can be viewed as *internal* and the anatomic angles as *external* joint angles.

The clinical reference system provides anatomically meaningful definitions of main segmental movements. The system is convenient when joint motion is performed from a standard anatomic, or neutral, position. However, when a joint rotation commences from a nonneutral configuration, the clinical system is not suitable. Also, it is not an appropriate format to describe complex movement patterns. This limitation can be illustrated with the following example. Imagine a subject who performs three shoulder joint motions in succession (a reader may perform this simple experiment). The joint movements begin with the shoulder in a dependent posture and are (1) flexion of 90°; (2) horizontal extension (abduction) of 90°; and (3) adduction. After such movements, the subject finds his or her hand in a pronated position (anatomic posture), although pronation was not performed. This phenomenon is called Codman's paradox (after E.A. Codman, who described it in 1934). This phenomenon occurs because rotations in the clinical system are not defined in accordance with the requirements of kinematics.

To define a joint rotation in three-dimensional space, the axis of rotation must be explicitly defined. In the clinical system, the rotations, especially abduction-adduction, are defined relative to the plane rather than the axis. When a rotation is defined relative to the plane, the axis of rotation may vary. For example, when an arm is flexed at the shoulder joint, the movement away from the midsagittal plane is definitely performed relative to a different axis than when the arm is oriented vertically. Additionally, even when the axes of rotation are indicated, they are described ambiguously, that is, without regard

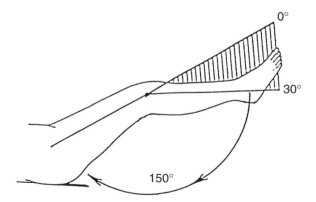

Figure 2.5   Included and anatomic angles used to designate joint configuration. The shaded area (= 30°) represents the anatomic angle. The included angle equals 150° in this example.

to the requirements of a valid kinematic description. For example, in the framework of the clinical reference system flexion is defined as the rotation about a joint frontal axis. There is no problem with such a definition when the motion begins with the arm at the side of the body. Yet, when the arm is adducted 90°, rotation around the frontal axis can barely be called flexion anymore; rather, it is internal or external rotation.

From the foregoing examples, we can see that three-dimensional joint motion cannot, in principle, be described unequivocally by such a triad as "flexion-extension, abduction-adduction, and internal-external rotation." To describe movement unambiguously, a reference system must be defined according to the requirements of kinematics.

2.1.3 Globographic Representation

With a globographic representation, an attitude of a body segment is given in spherical coordinates. This representation is used for descriptions of a spherical joint position; for example, the position of the upper arm with regard to the trunk. When a proximal coordinate system originating in the shoulder joint is fixed with the trunk, the longitudinal axis of the upper arm passes through the origin. The angular position of this line is shown on a globe with meridians and parallels and can be described by two angles (Figures 2.6 and 2.7). Such a representation is free from the artificiality of the cardinal anatomic planes. The angles are calculated along meridians (angle α, for motion in the vertical plane) and parallels (angle β, for horizontal circular movement of the representative point). A unit vector $\mathbf{V}$, given by its spherical coordinates, α and β, can be written in a Cartesian system as

$$\mathbf{V} = \begin{bmatrix} \sin \alpha \sin \beta \\ \cos \alpha \\ \sin \alpha \cos \beta \end{bmatrix} \qquad (2.4)$$

The globographic angles are different from Euler's angles because the latter are defined as consecutive rotations. Thus, the second and third rotations are performed about the new axes in the movable coordinate systems. Globographic angles are calculated about a fixed system of coordinates; α and β together define a single rotation about an oblique (usually) axis through the joint center. When a twist (third) angle is calculated, the reference (or zero) position is different when Euler's or globographic angles are used. When the globographic method is used to describe consecutive displacements, an *induced twist* may be seen; final values of the twist are different when the body segment moves directly from position A to position C compared with when it moves from A to B and then to C. In a globographic system, $A \rightarrow C \neq A \rightarrow B \rightarrow C$.

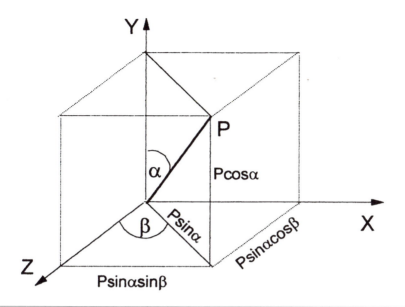

Figure 2.6   Angle convention used in a spherical (globographic) presentation.

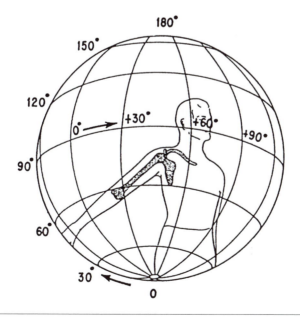

Figure 2.7   Globographic presentation of shoulder movement.

2.1.4 Segment Coordinate Systems

With segment coordinate systems, sagittal, frontal, and transverse planes are fixed within each segment. The result is a set of three reference planes for each segment that move with the segment in three-dimensional space. The intersection lines of the three planes make up the axes of the local Cartesian coordinate system. The attitude of one system, D, attached to a distal segment, relative to the second frame, P, fixed with the proximal segment, can easily be found. Although both segments may move, one segment is considered fixed and the second is considered to be moving. Because all reference axes are given, a joint angular position can be defined unambiguously with one of the described methods (rotation matrix, Euler's-Cardan's angles, or helical method).

If a set of Cardan's angles is chosen appropriately, it describes anatomic angles. For the majority of joints, the recommended sequence of Cardan's angles is (1) flexion-extension, (2) abduction-adduction, and (3) axial rotation. For the shoulder joint, the following sequence of the Euler's angles is advised: (1) rotation around the trunk-fixed vertical axis, which determines the plane of arm elevation; (2) arm elevation in the predecided plane; and (3) axial rotation. The helical method does not provide a clinical representation of three-dimensional joint motion.

Oftentimes, at least two reference frames are assigned to the body segment. One (mechanical) frame has an origin in the center of mass of the segment with the reference axes along the principal axes of inertia; this frame has no particular relation to any joint. The second frame is located at a joint and, with the joint in a neutral position, coincides with a joint reference system of the adjacent segment. With the exclusion of the so-called spin joint motion (see Chapter **4**), the axes of joint rotation do not pass through both segments forming the joint. Thus, the origin of at least one of the joint systems is physically outside the segment. The reference frame is only attached to, but not embedded into, the segment.

For two adjacent segments, four reference frames are introduced: two frames, P and D, are defined at the segments, and two, F and M, are defined at the joint (F is for "fixed" and M is for "moving"). Vectors defined in frame D can be readily presented in frame P if a homogeneous transformation matrix $[T_{PD}]$ is known:

$$P_T = [T_{PD}]P_D \qquad (2.5)$$

The transformation matrix $[T_{PD}]$ can be viewed as a composition of several displacements:

$$P \rightarrow F \rightarrow M \rightarrow D \qquad (2.6)$$

■ ■ ■ *FROM THE SCIENTIFIC LITERATURE* ■ ■ ■

Anatomic Frames for the Pelvis and Lower Limb Bones

Source: Cappozzo, A., Catani, F., Della Croce, U., & Leardini, A. (1995). Position and orientation of bones during movement: Anatomical frame definition and determination. *Clinical Biomechanics, 10,* 171-178.

Frames are proposed as a basis for standardization. The anatomic landmarks for the frames are presented in Table 2.1 and in Figures 2.8 and 2.9. The landmarks marked with (a) are used to define the anatomic frames, those indicated with (b) are used for determining (a)-type landmarks or to improve their estimation, and the (c) landmarks are recommended as supplementary.

The following reference frames are defined with respect to a subject standing in the anatomic position (Figure 2.9).

Pelvis

O_p – The origin is at the midpoint between anterior superior iliac spines (ASIS).

z_p – The z axis is oriented as the line passing through the ASISs with its positive direction from left to right.

x_p – The x axis lies in the quasitransverse plane defined by the ASISs and the midpoint between the posterior superior iliac spines (PSIS). Its positive direction is forward.

y_p – The y axis is orthogonal to the x-z plane and its positive direction is proximal.

Right and left thigh

O_t – The origin is the midpoint between the lateral and medial epicondyles (LE and ME).

y_t – The y axis joins the origin with the center of the femoral head (FH) and its positive direction is proximal.

z_t – The z axis lies in the quasifrontal plane defined by the y axis and the epicondyles and its positive direction is from left to right.

x_t – The x axis is orthogonal to the y-z plane and its positive direction is forward.

See **FRAMES,** *p. 92*

■ ■ ■ **FRAMES,** *continued from p. 91*

Table 2.1 Anatomic Landmarks

Hip bone

(a)	ASIS	Anterior superior iliac spine
(a)	PSIS	Posterior superior iliac spine
(b)	AC	Center of the acetabulum

Femur

(a)	FH	Center of the femoral head
(c)	GT	Prominence of the greater trochanter external surface
(a)	ME	Medial epicondyle
(a)	LE	Lateral epicondyle
(b) (c)	LP	Anterolateral apex of the patellar surface ridge
(b) (c)	MP	Anteromedial apex of the patellar surface ridge
(b) (c)	LC	Most distal point of the lateral condyle
(b) (c)	MC	Most distal point of the medial condyle

Tibia and fibula

(c)	IE	Intercondyle eminence
(a)	TT	Prominence of the tibial tuberosity
(a)	HF	Apex of the head of the fibula
(a)	MM	Distal apex of the medial malleolus
(a)	LM	Distal apex of the lateral malleolus
(b) (c)	MMP	Most medial point of the ridge of the medial tibial plateau
(b) (c)	MLP	Most lateral point of the ridge of the medial tibial plateau

Foot

(a)	CA	Upper ridge of the calcaneus posterior surface
(a)	FM	Dorsal aspect of the first metatarsal head
(a)	SM	Dorsal aspect of the second metatarsal head
(a)	VM	Dorsal aspect of the fifth metatarsal head

(a) Anatomic frames.

(b) Used to determine (a)-type landmarks.

(c) Supplementary landmarks.

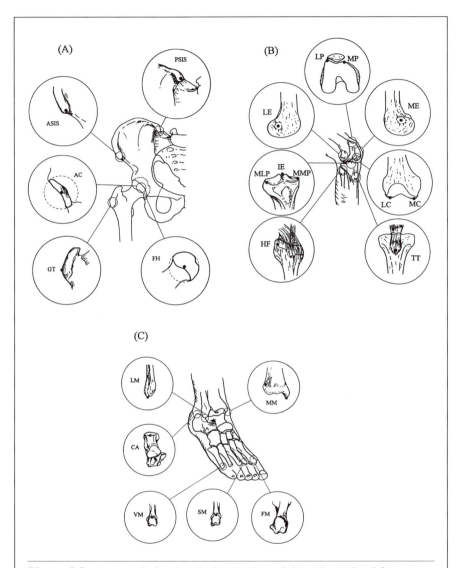

Figure 2.8 Anatomic landmarks in (A) the pelvis and proximal femur, (B) the distal femur and proximal tibia and fibula, and (C) the distal tibia and fibula and in the foot. See Table 2.1 for key to abbreviations.

Adapted from *Clinical Biomechanics, 10,* Cappozzo, A., et al., Position and orientation of bones during movement: Anatomical frame definition and determination, 171-178, 1995, with kind permission from Elsevier Science Ltd, The Boulevard, Langford Lane, Kidlington 0X5 1GB, UK.

See **FRAMES,** *p. 94*

■ ■ ■ **FRAMES,** *continued from p. 93*

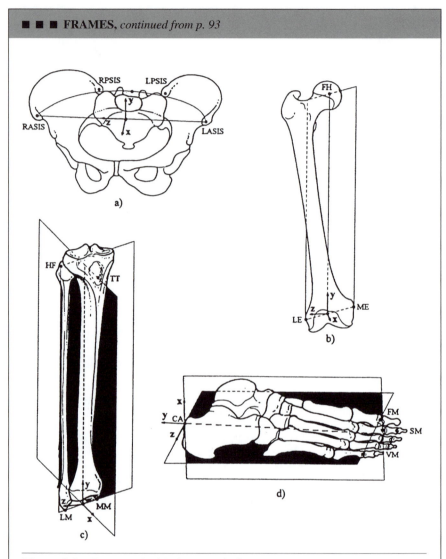

Figure 2.9 Bone-embedded anatomic frames.
Reprinted from *Clinical Biomechanics, 10,* Cappozzo, A., et al., Position and orientation of bones during movement: Anatomical frame definition and determination, 171-178, 1995, with kind permission from Elsevier Science Ltd, The Boulevard, Langford Lane, Kidlington 0X5 1GB, UK.

Right and left shank

O_s – The origin is at the midpoint of the line joining the lower ends of the medial and lateral malleoli (MM and LM).

y_s – The malleoli and the head of the fibula (HF) landmarks define a plane that is quasifrontal. A quasisagittal plane, orthogonal to the quasifrontal plane, is defined by the midpoint between the malleoli and the tibial tuberosity (TT). The y axis is defined by the intersection between the above-mentioned planes and its positive direction is proximal.

z_s – The z axis is in the quasifrontal plane and its positive direction is from left to right.

x_s – The x axis is orthogonal to the y-z plane and its positive direction is forward.

Right and left foot

O_f – The origin is at the calcaneus landmark (CA).

y_f – The calcaneus and the first and fifth metatarsal heads (FM and VM) define a quasitransverse plane. A quasisagittal plane, orthogonal to this latter plane, is defined by the CA and second metatarsal head (SM). The y axis is defined by the intersection of these two planes, and its positive direction is proximal.

z_f – The z axis lies in the quasitransverse plane and its positive direction is from left to right.

x_f – The x axis is orthogonal to the y-z plane and its positive direction is dorsal.

■ ■ ■ *FROM THE SCIENTIFIC LITERATURE* ■ ■ ■

Three-Dimensional Rotation of the Elbow Joint

Source: Chao, E.Y., & Morrey, B.F. (1978). Three-dimensional rotation of the elbow. *Journal of Biomechanics, 11,* 57-73.

The global, humeral, and forearm systems were defined (Figure 2.10, page 96). In the anatomic posture, the X and x axes corresponding to the humeral and the forearm systems were pointing in the anterior direction, the

See **ELBOW,** *p. 96*

■ ■ ■ **ELBOW,** *continued from p. 95*

Y and y axes were along the bones, and the Z and z axes were pointing laterally. The adopted rotational sequence was (1) flexion-extension, ϕ; (2) abduction-adduction, θ, or the carrying angle; and (3) axial rotation, ψ (Figure 2.10, page 97).

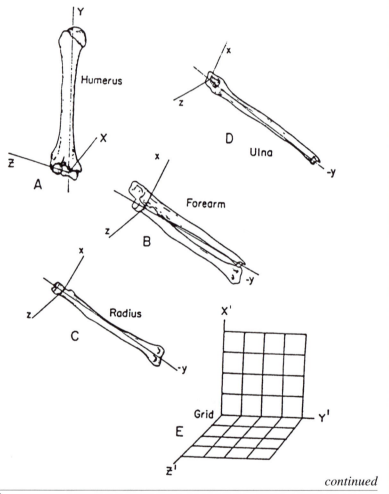

continued

Figure 2.10   Reference coordinate systems (this page) and the definition of the Euler's angles (next page).

Adapted from *Journal of Biomechanics, 11,* Chao, E.Y. & Morrey, B.F., Three-dimensional rotation of the elbow, 57-73, 1978, with kind permission from Elsevier Science Ltd, The Boulevard, Langford Lane, Kidlington 0X5 1GB, UK.

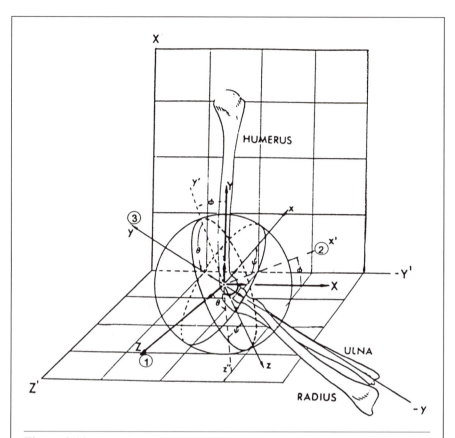

Figure 2.10 *(continued)*

The unit vectors of the forearm-fixed and the humerus-fixed systems are related by the rotation matrix

$$
\begin{bmatrix} \vec{x} \\ \vec{y} \\ \vec{z} \end{bmatrix} =
\begin{bmatrix}
\cos\phi\cos\psi - \sin\phi\sin\theta\sin\phi & \sin\phi\sin\psi - \sin\psi\sin\theta\cos\phi \\
-\cos\theta\sin\phi & \cos\theta\cos\phi \\
\sin\psi\cos\phi + \cos\psi\sin\theta\cos\phi & \sin\psi\sin\phi - \cos\psi\cos\theta\cos\phi
\end{bmatrix}
$$

$$
\begin{matrix} -\sin\psi\cos\theta \\ \sin\theta \\ \cos\psi\cos\theta \end{matrix}
\begin{bmatrix} \vec{X} \\ \vec{Y} \\ \vec{Z} \end{bmatrix}
$$

See **ELBOW,** *p. 98*

■ ■ ■ELBOW, *continued from p. 97*

where $\vec{x}, \vec{y}, \vec{z}$ and $\vec{X}, \vec{Y}, \vec{Z}$ are unit vectors along the forearm axes and the humeral axes, correspondingly. Note that in this equation the rows of the rotational matrix correspond to the axes of the system fixed at the distal segment, the forearm; the columns correspond to the axes of the proximal system, the humerus. The matrix is a transposition of the rotation matrices adopted in this book (see equation 1.2). Because of this difference and, also, because of the changed order of rotation, the rotational matrix is different from that presented in equation 1.32. In equation 1.32, the rotation sequence was Zy'x", in this example the sequence is Zx'y". If $\vec{x}, \vec{y}, \vec{z}$ and $\vec{X}, \vec{Y}, \vec{Z}$ are defined in the global reference system, the individual Euler's angles can be determined.

$$\text{Flexion angle, } \cos^{-1}\left(\frac{\vec{y} \cdot \vec{Y}}{\cos\theta}\right)$$

$$\text{Carrying angle, } \theta = \sin^{-1}\left(\vec{y} \cdot \vec{Z}\right)$$

$$\text{Axial rotation, } \psi = \left(\frac{\vec{z} \cdot \vec{Z}}{\cos\theta}\right)$$

Using equation 1.21 we can write in matrix form

$$[T_{PD}] = [T_{PF}][T_{FM}][T_{MD}] \tag{2.7}$$

The transformation matrices $[T_{PF}]$ and $[T_{MD}]$, as well as the transformations between the two-joint reference systems, proximal and distal, affixed to the same body segment are constant; they do not depend on the joint configuration. These transformations are known as *link transformations*. The transformation matrix $[T_{FM}]$ is called the *joint transformation*.

2.1.5 Joint Rotation Convention

Representative papers: Chao, 1980; Grood & Suntay, 1983; Woltring, 1994.

To identify the relative attitude of two body segments, three spatial axes should be specified. In the joint rotation convention (JRC), two axes are fixed with the body segments, proximal and distal, and one is a floating axis. The following procedure is recommended:

1. The first axis is the first fixed body axis and is perpendicular to the sagittal plane of the proximal segment.
2. The second axis is the floating axis (the cross product of the first and third axes).
3. The third, or the second fixed-body axis, is the long axis of the distal segment (Figure 2.11).

Also, two reference lines are defined. They are embedded in each body perpendicular to the fixed axis.

To visualize the JRC consider the human knee joint as an example (Figure 2.11). The flexion-extension axis (axis 1-1) is anchored to the distal femur. Consequently, the plane of flexion-extension does not change its orientation

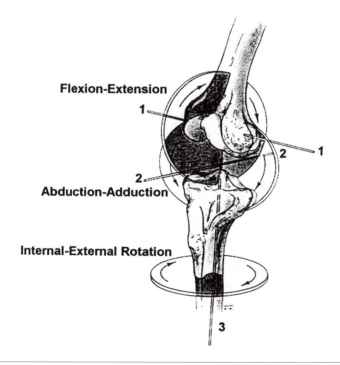

Figure 2.11   Joint rotation convention for a knee joint. Axis 1-1 is fixed to the femur and describes the flexion-extension motion. Axis 3-3 is fixed to the tibia along its longitudinal anatomic axis and defines internal-external rotation. The floating axis 2-2 is orthogonal to both axes 1-1 and 3-3 and is used to measure abduction and adduction.

From Chao, E.Y.S. (1990). Goniometry, accelerometry, and other methods. In N. Berme & A. Cappozzo (Eds.), *Biomechanics of human movement: Applications in rehabilitation, sports and ergonomics* (pp. 130-139). Worthington, Ohio: Bertec Corporation. Reprinted by permission.

throughout the motion history; it is always fixed with the femur. The axis of external-internal rotation (axis 3-3) changes its orientation with regard to the femur during the motion. This axis is along the long axis of the tibia. Consequently, the plane of the axial rotation always remains perpendicular to the longitudinal axis of the lower leg. The floating axis, which is the abduction-adduction axis in our example, is identical to the line of nodes. Although axis 2-2 is orthogonal to both axis 1-1 and axis 3-3, the abduction-adduction plane is fixed to neither the femur nor to the tibia. The orientation of this plane relies on the flexion-extension angle and does not depend on the internal-external rotation angle.

In all "traditional" angular conventions discussed in the preceding sections, the existence of two embedded reference frames was assumed. The relative position of the bodies was identified with the relative position of the two frames. In the JRC, however, one reference frame and two reference lines are defined. The magnitude of rotation around the fixed axis is measured by the angle formed between the reference line and the floating axis. For example, internal-external rotation in the knee joint is measured as the angle formed by the reference axis perpendicular to the long axis of the tibia and the floating, anteroposterior, axis.

Because the long axis of the distal segment is not perpendicular to the frontal axis of the proximal segment in all of the joints, the two terminal fixed axes are usually not orthogonal. Also, the fixed axes move with the body segments and their relative spatial orientation changes with motion. Thus, the JRC is not a Cartesian system of coordinates. The JRC angles are the Cardan's angles defined geometrically. The floating axis is parallel to the line of nodes and coincides with it when the three axes intersect at one point.

The JRC angles are anatomically meaningful. When the recommended procedure is followed, independent rotations around the defined axes match the clinical definitions for relative movement between segments. In addition, rotations about the JRC axes are time sequence–independent. The final displacement between the two bodies is independent of the temporal order in which the rotations are performed. It does depend, however, on the choice of the fixed and floating axes. The JRC angles should be defined in a certain sequence. After that, when the axes are defined, the rotations around these axes are sequence independent.

Because the JRC system is simply a variant of Euler's-Cardan's angles, the adverse effects seen in that system, particularly singularity, continue to occur. The JRC angles cannot be defined for some joint postures. When the longitudinal axis of a distal segment is collinear with the frontal axis of the proximal segment, the second axis of the reference frame is not defined (a singular position). The singularity appears because the cross product of two vectors cannot be calculated when the vectors are collinear. The magnitude of vector V, which is the cross product of vectors **P** and **Q,** equals PQ sin α. When α is $0°$ or $180°$, the cross product is sin $\alpha = 0$. Such a configuration can be seen when the arm is abducted $90°$ and the long axis of the arm is collinear with the shoulder frontal axis. The position of the second axis of the local frame is not defined for such a joint configuration. The JRC is convenient for identifying the instantaneous axis of rotation in the joint. However, as it was mentioned previously, the JRC is not an orthogonal system. Nonorthogonality can present a serious problem when joint forces and moments are going to be determined.

■ ■ ■ *FROM THE SCIENTIFIC LITERATURE* ■ ■ ■

Knee Movement During Walking

Source: Lafortune, M.A., Cavanagh, P.R., Sommer, H.J., III, & Kalenak, A. (1992). Three-dimensional kinematics of the human knee during walking. *Journal of Biomechanics, 25,* 347-357.

Metallic pins with the attached target markers were inserted into the tibia and femur of the subjects. Radiographs of the lower limbs were taken and then the subjects were filmed during walking at a mean speed of 1.2 m/s.

Six reference frames were defined: (1) the global reference frame for gait analysis; (2) the radiographic reference frame; (3) the tibial anatomic frame; (4) the tibial marker reference frame (the first technical frame); (5) the femoral anatomic reference frame; and (6) the femoral marker reference frame (the second technical frame). The 4×4 transformation matrices were used, and the following steps were taken:

1. The position vectors of target markers in the tibial and femoral anatomic frames were computed from values measured on the radiographs $[x_{TA}] = [T_{R/TA}][x_R]$ and $[x_{FA}] = [T_{R/FA}][x_R]$, where $[x_{TA}]$ and $[x_{FA}]$ are position vectors of point x in the tibial and femoral anatomic frames, $[x_R]$ is the position vector of point x in the radiographic reference frame, and $[T_{R/TA}]$ and $[T_{R/FA}]$ are corresponding transformation matrices.

2. The position vector of target markers in the tibial marker frame, $[x_{TM}]$, was computed from the known values of their position in the tibial anatomic frame, $[x_{TA}]$, $[x_{TM}] = [T_{TA/TM}][x_{TA}]$ and the position vector of target markers in the femoral anatomic frame, $[x_{FA}]$,

See **WALKING,** *p. 102*

■ ■ ■ WALKING, *continued from p. 101*

was computed from their location in the femoral marker reference frame $[x_{FA}] = [T_{FM/FA}][x_{FM}]$ where $[T_{TA/TM}]$ and $[T_{FM/FA}]$ are transformation matrices. Note that these two equations perform different transformations.

3. The two following transformations transform the position vector given in the tibial marker reference frame into the global reference frame $[x_G] = [T_{TM/G}][x_{TM}]$ and the position vector in the global frame into the vector in the femoral marker reference frame $[x_{FM}] = [T_{G/FM}][x_G]$.

4. The final computation of the location and attitude of the tibial anatomic reference frame with respect to the femoral anatomic reference frame was achieved according to the following equation:

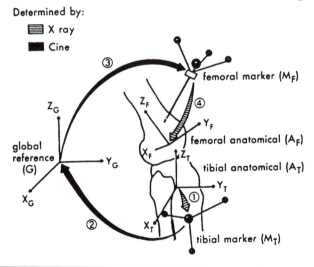

Figure 2.12 Transformations used to obtain the position of the tibial anatomic reference frame with respect to the femoral anatomic reference frame. Positions of the triads in the global reference frames were determined through cinematography, and their positions in the respective anatomic frames of reference were obtained through radiographs.
Reprinted from *Journal of Biomechanics, 25,* Lafortune, M.A., et al, Three-dimensional kinematics of the human knee during walking, 347-357, 1992, with kind permission from Elsevier Science Ltd, The Boulevard, Langford Lane, Kidlington 0X5 1GB, UK.

$[x_{FA}] = [T_{FM/FA}][T_{G/FM}][T_{TM/G}][T_{TA/TM}][x_{TA}]$. The sequence of matrix multiplication in the equation is shown in Figure 2.12.

Because $[x_{FA}] = [T_{TA/FA}][x_{TA}]$, it follows that

$$[T_{TA/FA}] = [T_{FM/FA}][T_{G/FM}][T_{TM/G}][T_{TA/TM}]$$

The matrix $[T_{TA/FA}]$ contains the kinematic information required to compute the locations of the markers in the femoral anatomic reference frame from their locations in the tibial anatomic reference frame. Although the general sequence of calculations does not depend on the chosen system of coordinates, the elements of the matrices do depend on them. In this study, the JRC system was used. Hence, the elements of matrix $[T_{TA/FA}]$ are the direction cosines of certain joint angles (flexion-extension, abduction-adduction, and internal-external rotation) and the translations along the lateromedial, longitudinal, and floating axes (shift, distraction, and drawer).

2.2 KINEMATIC CHAINS

A human body can be modeled as a multilink system comprising several body segments connected by joints. A linkage of rigid bodies is referred to as a *kinematic chain*. The simplest kinematic chain consists of two links connected by a single joint. A set of two adjacent links connected with one joint is called a *kinematic pair*. The fastening of bones in a kinematic pair is maintained either by their shape (a *form-closed pair*) or by externally applied forces (a *force-closed pair*). The hip joint comes nearest to the form-closed type, whereas the shoulder joint is considered force-closed because it is maintained by the surrounding muscles and ligaments.

Kinematic chains are either *serial* (simple) or *branched* (complex). In serial chains, each of the links is part of no more than two kinematic pairs. A branched chain contains at least one link that is part of more than two kinematic pairs. A human arm or leg can be considered a serial kinematic chain. When the trunk is also considered and articulations within the trunk are ignored, the chain is branched. The trunk enters five kinematic pairs: the two hip joints, the two shoulder joints, and the trunk-neck articulation. Kinematic chains are further classified as *open* or *closed*. The kinematic chain is referred to as open if one end of the chain (e.g., the distal segment) is free to move. In closed chains, constraints are imposed on both ends of the chain. Nonstationary (temporary) constraints are typical for many human movements (e.g., constraints imposed during the support periods in walking). Examples of open and closed kinematic chains in human movements are given in Figure 2.13.

2.2.1 Degrees of Freedom. Mobility of Kinematic Chains

The term "degrees of freedom" (DOF) refers to the independent coordinates required to completely characterize a body, or system, position. A single DOF can also be defined as an independent way the body can move. A rigid body, freely suspended in the air, has six DOF. It can translate along and rotate about three independent axes (longitudinal, vertical, frontal). When planar movement is considered, the body has three DOF: it can translate from one place to another in two directions and rotate.

The independent coordinates are any set of quantities that completely specify the state of a system. They are called *generalized coordinates*. The generalized coordinates are customarily written as $q_1, q_2, \ldots, q_n$, or simply as the q_i. The choice of a set of generalized coordinates to describe a system is not unique; there are many sets of quantities that completely specify the state of a given system.

A single point in space has three DOF. A system of n points has 3n DOF. However, if all the n points belong to a solid body they will maintain a constant distance from each other and the rigid body in space has only six DOF. A system of N rigid bodies, if not constrained, has 6N DOF. If some of the 6N coordinates are not independent (as would be the case if bodies were con-

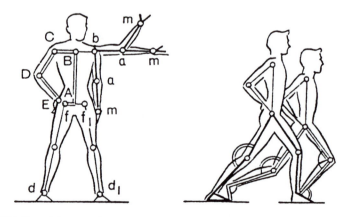

Figure 2.13 Kinematic chains in human movements. The left panel shows bam, an open chain, and ABCDEA and dff_1d_1, closed chains. The right panel shows a movement of the closed chain. In any closed chain, the joint angular motions are coupled.

nected by joints or if the motion were confined along some path or on a surface), then the number of DOF decreases. If there are m equations of constraint, then only (6N − m) coordinates are independent, and the system possesses (6N − m) DOF. The constraints may be *complete* or they may be *one-sided*. For example, during a stance period the foot is free to move up but it cannot move down. When the equations of constraint connect only the coordinates, the constraints are termed *holonomic*. Constraints imposed on the velocities, accelerations, and so on are called *nonholonomic*. If the equations do not explicitly contain the time, the constraints are said to be *fixed* or *scleronomic*. Constraints changing in time are *rheonomic*.

The maximum number of DOF for each joint is six. If the translational motion within a joint (normally assumed to be small) is ignored, the maximal number of DOF is three. The number of DOF of a joint can be represented as a contrast; i.e., *six minus the number of constraints imposed*. Instead of the DOF, the *class of joint* may be defined. The class of joint is determined by the number of imposed constraints. Human joint geometry is complex. Joint surfaces are innately irregular in shape, or idiosyncratic, and therefore cannot be described as simple geometric surfaces. Consequently, when modeling joints there is a trade-off between accuracy and simplicity.

To simplify the study of human motion, the human body joints are classified as having one, two, or three rotational DOF. The joints with one DOF are called *hinge*, or *revolute*, joints. Joints with three DOF are referred to as *ball-and-socket*, or *spherical*, joints. Finger (interphalangeal) joints are examples of hinge joints; the hip and shoulder (glenohumeral) joints are ball-and-socket joints. Some joints (such as the sternoclavicular, which is used to shrug the shoulders, or the temporomandibular) have two DOF.

In mechanical models, when a twisting motion (e.g., pronation-supination or internal-external rotation) is considered, the twist is usually assigned to one of two joints, proximal or distal. For example, pronation-supination can be associated with either the elbow or the wrist. Depending on whether pronation-supination is assigned to either the elbow or the wrist joint, the elbow joint has one or two DOF. Note that joints having one DOF can also be called joints of the 5th class (because five constraints are imposed); joints with three DOF are joints of the 3rd class, and so on.

The total number of DOF in a kinematic chain is called the *mobility* of the chain. To calculate the mobility of a kinematic chain, the following convention is usually adopted: the external reference frame, which is fixed with the immovable environment, is counted as an additional link of the chain. With this convention, the number of rigid bodies included in the chain increases to N + 1. "Joint" 1 (between the external reference and the first link of the system) may have up to six DOF. The same is accepted for the last link of a closed kinematic chain. When using this convention, the mobility of a kinematic chain in space is described by the following formula:

$$F = 6(N - k) + \sum_{i=1}^{k} f_i \qquad (2.8)$$

where N is the number of links, k is the number of joints, and f_i is the number of DOF in the *i*th joint. This equation is known as *Gruebler's formula*. For an open chain, the number of joints equals the number of links. This equality allows equation 2.8 to be simplified to

$$F \text{ (for an open chain)} = \sum_{i=1}^{k} f_i \qquad (2.9)$$

If the chain is closed, it has one more joint than links, k = N + 1. The mobility of the chain is

$$F \text{ (for a closed chain)} = \sum_{i=1}^{k} f_i - 6 \qquad (2.10)$$

When the kinetic chain is described using the planar system, Gruebler's formula is modified as

$$F = 3(N - k) + \sum_{i=1}^{k} f_i \qquad (2.11)$$

For a closed chain in a planar system, the mobility is calculated as

$$F \text{ (for a closed planar chain)} = \sum_{i=1}^{k} f_i - 3 \qquad (2.12)$$

According to equation 2.12, a closed chain with four revolute joints in a planar system has four DOF with regard to the environment; three DOF are due to the mobility of the fictitious "joint" 1 and one DOF is due to the relative movement of the links. The chain has only one DOF when the relative movement of the links is considered exclusively.

Kinematic chains having only one DOF are called *mechanisms*. In mechanisms, at least one of the members is attached to an unmovable base (frame), which is considered an integral part of the mechanism. The position or movement of any one link of the mechanism prescribes the position or movement of all the other links. When such a situation takes place in human or animal movements the chain is called the *biomechanism*. For example, the rib cage during breathing and the legs of a fencer in the lunge position (if both the feet are on the support and the posterior leg is extended) have only one DOF; the chains are considered biomechanisms.

2.2.2 Open Kinematic Chains: The End-Effector Mobility

In many human movements, the last link of an open kinematic chain, typically the hand (a working tool) or the foot, needs to be positioned in a specified place with a specific orientation. This link is termed the *end effector.* The end effector may have its own mobility, for example, the human hand. The gripping function of the human hand and the positioning function of the human arm can be regarded as separate functions. To be positioned at an arbitrary point with an arbitrary orientation in space, the end effector must have at least six DOF. The number of DOF of the end effector equals the mobility of the kinematic chain. If the chain has less than six DOF, the end effector cannot be arbitrarily positioned within the reach area. When the chain has exactly six DOF, there exists exactly one joint configuration that places the end effector in a required position, i.e., in a required place with a required orientation. When the chain has more than six DOF, the end effector can be positioned at a required point with a required orientation in an infinite number of ways. Open kinematic chains with more than six DOF are termed *redundant chains.* For redundant chains, an infinite set of joint positions leads the end effector to the same location.

Maneuverability of the chain is described as $M = F - 6$, where F is the number of DOF of the chain. The maneuverability, M, equals the number of extra parameters that can be used to guide the end effector in a desired way; for example, to avoid obstacles or to minimize energy expenditure.

When studying the position of a kinematic chain, two main problems exist:

1. The joint coordinates are known and the end effector position is sought. This is called the *direct kinematic problem.*

2. The position of the end effector is known and the joint coordinates are sought. This problem is termed the *inverse kinematic problem.*

Note that the end-effector position is sought, or given, in absolute coordinates. The joint coordinates are the joint angles between adjacent segments.

2.2.3 Kinematics Models and Mobility of the Human Body

Kinematic models of the human body are those that represent its mobility and neglect all other aspects (e.g., the mass distribution). The models are classified as *anthropomorphic*, also called *skeletal,* or *functional.*

Skeletal models visually resemble the construction of the human body; the body segments are (typically) modeled as solid links and the human joints as the joints of the model. In functional models, the body segments are modeled as nodes of a graph (of a tree) and the joints as arcs connecting the nodes

(Figure 2.14). This representation has certain advantages for computer modeling and data structures; the joints are innately binary (they connect exactly two segments). Those connections are conveniently represented by the arcs having, not surprisingly, two ends. The segments may have several joints. For example, the trunk, if considered as a whole, provides a relationship between the positions of the two hip joints, the two shoulder joints, and the trunk-neck articulation.

In this book, only anthropomorphic models are used. One of them, in which the body is rendered as a stick figure, is shown in Figure 2.15.

The model represented in Figure 2.15 consists maximally of 18 rigid segments and 17 joints and possesses 41 DOF. If the sternoclavicular joints are ignored and the spine is considered one solid segment, the number of DOF decreases to 31. This is called a *gross body model,* in which many small joints (e.g., the interphalangeal joints) are not included. In the gross body model, the arm has seven DOF. Thus, its kinematic chain is redundant. This is easily demonstrated. When the hand is fixed on a table and the shoulder is fixed too, one can move the elbow without changing the hand or the shoulder position. The serious problem with gross body models is how to model the trunk segment. To consider the trunk as one rigid body is too unrealistic. Incorporation of fictitious trunk joints improves the quality of modeling but does not completely solve the problem. For example, when the model is presented on a screen, the trunk segments are separate from each other at some body angles. Although it is more realistic to represent the torso as a nonrigid (bending and twisting)

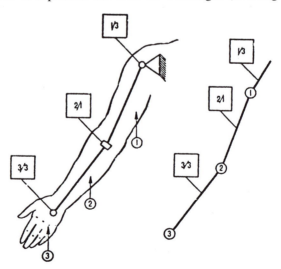

Figure 2.14 Two models of the arm: left, an anthropomorphic model, and right, a functional model. The "ratios" in the squares are the joint number over the number of DOF in the joint. The segment numbers are shown in the circles.

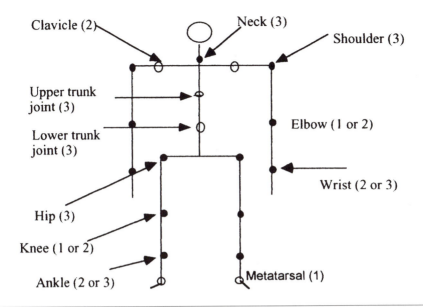

Figure 2.15   Kinematic model of human body. Filled circles designate the joints that are usually included in the model. Open circles are for the joints that are included only in some models.

segment, mathematical difficulties associated with this approach are, as of yet, unsolved.

To estimate the total mobility of the body all joints and body segments must be considered. According to estimations, there are 148 movable bones and 147 joints in the human body. Such a model is shown in Figure 2.16. With this approach the arm has not 7 but 30 DOF.

The total mobility can be estimated using a slightly modified Gruebler's formula, in which classes of joints instead of the number of DOF are used:

$$F = 6N - \sum_{i=3}^{5} i \cdot j_i \qquad (2.13)$$

where F is the mobility of the body (the total number of DOF), N is the number of movable bones, i is the class of the joint (based on the number of imposed constraints, $i = 6 - f$, where f is the number of DOF), and j_i is the number of joints of the class i. It has been estimated that the human body has 148 movable bones connected by the joints, 29 joints of the 3rd class (with three DOF), 33 joints of the 4th class (with two DOF), and 85 joints of the 5th class (with one DOF). The total mobility of the human body is

$$F = (6 \cdot 148) - (3 \cdot 29) - (4 \cdot 33) - (5 \cdot 85) = 888 - 87 - 132 - 425 = 244 \qquad (2.14)$$

   Thus, the human skeletal system is highly redundant. It has 244 DOF and its maneuverability is 238. To position an end effector in space, the brain must specify not 6 but 244 variables, of which 238 are redundant and may be used to perform the motor task in an optimal way. In comparison, the arms of contem-

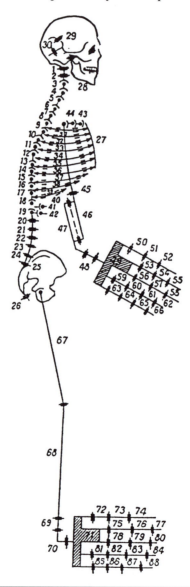

Figure 2.16   Mobility of the human body. The numbers correspond to the joints. Note that only one side of the body is shown.
From Morecki, A., Ekiel, J., & Fidelus, K. (1971). *Bionika ruchu*, Warsaw. (In Polish.) Reprinted by permission.

porary robots usually have no more than five or six DOF. In effect, only nonredundant robotic manipulators are used today.

2.2.4 Constraints on Human Movements

Representative paper: Saltzman, 1979.

Holonomic (geometric) constraints in human movements can be classified into several groups: anatomic, actual, mechanical, and motor task constraints.

Anatomic constraints are those imposed by the structure of the musculo-skeletal system. All joints are constrained in at least in two respects: first, body segments forming the joint contact each other (the requirement of the joint integrity), and second, the range of joint motion is limited. Constraints can also be put on movements about different joint axes. The movements about different axes are called *adjunct* movements when they can occur independently, and they are called *conjunct*, or *coupled*, movements when they have to occur together. If the joint movements are coupled, the number of axes of rotation in the joint exceeds the number of DOF. An example of this is the first carpometacarpal joint in which thumb flexion-extension occurs coupled with pronation-supination. Other examples of the coupled joint movements will be discussed in Chapter **5**.

Actual constraints are tangible physical obstacles to movement, such as supporting surfaces or elements of the surrounding environment. These constraints decrease human mobility. For example, during pedaling the leg can be modeled as a planar kinematic chain with three revolute joints (the hip, knee, and ankle; the location of the hip is assumed stationary). However during bicycling, one additional constraint is imposed by the constant contact with the pedal. Thus, only two DOF are left. One DOF is used to rotate the pedal; thus, the last unconstrained DOF is used to vary the pedaling technique. The technique variants are rather limited and only two exist: fixed ankle (or "toe" pedaling) and movable ankle (or "heel" pedaling).

Both anatomic and actual constraints are real; they are caused by material bodies that prevent motion beyond a given border. In human movements, however, other requirements must also be satisfied to perform a motor task. It is convenient to consider these requirements as constraints.

Mechanical constraints are those that define movement geometry in an indirect way; a human must rely on them to prevent an accident, such as falling down. Two mechanical constraints are most typical: constraints necessitated by limited friction and constraints necessitated by balance demands. An example of the first is that to prevent slipping accidents people should perform takeoffs at angles that are larger than the so-called angle of friction, because small takeoff angles lead to slipping and falling. (The takeoff angle is the angle formed by the ground force vector and the support surface.) An example of the second is seen if we again consider the legs as chains with three rotational DOF. In a standing posture, one constraint is imposed because the vertical projection of the general center of mass must be inside a narrow area. Thus, just two DOF are left. As a result, during landing only two distinct strategies are possible: "trunk bending" and "knee bending" (Figure 2.17). When knee bending is prohibited, as in a pure "trunk-bending" strategy, the system has only one DOF; movements in the hip and the ankle joints are coupled. Mechanically unconstrained DOF are called the *permitted DOF*, or *mechanically permitted DOF*.

Motor task constraints are imposed both voluntarily and involuntarily by the performer to execute the desired movement or to fulfill the planned motor task. These constraints are classified as *instructional* (defined by an instruction of the experimenter, competition rules, or the like) and *intentional* (imposed by the performer himself or herself). Because the human motor system is extremely redundant, the number of these constraints is very high. The control of human and animal movements can be seen as the elimination of the redundant DOF. To accomplish a motor task, excessive mechanical DOF must somehow be conquered. This is called *Bernstein's problem*, after Nicholai A. Bernstein (1896-1966), a Russian biomechanist and physiologist.

The CNS reduces the number of DOF in several ways. For example, some joints are *frozen* during the fulfillment of a motor task; that is, their angular values do not change. Also, the movement of some joints is coupled over the duration of some actions. These couplings are called *functional synergies*. The

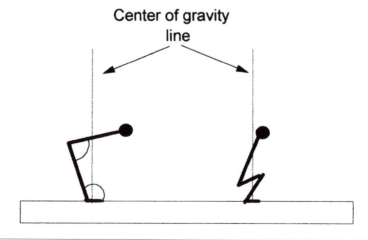

Figure 2.17   Planar kinematic chains with one permitted DOF (left, "trunk-bending" strategy) and two permitted DOF (right, "knee-bending" strategy). In the figure on the left, the hip and ankle angles are coupled. In the figure on the right, three angles (the hip, knee, and ankle) are interrelated; their values cannot be chosen arbitrarily.

concept of joint synergy suggests that several joints are controlled as a unit. Functional synergies and frozen joints decrease the number of DOF controlled by the nervous system. Also, only some DOF, called *essential*, are controlled unconditionally throughout the performance. Other variables, called *nonessential,* are under rather loose control. If a movement is repeated several times, the essential variables show small variability. The variability of nonessential movement parameters is much greater. For example, when hammering a nail, the variability of the hammer position is much less at the instant when the hammer touches the nail than during the wind-up phase. As it happens, this observation, made by Bernstein as early as 1920, was the starting point of his studies in biomechanics and motor control.

■ ■ ■ FROM THE SCIENTIFIC LITERATURE ■ ■ ■

Classification of Closed Kinematic Chains in Human Movement

Source: Vaughan, C.L., Hay, J.G., & Andrews, J.G. (1982). Closed loop problems in biomechanics. Part 1. A classification system. *Journal of Biomechanics, 15,* 197-200.

Kinematic analysis of human motion has often been used as a starting point for solving the *inverse dynamics problem*—calculating the forces and torques acting on the joints from the kinematic recordings. As in most mathematics, a solution is feasible when the number of unknowns does not exceed the number of equations; when one equation contains two unknowns it cannot be solved in a unique way. In the framework of the inverse dynamics problem, the number of equations is equal to the number of DOF in the chain. Closed chains have a decreased number of DOF, e.g., a planar closed four-link chain with revolute joints has only one DOF. Four unknown joint torques cannot be calculated from a single equation. Hence, the difference between the number of equations and the number of unknowns, Δ, is important. The difference defines whether the system of equations is overdetermined, determined, or underdetermined.

Body positions are classified according to the number of

- extremities in contact with a fixed external reference system: nonsupport (NS), single support (SS), double support (DS), triple support (TS), and quadruple support (QS), and

See **CHAINS,** *p. 114*

■ ■ ■ **CHAINS,** *continued from p. 113*

- closed loops formed by the extremities that are not in contact with the fixed external system: open loop (OL), single closed loop (SCL), and double closed loop (DCL).

As a result, five groups of body positions are defined (Figure 2.18 and Table 2.2). Among the groups, Δ varies from two to negative six. In the

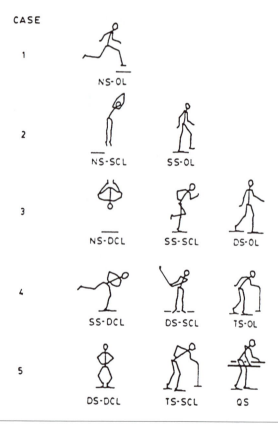

Figure 2.18 Classification system of whole body activities. See Table 2.2 for details.

Reprinted from *Journal of Biomechanics, 15,* Vaughan, C.L., Hay, J.G., & Andrews, J.G., Closed loop problems in biomechanics. Part 1. A classification system, 197-200, 1982, with kind permission from Elsevier Science Ltd, The Boulevard, Langford Lane, Kidlington 0X5 1GB, UK.

first group, the number of equations exceeds the number of unknowns by two—the system is overdetermined. The last three groups are underdetermined—the number of unknowns surpasses the number of equations.

Indeterminate problems can be made determinate by the use of appropriately placed force measuring devices, e.g., using two force platforms instead of one.

Table 2.2   Classification of Body Positions

Case	Δ	System of equations	Names[1]	Examples
1	2	Overdetermined	NS-OL	Airborne phase in running
2	0	Determined	NS-SCL	Diving, hands clasped together
			SS-OL	Walking, single support period
3	−2	Underdetermined	NS-DCL	Tucked position in somersaulting
			SS-SCL	Sprint relay exchange
			DS-OL	Walking, double support period
4	−4	Underdetermined	SS-DCL	Ice skating, hands on hips
			DS-SCL	Golf swing
			TS-OL	Walking with a cane
5	−6	Underdetermined	DS-DCL	Takeoff, hands on hips
			TS-SCL	Walking with cane, a hand on hip
			QS	Walking, hands holding parallel bars

[1]NS, nonsupport; OL, open loop; SCL, single closed loop; SS, single support; DCL, double closed loop; DS, double support; TS, triple support; QS, quadruple support.

2.2.5 Position Analysis of Kinematic Chains

In this section, unless stated otherwise, I am referring to open chains.

2.2.5.1 Two Simple Chains

Consider two simple chains, a two-link and a three-link. The problems to be addressed are direct kinematics, inverse kinematics, and a way of representing the chains. These chains will be analyzed repeatedly throughout the textbook from different perspectives.

2.2.5.1.1 Two-Link Planar Chain

The link lengths are ℓ_1 and ℓ_2 (Figure 2.19). The X axis is counted as an additional link of the chain, ℓ_0. The proximal end of link ℓ is constrained to the origin and, thus, only rotations in the joints are permitted. The external (or anatomic) angles in joints 1 and 2 are α_1 and α_2. The terminal point of the distal segment is designated P. The two-link planar chain is the simplest possible model of a human extremity. Oftentimes, the proximal link of this chain will be further referred to as the "upper arm" or "thigh" and the distal link as the "forearm" or "shank." Joint 2 in this case is the "elbow" or "knee" and joint 1, at the origin O, is the "shoulder" or "hip." The chain represents a human arm or leg without motion at the wrist or ankle.

Direct kinematics. When angles α_1 and α_2 are known, the position of point P, the tip of the chain, can be easily found. The projections of joint 2 and point P on the X and Y axes are

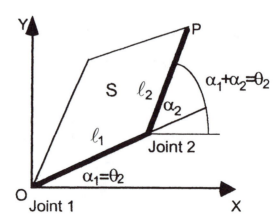

Figure 2.19   Model of a two-link planar chain.

$$X_2 = \ell_1 \cos \alpha_1$$

$$Y_2 = \ell_1 \sin \alpha_1$$

$$X_P = \ell_1 \cos \alpha_1 + \ell_2 \cos(\alpha_1 + \alpha_2) \qquad (2.15)$$

$$Y_P = \ell_1 \sin \alpha_1 + \ell_2 \sin(\alpha_1 + \alpha_2)$$

Thus, the direct kinematics problem has a unique solution. The solution can also be written in a matrix form:

$$\begin{bmatrix} X_P \\ Y_P \end{bmatrix} = \begin{bmatrix} \cos \alpha_1 & \cos(\alpha_1 + \alpha_2) \\ \sin \alpha_1 & \sin(\alpha_1 + \alpha_2) \end{bmatrix} \cdot \begin{bmatrix} \ell_1 \\ \ell_2 \end{bmatrix} \qquad (2.16)$$

Inverse kinematics. The set of points that can be reached by the terminal point of the distal segment when the terminal point of the proximal link is affixed to an unmovable body is called the *reach area*. If the position of point P is given within the reach area, two configurations of the chain can be used to locate the tip P in the position X_P, Y_P. The second configuration (the "elbow-up" position; thin lines) is a reflection of the first one in the line through O and P. So, even in the simplest case, the inverse solution can be not unique.

Representation. A two-link chain can be conveniently described by the position and length of the radius-vector drawn from the origin O to the terminal point of the distal segment P. The distance between the origin of system O and the tip position $P(X_P, Y_P)$ can readily be obtained from the law of cosines. The elbow, or knee, angle is $(\pi - \alpha_2)$ and the side OP of the triangle is

$$OP = \left[\ell_1^2 + \ell_2^2 - 2\ell_1\ell_2 \cos(\pi - \alpha_2) \right]^{0.5} \qquad (2.17)$$

This distance will be called the *extremity stretch*. To designate the stretch, the symbol S will be used. A movement that decreases or increases the stretch is called extremity flexion or extension. A change in orientation of the radius-vector is called extremity rotation. Planar movement of the extremities is a combination of rotation and flexion-extension. The described representation of the two-link chain is equivalent to its representation through joint angles. In both cases the system with two DOF is explicitly defined by two independent parameters. Note that arm or leg flexion-extension depends on the changes in the second joint only; extremity rotation is a function of two joint angles. When a pure flexion-extension takes place (without extremity rotation), the terminal point moves along the radius, which maintains the same orientation in space. During such a pushing or pulling motion, links 1 and 2 (the upper arm and forearm or the thigh and shank) concurrently rotate in opposite directions in a coupled way. When planar arm movements are discussed, simultaneous flexion-extension of the elbow and shoulder joints in the same direction is oftentimes called the *whipping* action. A *reaching* action involves flexion-extension of the elbow and shoulder joints in opposite directions.

2.2.5.1.2 Three-Link Planar Chain

The link lengths are ℓ_1, ℓ_2, and ℓ_3 (Figure 2.20). The angles in the joints 1, 2, and 3 (the numbers are not shown in the picture) are α_1, α_2, and α_3. The distal segment, ℓ_3, is the end effector and should be placed in the plane in such a way that its distal terminal P is located at a point with the coordinates (X_P, Y_P). Let us try first to solve the direct problem of kinematics, and then the inverse problem. After that, representation of the three-link planar models will be discussed.

Direct kinematics. The joint coordinates, α_1, α_2 and α_3, are given; the position of the end effector, ℓ_3, or its terminal point $P(X_P, Y_P)$, need to be found. It is easily seen from the figure that orientation of link ℓ_i in the absolute reference frame, θ_i, is the sum of the more proximal joint angles: $\theta_1 = \alpha_1$, $\theta_2 = \alpha_1 + \alpha_2$, and $\theta_3 = \alpha_1 + \alpha_2 + \alpha_3$. Note that such a simple representation is possible because the external (or anatomic) rather than internal (or included) joint angles were used. The projections of joints 2 and 3 and of the terminal point P on the X and Y axes are

$$X_2 = \ell_1 \cos \theta_1 = \ell_1 \cos \alpha_1$$
$$Y_2 = \ell_1 \sin \theta_1 = \ell_1 \sin \alpha_1$$
$$X_3 = \ell_1 \cos \theta_1 + \ell_2 \cos \theta_2 = \ell_1 \cos \alpha_1 + \ell_2 \cos(\alpha_1 + \alpha_2)$$
$$Y_3 = \ell_1 \sin \theta_1 + \ell_2 \sin \theta_2 = \ell_1 \sin \alpha_1 + \ell_2 \sin(\alpha_1 + \alpha_2)$$
$$X_P = \ell_1 \cos \theta_1 + \ell_2 \cos \theta_2 + \ell_3 \cos \theta_3$$
$$= \ell_1 \cos \alpha_1 + \ell_2 \cos(\alpha_1 + \alpha_2) + \ell_3 \cos(\alpha_1 + \alpha_2 + \alpha_3) \qquad (2.18)$$
$$X_P = \ell_1 \sin \theta_1 + \ell_2 \sin \theta_2 + \ell_3 \sin \theta_3$$
$$= \ell_1 \sin \alpha_1 + \ell_2 \sin(\alpha_1 + \alpha_2) + \ell_3 \sin(\alpha_1 + \alpha_2 + \alpha_3)$$

Again, the direct problem is easily solved. In matrix form the equation for the point P is

$$\begin{bmatrix} X_P \\ Y_P \end{bmatrix} = \begin{bmatrix} \cos \theta_1 & \cos \theta_2 & \cos \theta_3 \\ \sin \theta_1 & \sin \theta_2 & \sin \theta_3 \end{bmatrix} \cdot \begin{bmatrix} \ell_1 \\ \ell_2 \\ \ell_3 \end{bmatrix} \qquad (2.19)$$

Inverse kinematics. The coordinates of the end effector point, X_P, Y_P, are given. The joint coordinates, α_1, α_2, and α_3, need to be found. The problem has an

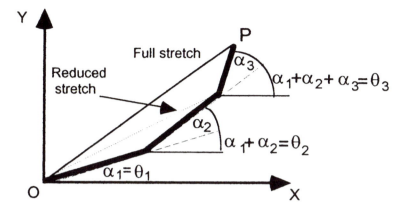

Figure 2.20 Model of a three-link planar chain. The terminal point of the distal segment P has the coordinates (X_p, Y_p) in the absolute reference frame.

infinite number of solutions: two equations in three unknowns cannot be solved in a unique way. Because of the redundancy of the kinematic chain, the same pointing task may be performed in many different ways.

Representation. When radius-vectors are drawn from the origin O to the tip P or to joint 3 (i.e., the wrist or ankle), the corresponding distances are called *full stretch* and *reduced stretch* (see Figure 2.20). The arm or leg rotation and flexion-extension are as previously described. The three-link extremity defined by its radius-vector (stretch and position) has one redundant DOF. When the terminal point is fixed, the chain can still change its configuration; however, all its joint angles are coupled.

2.2.5.2 Multilink Chains and Transformation Analysis

Even a simple three-link chain, when analyzed in space, is a rather complex object, almost impossible to visualize and geometrically analyze. Geometric representation is becoming useless for complex systems; analytic methods should be used. Consider an open kinematic chain of N rigid bodies where N > 2 (Figure 2.21). The links are numbered from zero (for an imaginary unmovable object connected with link 1) to N. Except for the last link, each link of the chain has two joints connecting it to adjacent bodies. The joints are numbered from 1, between the axis X and link 1, to N, between links (N – 1) and N. Local coordinate systems, x_i, y_i, z_i, are attached to each link i. The position of each link can also be represented in the global frame X,Y,Z. In the ensuing paragraphs, the direct kinematic problem is discussed. A single solution to the inverse problem does not exist because of the chain's redundancy.

The position of frame i relative to the previous frame (i − 1) can be described by the 4×4 transformation matrix $[T_{(i-1),i}]$, where (i − 1) and i stand

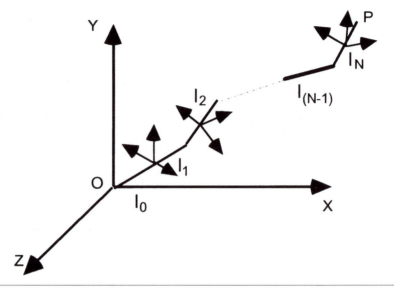

Figure 2.21   Model of a multilink open chain in space.

for the link numbers. The matrices of local transformation are assumed to be known. The end-effector position is then defined as the composition of sequential coordinate transformations $[T_{(i-1),i}]$, where i runs from 1 to N. Because the composition of several displacements is given by matrix multiplication of the corresponding transformation matrices (see equation 1.22), the end-effector position in the global space is defined by the homogeneous transformation

$$\left[T_{0,N}\right]=\left[T_{0,1}\right]\left[T_{1,2}\right], \ldots, \left[T_{(N-1),N}\right] \qquad (2.20)$$

where $[T_{0,N}]$ is the homogeneous transform. Equation 2.20 is termed the *structure equation* of the chain. It defines the position of the end effector, N, in terms of the relative positions of each link in the chain. The 4×4 matrix $[T_{0,N}]$ comprises a position vector of the origin of the local end-effector coordinate frame as well as a 3×3 rotation matrix.

The matrices $[T_{(i-1),i}]$ define both the *link* transformations, prescribed by the link anatomy, and the *joint* transformations determined by the joint position. To clarify an origin of transformation, the joint coordinate systems should be used in addition to the local frames fixed with the links (similar to the local systems described in Section **1.3.3**). After defining the link transformations as $[L_i]$ and the joint transformations as $[J_{(i-1),i}]$, the position of the end effector is given as

$$\left[T_{0,N}\right]=\left[J_{0,1}\right]\left[L_1\right]\left[J_{1,2}\right]\left[L_2\right], \ldots, \left[J_{(N-1),N}\right] \qquad (2.21)$$

and the position of the tip of the end effector as

$$\left[T_{0,N} \right] = \left[J_{0,1} \right] \left[L_1 \right] \left[J_{1,2} \right] \left[L_2 \right], \ldots, \left[J_{(N-1),N} \right] \left[L_N \right] \qquad (2.22)$$

where $[L_N]$ is the position of the tip in the joint coordinate system of joint N. The transformations $[L_i]$ are constant; they are defined by link geometry. The transformations $[J_{(i-1),i}]$ contain the joint coordinates.

When a closed rather than an open chain is analyzed, the position of the end effector and the starting point is the same. Hence, for the closed chain, the

---

■ ■ ■ *From the Scientific Literature* ■ ■ ■

Describing Arm Position With Respect to the Trunk

Source: van der Helm, F.C.T. & Pronk, G.M. (1995). Three-dimensional recording and description of motions of the shoulder mechanism. *Journal of Biomechanical Engineering, 117,* 27-40.

When reading this example from the literature the reader is advised to refresh the knowledge from Section **1.2.5.1.3,** in addition to that in Section **2.2.5.2**.

The arm is connected to the trunk through the shoulder mechanism, a constellation of four bones—the humerus, scapula, clavicle, and thorax. Upper arm motion involves three joints: (a) the sternoclavicular, between the thorax and clavicle; (b) the acromioclavicular, between the clavicle and scapula; and (c) the glenohumeral, between the scapula and humerus. Also, the thorax rotates with respect to the environment at the utmost humeral elevation angles. The following procedure was used to describe an arm movement.

1. A global reference system was selected with the origin at the suprasternal notch, with the X axis pointing laterally, the Y axis pointing cranially, and the Z axis pointing dorsally. Only right shoulders were analyzed. The local reference systems were fixed with the thorax, clavicle, scapula, and humerus (Figure 2.22).

2. The following matrices defining orientation of the local coordinate systems with respect to the global system were introduced: [Th] for the thorax, [C] for the clavicle, [S] for the scapula, and [H] for the humerus. If the notation system used in this book were employed, these matrices would have been labeled $[J_{01}]$, $[J_{02}]$, $[J_{03}]$,

See **ARM,** *p. 122*

■ ■ ■ **ARM,** *continued from p. 121*

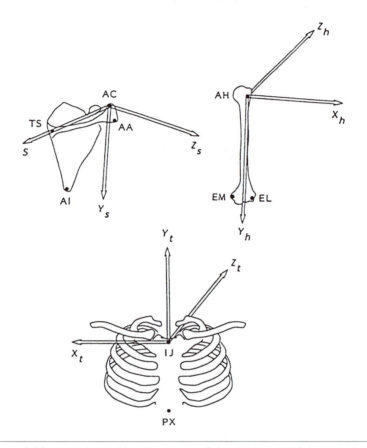

Figure 2.22   Local coordinate systems of the thorax, the right scapula, and the humerus. The scapula and humerus are drawn from the rear. AC, acromion; TS, trigonum spinae; AI, angulus inferior; AA, angulus acromialis; AH, the gap between acromion and humerus; EM and EL, medial and lateral epicondyli; IJ, incisura jugularis; PX, processus xyphoideus.

Reprinted from *Clinical Biomechanics, 8,* Veeger, H.E.J., van der Helm, F.C.T., & Rozendal, R.H., Orientation of the scapula in a simulated wheelchair push, 81-90, 1993, with kind permission of Elsevier Science Ltd, The Boulevard, Langford Lane, Kidlington 0X5 1GB, UK.

and $[J_{04}]$ or, using the notation from Chapter **1**, $[R_{G1}]$, $[R_{G2}]$, $[R_{G3}]$, and $[R_{G4}]$.

3. The rest positions were labeled with subscript 0 and the final position with subscript i. Hence, $[H_0]$ means "the rest position of the humerus with respect to the global system of coordinates," and $[H_i]$ indicates "the final position of the humerus in the global coordinate system."

4. The following matrices define *rotation* of the bone from its rest position to position i: $[Rt_{0i}]$ for the thorax, $[Rc_{0i}]$ for the clavicle, $[Rs_{0i}]$ for the scapula, and $[Rh_{0i}]$ for the humerus. With the notation system employed in Chapter **1**, these matrices would be labeled as $[R_{12}]$ or $[D_{12}]$, with an additional symbol for the thorax, clavicle, scapula, and humerus. Note that the authors of this paper did not use specific symbols for $[R_{12}]$ or $[D_{12}]$. Rather, they simply indicate whether the rotation is defined with respect to the global or local axes and use the same symbol for the different matrices.

5. The final positions of the bones as well as the global and local rotation matrices were represented as shown in Table 2.3. Note that because rotations only are discussed, the matrices are orthogonal, that is, their transposes are equal to their inverses.

6. The shoulder mechanism, as compared with other kinematic chains, is specific in that (a) movements in the sternoclavicular, acromioclavicular, and glenohumeral joints are not precisely defined with respect to the proximal bones, e.g., arm flexion is usually described as joint movement around the frontal axis of the trunk but not as a rotation around an axis fixed with the scapula, and (b) the clavicle and scapula are hidden underneath the skin and their instant positions are difficult to record. As a result, the joint rotations were defined with respect to the global axes (Table 2.4).

The left column of Table 2.4 gives the final position of the bone as a function of its initial position and joint angular displacements. The right column represents the joint angular displacements (rotation matrices) as a function of the initial position of the bone in the global frame, the final position of the bone, and the rotation matrices describing joint angular displacement in the proximal joints. Note how troublesome it is to describe the distal bone orientation when all the joint displacements (rotation matrices) are taken with respect to the global axes (compare with equation 2.20).

See **ARM**, *p. 124*

■ ■ ■ ARM, *continued from p. 123*

Table 2.3 Orientational Matrices Describing a Bone Orientation With Regard to Bone Reference Position

Bone	Bone rotation with regard to local axes		Bone rotation with regard to global axes	
	Final bone orientation	Local rotation matrix (transformation matrix)	Final bone orientation	Global rotation matrix (displacement matrix)
Thorax, $[Th_i]$	$[Th_i] = [Th_0][Rt_{0i}]$	$[Rt_{0i}] = [Th_0]^T[T_i]$	$[Th_i] = [Rt_{0i}][Th_0]$	$[Rt_{0i}] = [T_i][Th_0]^T$
Clavicle, $[C_i]$	$[C_i] = [C_0][Rc_{0i}]$	$[Rc_{0i}] = [C_0]^T[C_i]$	$[C_i] = [Rc_{0i}][C_0]$	$[Rc_{0i}] = [C_i][C_0]^T$
Scapula, $[S_i]$	$[S_i] = [S_0][Rs_{0i}]$	$[Rs_{0i}] = [S_0]^T[S_i]$	$[S_i] = [Rs_{0i}][S_0]$	$[Rs_{0i}] = [S_i][S_0]^T$
Humerus, $[H_i]$	$[H_i] = [H_0][Rh_{0i}]$	$[Rh_{0i}] = [H_0]^T[H_i]$	$[H_i] = [Rh_{0i}][H_0]$	$[Rh_{0i}] = [H_i][H_0]^T$

Table 2.4   Bone Orientation as a Result of Joint Rotations With Regard to Global Axes

Bone	Bone Orientation	Global Rotation Matrices
Thorax	$[Th_i] = [Rt_{0i}][Th_0]$	$[Rt_{0i}] = [T_i][Th_0]^T$
Clavicle	$[C_i] = [Rt_{0i}][Rc_{0i}][C_0]$	$[Rc_{0i}] = [Rt_{0i}]^T[C_i][C_0]^T$
Scapula	$[S_i] = [Rt_{0i}][Rc_{0i}][Rs_{0i}][S_0]$	$[Rs_{0i}] = [Rc_{0i}]^T[Rt_{0i}]^T[S_i][S_0]^T$
Humerus	$[H_i] = [Rt_{0i}][Rc_{0i}][Rs_{0i}][Rh_{0i}][H_0]$	$[Rh_{0i}] = [Rs_{0i}][Rc_{0i}]^T[Rt_{0i}]^T[H_i][H_0]^T$

successive transformation given by equation 2.22 is an identity transformation, $[T_{0,N}] = [I]$.

$$\left[J_{0,1}\right]\left[L_1\right]\left[J_{1,2}\right]\left[L_2\right], \ldots, \left[J_{(N-1),N}\right]\left[L_N\right] = [I] \qquad (2.23)$$

This is a matrix loop equation. A vector loop approach can also be employed. The vector technique includes several consecutive steps:

1. Attach vectors to the links forming a closed loop. The magnitude and the direction of the vectors should correspond to the length and orientation of the links.
2. Write a vector loop position equation. The equation states that the sum of the vectors in the loop is zero.
3. Break the vector equation into scalar component equations. Solve the equations for the position unknowns.

Unknown velocity and acceleration can also be found by differentiating the equations with regard to time.

2.2.5.3 Denavit-Hartenberg Convention

Representative papers: Chao, Rim, Smidt, & Johnston, 1970; Denavit & Hartenberg, 1955; Raikova, 1992.

Structure equations are the most universal tools for defining positions of serial link chains. The equations, however, are redundant and their parameters are difficult to visualize. This is because specific information about the chain (for example, joint constraints) is not used in the definition. The method called the

■ ■ ■ *FROM THE SCIENTIFIC LITERATURE* ■ ■ ■

Vector Loop Technique in Analysis of Closed Kinematic Chains

Source: You, H. (1996). A study on the determination of an optimal seat-pedal relationship using kinematic simulation. A student research project in the Advanced Biomechanics of Human Motion class. Department of Kinesiology, Penn State University.

The goal of the study was to find an optimal seat-pedal location for drivers of various body dimensions. The seat-pedal system and its vector model are shown in Figure 2.23. The system includes two closed kinematic chains. To show how the method works, we apply it to the first closed chain, which comprises five segments. The corresponding vectors represent the hip joint height ($\mathbf{r}_1$), the femur ($\mathbf{r}_2$), the shank ($\mathbf{r}_3$), the segment from the ankle joint to the heel contact point ($\mathbf{r}_4$), and the vector from the heel point to the vertical projection of the hip joint ($\mathbf{r}_5$). The following vector loop equation is valid

$$\mathbf{r}_1 + \mathbf{r}_2 + \mathbf{r}_3 + \mathbf{r}_4 + \mathbf{r}_5 = 0$$

For the horizontal and vertical components the equations are

$$\mathbf{r}_1 \sin \theta_1 + \mathbf{r}_2 \sin \theta_2 + \mathbf{r}_3 \sin \theta_3 + \mathbf{r}_4 \sin \theta_4 + \mathbf{r}_5 \sin \theta_5 = 0$$
$$\mathbf{r}_1 \cos \theta_1 + \mathbf{r}_2 \cos \theta_2 + \mathbf{r}_3 \cos \theta_3 + \mathbf{r}_4 \cos \theta_4 + \mathbf{r}_5 \cos \theta_5 = 0$$

The length of the femur, $\mathbf{r}_2$, the shank, $\mathbf{r}_3$, and the ankle-to-heel distance, $\mathbf{r}_4$, are constant. Also, $\theta_1 = \pi/2$ and $\theta_5 = 0$. The distances $\mathbf{r}_1$ and $\mathbf{r}_5$ are unknown or controlled variables. The angles θ_2, θ_3, and θ_4 are also unknown or controlled variables. The angles depend—among other parameters—on the knee included angle, α_1, and the ankle included angle, α_2. The chain consists of five links and five revolute joints. Hence, according to Gruebler's formula for closed chains in a planar system (equation 2.12) the chain possesses two DOF.

The loop vector equations of the chain were solved and the solutions were tabulated. Using the tables, a designer interested in an adjustment of a car seat to a driver of certain body dimensions (the lengths $\mathbf{r}_2$, $\mathbf{r}_3$, and $\mathbf{r}_4$ are known) can use either one of two strategies: (1) select the values of $\mathbf{r}_1$ and $\mathbf{r}_5$ and then determine the joint angles, or (2) select the values of any two angles and determine the values of $\mathbf{r}_1$ and $\mathbf{r}_5$ and the third angle.

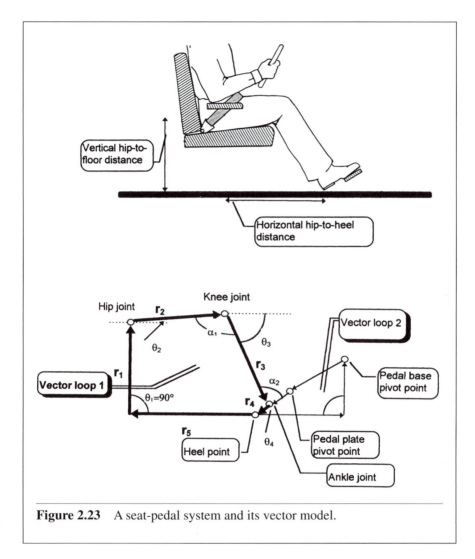

Figure 2.23 A seat-pedal system and its vector model.

Denavit-Hartenberg convention provides a de facto standard procedure for describing multilink chains, especially in robotics. The convention is based on local reference systems attached to the links but defined in the joints. The *i*th frame moves with the *i*th link. The Denavit-Hartenberg notation is introduced for chains with one-DOF joints, both revolute and prismatic (telescopic). We confine the discussion to the revolute joints only. The convention is further explained through an example (Figure 2.24).

Suppose one is interested in a two-link chain formed by the link (i − 1) and the link i. The motion of interest is rotation in joint i. The motion is performed

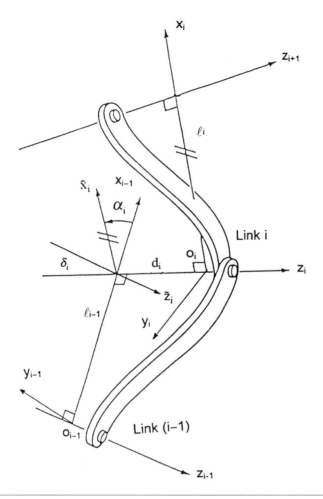

Figure 2.24 Parameters used in the Denavit-Hartenberg notation. The two back slashes (\\) through the lines x_i and $\hat{x}_i$ indicate that the lines are parallel. Rotation in the joint i is defined by the angle α_i formed by the axes x_{i-1} and $\hat{x}_i$.

about the axis z_i. For the sake of illustration, the axes of rotation in the joints $(i - 1)$, i, and $(i + 1)$ are not assumed to lie in one plane. In general, for a link i, the proximal joint is labeled joint i and the distal joint is joint $(i + 1)$. The axes z_i and z_{i+1} define two lines in the link i. In a similar way, two lines are defined for each (intermediate) body segment. The problem is to assign local coordinate frames that permit the compact description of joint transformations. We define, as an example, the frame for the link i. Frames are determined by two rules:

1. The z_i axis lies along the axis of rotation of the ith joint.

2. The x_i axis is perpendicular to both the z_i axis and the z_{i+1} axis.

Rule 2 means that the x_i axis is a common normal (perpendicular) to the axes of rotation in the two neighboring joints, i and (i + 1). Trace out this relationship in Figure 2.19 to make sure this point is clear. The interception of the z_i and x_i axes defines the origin of the ith local frame, o_i, and y_i completes the right-handed coordinate system. Whenever z_i and z_{i+1} are not parallel, the common perpendicular is unique and o_i is uniquely specified. Note that the position of o_i along z_i might be outside of the joint i. Selection of the o_i location for the parallel axes of rotation in joints will be discussed later in the text.

Assuming that local coordinate systems in other joints are defined in the same way, the following parameters are used to determine the relative location of the two frames:

- The segment length, ℓ, which is defined as the shortest distance between the axes of flexion-extension in the two joints, i and (i + 1), i.e., the elbow and the wrist. The segment length defined in such a way is called the *biomechanical length of the body segment*. Recall that in anatomy and anthropology the segment length is defined differently, as a projected distance between anatomic points rather than the axes of rotation. Contrary to the *anatomic length*, the biomechanical length is measured along the common normal of the two axes of rotation.

- The interjoint offset, d_i, is defined as the distance between the points o_{i-1} and o_i measured along the elbow joint axis, z_i (in joint i). In the elbow joint, because of the valgus angulation of the forearm with the elbow fully extended and the forearm supinated, the joint offset is easily observed.

- The twist, or offset, angle, δ_i, which differs from zero when the axes of flexion-extension in two joints of the same link, for example, in the elbow and the wrist, are not exactly in the same (frontal) plane. For the elbow and wrist, this angle is small. In the ankle joint, the axis of flexion-extension does not rest in the frontal plane. Therefore, the twist angle formed by the axes of rotation in the knee joint and the ankle joint is rather large.

- The angle of rotation (flexion-extension) in the joint i, α_i.

The *segment length*, *joint offset*, and *twist angle* are anatomic parameters that are invariable. Joint configuration depends on the angle of rotation only. The four parameters of the Denavit-Hartenberg convention are used to define the local coordinate systems and the transformation matrices. The transformation matrices are homogeneous: they define both the rotation and translation of one local joint system with regard to another. A 4 × 4 matrix associated with joint (i − 1) transforms coordinates given in the local frame i into the coordinates in the frame i − 1. This transformation can be seen as the following sequence of operations:

1. rotation by $-\delta_i$ to align z_i and z_{i-1};
2. translation along x_{i-1} by ℓ_{i-1}, bringing z_i and z_{i-1} into coincidence;
3. translation along z_i by d_i aligning the xy_i plane (defined in the joint i) with the xy_{i-1} plane, defined in the joint (i − 1); and
4. rotation by $-\alpha_i$, bringing the two frames into the full alignment.

In matrix form this is expressed as

$$
[T_{i-1,i}] =
\underbrace{\begin{bmatrix} \cos\theta_i & -\sin\theta_i & 0 & 0 \\ \sin\theta_i & \cos\theta_i & 0 & 0 \\ 0 & 0 & 1 & 0 \\ 0 & 0 & 0 & 1 \end{bmatrix}}_{\substack{\text{2nd rotation} \\ \text{(flexion-extension)}}}
\underbrace{\begin{bmatrix} 1 & 0 & 0 & 0 \\ 0 & 1 & 0 & 0 \\ 0 & 0 & 1 & d_i \\ 0 & 0 & 0 & 1 \end{bmatrix}}_{\substack{\text{2nd translation} \\ \text{(joint offset} \\ \text{alignment)}}}
\underbrace{\begin{bmatrix} 1 & 0 & 0 & \ell_i \\ 0 & 1 & 0 & 0 \\ 0 & 0 & 0 & 0 \\ 0 & 0 & 0 & 1 \end{bmatrix}}_{\substack{\text{1st translation} \\ \text{(along the} \\ \text{common normal)}}}
$$

$$
\times \underbrace{\begin{bmatrix} 1 & 0 & 0 & 0 \\ 0 & \cos\alpha_i & -\sin\alpha_i & 0 \\ 0 & \sin\alpha_i & \cos\alpha_i & 0 \\ 0 & 0 & 0 & 1 \end{bmatrix}}_{\substack{\text{1st rotation} \\ \text{(twist angle)}}}
= \begin{bmatrix} \cos\theta_i & -\cos\alpha_i\,\sin\theta_i & \sin\alpha_i\,\sin\theta_i & \ell_i\,\cos\theta_i \\ \sin\theta_i & \cos\alpha_i\,\cos\theta_i & -\sin\alpha_i\,\cos\theta_i & \ell_i\,\sin\theta_i \\ 0 & \sin\alpha_i & \cos\alpha_i & d_i \\ 0 & 0 & 0 & 1 \end{bmatrix}
$$

$$(2.24)$$

Equation 2.24 represents a standard Denavit-Hartenberg transformation matrix. The last column of this matrix locates the origin of frame i in terms of the reference system (i − 1). The upper left 3 × 3 submatrix represents rotation of the *i*th frame relative to the frame (i − 1). There are n + 1 frames on an n-joint kinematic chain.

Several additional rules apply in the Denavit-Hartenberg notation. Ball-and-socket joints are described as three joints with the twist angle 90° and the zero values of the segment length and the joint offset. When two axes of an intermediate link are parallel (making the twist angle 0°), the common normal is usually chosen in such a way that the joint offset d_i also becomes zero. The first and the last bodies of the chain have no associated joints. To define the local frames for them, two fictitious "joints" are introduced: joint 1 is between the global frame counted as the zero link of the chain and the first link of the chain, and the joint (n + 1) is arbitrarily specified at the last link. Once the axes of rotation in these "joints" are defined, other parameters of the Denavit-

■ ■ ■ *FROM THE SCIENTIFIC LITERATURE* ■ ■ ■

Using Denavit-Hartenberg Convention for Studying Finger Movement

Source: Casolo, F., & Lorenzi, V. (1994). Finger mathematical modeling and rehabilitation. In: F. Schuind, K.N. An., W.P. Cooney III, & M. Garcia-Elias (Eds.). *Advances in the Biomechanics of the Hand and Wrist.* NATO ASI Series. Series A: Life Sciences Vol. 256. New York: Plenum Press, pp. 197-223.

The finger was described as an open chain of three rigid bodies with the metacarpal bone as a reference frame (Figure 2.25). The interphalangeal

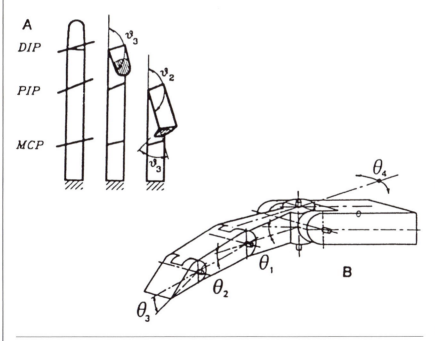

Figure 2.25 The Denavit-Hartenberg convention can be used for representing finger movement. A. Joint axes of rotation are generally not parallel. B. Scheme of the finger joint kinematics.

Adapted from Casolo, F. & Lorenzi, V. (1994). Finger mathematical modeling and rehabilitation. In F. Schuind, K.N. An., W.P. Cooney III, & M. Garcia-Elias (Eds.), *Advances in the biomechanics of the hand and wrist.* (pp. 197-223). NATO ASI Series. Series A: Life Sciences Vol. 256. New York: Plenum Press.

See **DENAVIT-HARTENBERG,** *p. 132*

■ ■ ■ **DENAVIT-HARTENBERG,** *continued from p. 131*

joints were modeled as hinges, i.e., revolute joints with one DOF. In accordance with experimental observations, the axes of the interphalangeal joints were not assumed to be parallel (see Figure 2.25A). The metacarpophalangeal joint was represented as a joint with two DOF, flexion-extension and abduction-adduction. Hence, the chain had four DOF (Figure 2.25B). To define the chain configuration, the following parameters should be known: (a) segment length, in total 3; (b) angles defining axes of orientation, e.g. Euler's angles, in total 9; and (c) angles associated with the finger movement, in total 4, flexion-extension angle in the distal interphalangeal joint, proximal interphalangeal joint, and metacarpophalangeal joint, and abduction-adduction angle in the metacarpophalangeal joint.

The coordinate systems were selected according to the Denavit-Hartenberg convention. Each phalanx and joint, from proximal to distal, was labeled with k = 1, 2, 3 (0 for the metacarpal bone). A reference frame k was assigned to the center of joint k. The reference axes were oriented as follows: x_k along the joint axis of rotation, directed from the ulna to the radius; y_k perpendicular to a plane defined by x_k and k + 1 joint center, directed dorsally; and z_k (for the right hand) coherent with the right-hand thumb rule.

With such a convention, the position matrix of the distal phalanx with respect to the metacarpal frame can be calculated by means of the simple expression:

$$[T_{0,3}] = [T_{0,1}] [T_{1,2}] [T_{2,3}]$$

Hartenberg notation can be measured in the manner discussed in the previous paragraphs.

With Denavit-Hartenberg matrices, the structure equation of the chain is defined as usual (see equation 2.20).

2.3 BIOLOGICAL SOLUTIONS TO KINEMATIC PROBLEMS

Representative papers: Abeele et al. 1993; Soechting & Flanders, 1989.

Fit humans move easily in an immediate environment, catch and manipulate objects, avoid obstacles, and adjust their motion to the changes in the extrapersonal space. Purposeful movement toward an object can be performed at a very early age. Four-month-old babies can grasp a toy. To move about efficiently in the physical environment and to manipulate objects, the CNS needs to integrate the sensory information about the location of the object relative to the body (or the body with regard to the object) on the one hand, and the body posture (joint configuration) on the other.

The methods of kinematic geometry described previously are commonly used to define the position of body parts with regard to each other and the environment. Together, they provide a powerful tool for describing body position, including its location, orientation, and posture. These tools are used efficiently by external observers, both biomechanists and engineers. This does not imply, however, that the CNS uses similar approaches to control body position. Most probably, it does not. The internal representation of body position is executed in another way, which still remains unsolved. In what follows, a brief description of the problem is provided.

2.3.1 Internal Representation of the Immediate Extrapersonal Space

Representative paper: Kosslyn et al., 1992.

Human beings use two approaches to judge the location of objects in space. In the first, which is called *categorical*, spatial relations of objects are described verbally by such expressions as "to the left" or "to the right" or by prepositions "below," "on," and "in front of." The precise location of the object is not made explicit in such terms. For example, when the object is located to the left of another, it can be at any place with regard to the horizontal and vertical axes. Some spatial categories, for instance, laterality, develop relatively late in life. Children begin to discern left and right at the age of 7 or 8. Neural structures underlying the perception of categorical relations are contained mainly within the left hemisphere of the brain. In general, the left hemisphere is important for conceptual thinking, including any category formation, and language. Patients with a left-hemisphere stroke have decreased ability to recognize categorical spatial transformations.

Representation of the precise location of objects is called *coordinate representation*. The neural substrate of the perception of coordinate relations is mainly within the right hemisphere of the brain. Patients with right-hemisphere strokes mistakenly estimate the coordinates of the object. Although the metric aspects, such as distances, angles, or volume, can be put into language, this is typically done only in technology and science and requires special education. When asked to estimate metric aspects verbally, most people do it rather imprecisely. However, to move successfully, the metric aspects must be perceived accurately.

To discuss the internal coordinate representation, it seems natural to preserve the main terminology used in kinematics and to speak about spatial reference frames used by the CNS. These (internal) reference frames, however, may be of any nature and type. They can be formed in the sensors (e.g., location of a point on the retina, direction of velocity sensed by semicircular canals in the vestibular organ), in the motor system (e.g., the location of the head with regard to the ground), or they can be constructed by the brain. Sensory messages initially expressed in various frames must be somehow integrated by the CNS. This is believed to be a multistage (hierarchical) process, a sequence of transformations (mappings) between successive reference frames. For example, to locate a visual target with regard to the performer's body, the following neural transformations should occur:

1. The visual object produces an image on the retina; the object's location is presented in the retinal reference frame.

2. The eye orientation within the head should be sensed. The object's coordinates are transformed from the retina-centered reference frame into the head-centered reference frame.

3. The head position with regard to the trunk should be taken into account. The transformation of the object's coordinates from the head-centered to the trunk-centered frame takes place. As a result, the position of the target with regard to the body (if the trunk and the body are synonymous) is established.

If the matrix method were used, it would result in multiplication of retinal coordinates by the two matrices describing the transformation "from the eyes to the head" and "from the head to the trunk." If the target position needs to be determined in relation to a particular body part (e.g., the shoulder, hand, finger) rather than to the trunk, the computational complexity increases.

2.3.2 Internal Representation of the Body Posture

For an illustration, first consider how the problem of end-effector positioning has been solved in robotics. This is usually done in two steps:

1. the (real or desired) position of the end effector is given in an external reference system, and then

2. the inverse kinematics solution is sought to define the appropriate joint angular coordinates.

Hence, the position of the robot's hand is given first in the extrinsic coordinates, and then the intrinsic coordinates of the joint angles are found. Evidently, for multilink chains with many DOF, there is no unique solution to the inverse kinematics problem. Many joint configurations can bring the end effector to the same position.

2.3.2.1 Frames of Reference and the "Body Scheme"

Representative papers: Berkenblit et al., 1986; Soechting & Flanders, 1992.

The frames used by the CNS for spatial representation of body posture (e.g., joint angles, limb angular orientation in the external space) are still objects of discussion and intensive research. The possibility also exists that various reference frames are used concurrently or in sequence. In the latter case, the movement is broken into parts for which different frames are used. In what follows, some considerations regarding the internal representation of body posture are provided.

The "robotics" approach, in which the end effector position is described by a composition of several joint transformations, inevitably leads to the accumulation of errors from various joints. The positioning accuracy of the terminal point in the chain is less than the accuracy of individual joint positioning. Surprisingly, joint errors are not accumulated in human movements: the accuracy of the end-effector positioning does not depend on the length of the chain. To illustrate this you can conduct a simple experiment. With any finger of one hand touch the same finger of the opposite hand behind your back. Most people can do this immediately or during a second attempt. Let us formally describe the motor task. Suppose that an internal global system is fixed somewhere in the shoulder girdle. In the experiment, each arm chain includes at least six joints (the glenohumeral, elbow, wrist, and the three finger joints). Therefore, for each arm the following structure equation can be written:

$$P = [T_{sh}] [T_{elb}] [T_{wrist}] [T_{carpometacarpal}] [T_{metacarpophalangeal}] [T_{interphalangeal}] [L_N] \quad (2.25)$$

where **P** is a three-dimensional vector of the position of the terminal point in the global frame, the [T]s are transformation matrices, and $[L_N]$ is the position of the tip in the joint coordinate system of joint N. Each transformation matrix contains certain "errors of measurement" (people reproduce a joint angular position with some variation). Despite these errors, and the involvement of 12 joints totally, the final accuracy of the two chains is very high. When the same movement is performed several times, the variability of the end-effector trajectory is less than that of the individual joints.

The complexity and robustness of strategies used by animal species for positioning their limbs are confirmed by classic physiological experiments with spinal frogs (frogs with dissected brains and intact spinal cords). When a piece of paper soaked in acid is applied to a frog's skin, the frog wipes the paper away from the body. This multijoint movement is performed in a coordinated manner, disregarding almost all attempts to fool the spinal frog (e.g., to place the stimulus at different locations, like on the ipsilateral forelimb, to change the limb position, to load the limb, to introduce mechanical obstacles to the movement). If the task is solvable, the spinal frogs solve it, and usually on the first trial.

These experiments have led to the conclusion that a *body scheme* is somehow represented in the CNS. The body scheme can be viewed as a specific internal representation of the body's kinematic geometry. At present we know little about the internal representation. It is postulated that a body-centered frame of reference is somehow constructed and used.

It seems that the orientation of individual body segments in the gravity field is one of the parameters used in the body scheme. Gravity provides a stable vertical reference axis (a *geocentric reference frame*). The inclination of a body limb to the gravitational vertical is perceived with a higher accuracy than joint angles. It has been hypothesized that the CNS employs a geocentric reference system rather than a joint-based system. This hypothesis stems, in part, from the fact that people can easily maintain a joint angular position, regardless of the whole body's positioning in the gravity field. For instance, during trunk bending the head is easily maintained in the same position with regard to the trunk even though the head is differently oriented in the field of gravity. This is possible only because the activities of the neck muscles at each trunk position are adjusted to counterbalance the gravity force. Likewise, when a horizontally placed forearm is supinated or pronated, the muscles maintaining the forearm horizontally vary according to the orientation of the forearm relative to the upper arm. Thus, by assumption, the position of the body segments relative to the vertical, rather than to the neighboring segment only, is used in the internal representation of the body posture. This approach is used when studying the internal representation of kinematic geometry in human movements (Figure 2.26). The geocentric system, however, is not the only frame of reference used by the CNS; astronauts perform skilled maneuvers in the absence of gravity.

Evidence has been provided to show that errors in movement execution are due to mistakes in sensorimotor transformation from the extrinsic to the intrinsic coordinates: the subjects readily recognize small changes in both the location of an external target (thus, extrinsic representation was very accurate) and in limb position (intrinsic representation was accurate too), but they fail to precisely point out the target without visual control.

When a tool is used by a person, that person should locate and orient the tool in an extrinsic coordinate system (also called a *task space*). To do that, the person should construct a local frame centered on the object being used. He or she should sense (or know) the dimensions of the instrument, its functional axes, inertial properties, and so on. The tool becomes an extension of oneself, a new end effector of the human body's kinematic chain.

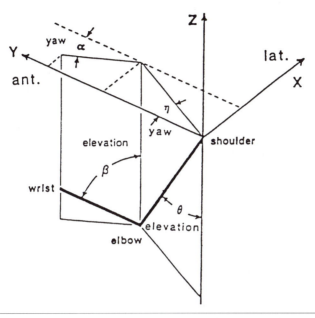

Figure 2.26 The relationship between target location (extrinsic coordinates) and arm orientation (intrinsic coordinates). Arm and forearm inclination are given with respect to the vertical, rather than as joint angles. Two angles are used: (a) elevation (pitch), which is the angle between the limb segment and the vertical axis, measured in the vertical plane; and (b) the yaw angle between the limb segment and the anterior axis, measured in the horizontal plane. The target location is described in Cartesian coordinates X,Y,Z. The relations between extrinsic and intrinsic coordinates are:

$$X = \ell_a \sin \theta \sin \eta + \ell_f \sin \beta \sin \alpha$$
$$Y = \ell_a \sin \theta \cos \eta + \ell_f \sin \beta \cos \alpha$$
$$Z = \ell_a \cos \theta + \ell_f \cos \beta$$

where ℓ_a is the length of the upper arm, ℓ_f is the length of the forearm, the angles θ and η are the vertical and horizontal angles of the upper arm, and β and α are the vertical and horizontal angles of the forearm. Note that the chain is redundant (three equations with four unknowns) and the equations are nonlinear.

From Soechting, J.F. & Flanders, M. (1989). Errors in pointing are due to approximations in sensorimotor transformations. *Journal of Neurophysiology, 62,* 595-608. Reprinted by permission.

2.3.2.2 Vector Field Representation of the Movement Geometry

Representative papers: Bizzi et al., 1991; Georgopulos et al., 1993; Mussa-Ivaldi and Giszter, 1992.

From neurophysiological experiments it follows that some populations of neural cells are tuned to a certain direction of movement. Two examples follow.

In the experiments with spinal frogs, a certain region of the spinal cord (the premotor layers in the lumbar gray matter) was stimulated with microelectrodes and the forces at the ipsilateral ankle were registered. The position of the limb was systematically changed (Figure 2.27A) while the same site was stimulated. The registered forces were directed to a single point within the limb reach. Considered together, all of the forces formed a field that converged at the equilibrium point (Figure 2.27B). When the limb was at the equilibrium point, the electrostimulation did not elicit force. Microstimulation of various sites generated different fields with different zero-force points. The fields were added vectorially. Two simultaneous microstimulations to two spinal regions elicited a vector field that was proportional to the vector sum of the fields produced by the independent stimulation of each region. These experiments suggested that the limb postures are represented in the form of convergent force fields acting on the limb's end point.

In another group of experiments, neural cells in motor and premotor cortices, whose activity depends on movement direction, have been discovered. Individual neurons from this population are tuned to the direction of move-

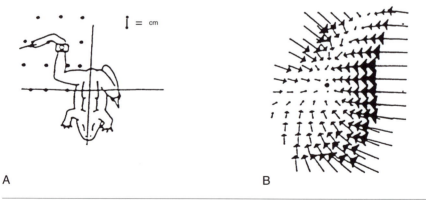

A   B

Figure 2.27   Force fields obtained from microstimulation of the frog's spinal cord. A. The positioning of the frog's extremity. The black dots indicate the locations at which the forces were measured. B. Vector field of forces (interpolated). The zero-force (equilibrium) point is indicated by a filled circle.
Adapted by permission from Mussa-Ivaldi, F.A. & Giszter, S.F. (1992). Vector field approximation: A computational paradigm for motor control and learning. *Biological Cybernetics, 67*, 491-500.

ment and have a preferred direction. When movement is performed in the preferred direction, the activity of a specific neuron is maximal (Figure 2.28, top panel). Otherwise, the activity of the neuron decreases progressively with movement farther away from the preferred direction. The cell's discharge rate changes as a linear function of the cosine of the angle between the preferred direction of the cell and the movement direction (Figure 2.28, bottom panel):

$$E_i = b + k \cos \theta_i \tag{2.26}$$

where E_i is a firing rate of the ith neuron, θ_i is the angle between the movement direction and the preferred direction for the ith neuron, and b and k are parameters.

When movement is performed in a certain direction, the large ensemble of directionally tuned cells is activated (Figure 2.29). The activity of the ensemble is represented by the neuronal population vector, which is calculated as a weighted sum of vector contributions of single cells:

$$\mathbf{P}(t) = \sum_i V_i(t)\mathbf{C}_i \tag{2.27}$$

where $\mathbf{P}(t)$ is the neuronal population vector, $\mathbf{C}_i$ is the unit preferred direction vector of the ith neuron, V_i is the weight proportional to the firing rate of the ith neuron, and t is time.

Neuronal population vectors accurately predict the direction of movement both during the periods at which the movement is being planned and when the movement is executed. The neuronal population vectors do not differ for movements of various amplitudes. It is important to mention that the movement direction in this set of experiments have been determined relative to the initial position of the hand. Hence, the direction angle θ is represented in the reference frame fixed with the hand. The preferred direction of the individual cells has not been affected by the changes in initial and intended arm position. This fact implies that the cells do not encode the location of the target itself but rather they encode the vectorial difference between the two hand positions, initial and intended.

The described findings suggest that the CNS represents movement geometry in a manner that hardly resembles the methods that have been developed in theoretical kinematics and realized in engineering.

2.4 SUMMARY

Various coordinate systems are used to describe a joint configuration, i.e., body posture. The following systems were discussed in the chapter: the clinical

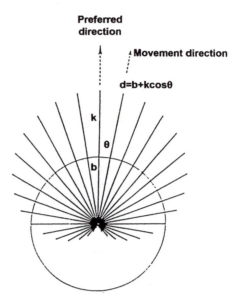

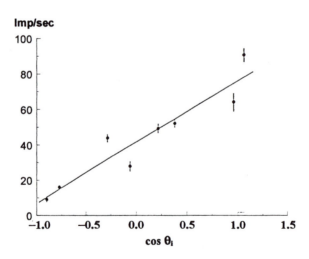

Figure 2.28 Relationship between single cell discharge and direction of movement. On the top panel, the relationship between the movement direction and the preferred direction of a cell is shown. The length of a line indicates the discharge rate of the cell with movement in the direction of the line. The bottom panel shows the mean discharge rate of the cell versus $\cos \theta_i$. Vertical bars are ± 1 standard deviation.

Adapted by permission from Schwartz, A.B., Kettner, R.E., & Georgopulos, A.P. (1988). Primate motor cortex and free arm movements to visual targets in three-dimensional space. 1. Relations between single cell discharge and direction of movement. *Journal of Neuroscience, 8,* 2913-2927.

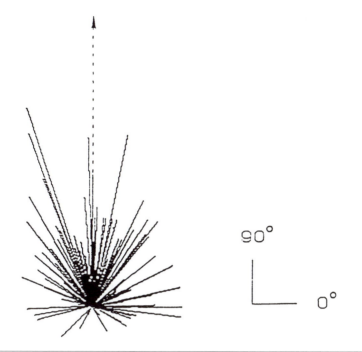

Figure 2.29 Vector contributions of 151 directionally tuned cells for movements directed at 90°. The spatial congruence between the movement direction and the direction of the population vector (dashed arrow) is evident.
Adapted by permission from Kalaska, J.F., Gaminiti, R., & Georgopulos, A.P. (1983). Cortical mechanisms related to the direction of two-dimensional arm movements: Relations in parietal area 5 and comparison with motor cortex. *Experimental Brain Research, 51,* 247-260.

system, globographic presentation, segment coordinate system, and the joint rotation convention.

Human body segments are combined in kinematic chains. The kinematic chain is open if one end of the chain (e.g., distal segment) is free to move. In closed chains, constraints are imposed on both ends of the chain. The term "degree of freedom" refers to an independent way in which the body can move. A free rigid body has six DOF, three translatory and three rotatory. The total number of DOF of a kinematic chain is called the mobility of the chain. Interest in the study of kinematic chains primarily revolves around two issues: (a) direct kinematics, when the joint coordinates are known and the end-effector position is sought; and (b) inverse kinematics, when the end-effector position is known and the joint coordinates are sought. The human skeletal system has many (about 245) DOF and is highly redundant. To perform a coordinated movement, excessive mechanical DOF should be overcome (Bernstein's problem).

The structure equation defines the position of the end effector (in the global system of coordinates) in terms of the relative position of each link in the chain. The Denavit-Hartenberg convention is often used when describing multilink chains.

The manner in which the geometries of the body and the immediate environment are represented in the CNS is currently the object of intensive research. At least three reference frames are used by the CNS to control body posture and movement: (1) the exocentric frame, in which immediate extrapersonal space is represented; (2) the geocentric frame, based on the gravity vector; and (3) the egocentric frame, representing the relative position of various body parts. Recent findings suggest that body posture, as well as movement geometry, is represented in the CNS in a vector form.

2.5 QUESTIONS FOR REVIEW

1. Give an example of three different technical coordinate systems.

2. What are the advantages and disadvantages of the clinical reference system?

3. Discuss the difference between the joint angles used in the globographic convention and Euler's angles.

4. Write down an equation relating coordinates of a point measured in two different segmental coordinate systems, one of which is fixed with a distal body segment and the second with the proximal segment. Use a homogeneous transform.

5. Using the joint rotation convention, define the coordinate axes for the hip joint. Which of the axes is the floating axis?

6. Give an example of singular joint configuration.

7. How many DOF does a joint of the 4th class have? Do rotational joints of the 1st class exist? What is the class of a hinge joint? How many DOF does a ball-and-socket joint have?

8. Define mobility and maneuverability of a kinematic chain.

9. Write down and explain Gruebler's formula for an open and closed kinematic chain.

10. How many DOF does a planar four-link model consisting of the foot, shank, thigh, and torso (the trunk, head and neck, and arms combined) have?

11. A driver pushes a pedal that rotates about a fixed axis. By assumption, the driver's pelvis does not move. How many DOF does a planar kinematic chain consisting of the hip, shank, foot, and the pedal have? Is

this chain open or closed? How do various assumptions about relative motion of the foot with regard to the pedal influence the answer?

12. Discuss the direct and inverse problem of kinematics.

13. An end point of the two-link kinematic chain has coordinates X and Y (see Figure 2.19). The length of the links is ℓ_1 and ℓ_2. Determine the angles θ_1 and θ_2.

14. A four-link planar chain is either open or closed. Write down the structure equations for the chain assuming the two configurations.

15. Discuss the Denavit-Hartenberg convention.

16. Discuss the problem of the external representation of extrapersonal space and body posture.

Answer to Question 13

$$\theta_2 = \cos^{-1}\left(\frac{X^2 + Y^2 - \ell_1^2 - \ell_2^2}{2\ell_1\ell_2}\right) \quad \theta_1 = \tan^{-1}\left(\frac{Y}{X}\right) - \tan^{-1}\left(\frac{\ell_2\sin\theta_2}{\ell_1 + \ell_2\cos\theta_2}\right)$$

BIBLIOGRAPHY

Abeele, S.V., Crommelinck, M., & Roucoux, A. (1993). Frames of reference used in goal-directed arm movement. In A. Berhoz (Ed.), *Multisensory control of movement* (pp. 362-378). Oxford, United Kingdom: Oxford Scientific Publications.

American Academy of Orthopaedic Surgeons. (1965). *Joint motion. Method of measuring and recording*. Chicago.

Andrews, J.G. (1984). On the specification of joint configurations and motions. *Journal of Biomechanics, 17,* 155-158.

Andrews, J.G. (1995). Euler's and Lagrange's equations for linked rigid-body models of three-dimensional human motion. In P. Allard, I.A.F. Stokes, & J.-P. Blanchi (Eds.), *Three-dimensional analysis of human movement* (pp. 145-176). Champaign, IL: Human Kinetics.

Berkenblit, M.B., Feldman, A.G., & Fukson, O.I. (1986). Adaptability of innate motor patterns and motor control mechanisms. *Behavioral and Brain Sciences, 9,* 585-638.

Berme, N., Cappozzo, A., & Meglan, J. (1990). Kinematics. In N. Berme & A. Cappozzo (Eds.), *Biomechanics of human movement: Applications in rehabilitation, sports and ergonomics* (pp. 89-102). Worthington OH; Bertec Corporation.

Bizzi, E., Mussa-Ivaldi, F.A., & Giszter, S.F. (1991). Computations underlying the execution of movement: A biological perspective. *Science, 253,* 287-291.

Bottema, O. (1950). On Gruebler's formulae for mechanisms. *Applied Scientific Research, A-2,* 162-164.

Cappozzo, A., Catani, F., Della Croce, U., & Leardini, A. (1995). Position and orientation of bones during movement: anatomical frame definition and determination. *Clinical Biomechanics, 10,* 171-178.

Cappozzo, A., Catani, F., Leardini, A., Benedetti, M.G., & Della Croce, U. (1996). Position and orientation of bones during movement: Experimental artifacts. *Clinical Biomechanics, 11,* 90-100.

Cappozzo, A. & Gazzani, F. (1990). Joint kinematic assessment during physical exercise. In N. Berme & A. Cappozzo (Eds.), *Biomechanics of human movement: Applications in rehabilitation, sports and ergonomics* (pp. 263-273). Worthington OH: Bertec Corporation.

Casolo, F. & Lorenzi, V. (1994). Finger mathematical modeling and rehabilitation. In F. Schuind, K.N. An., W.P. Cooney III, & M. Garcia-Elias (Eds.), *Advances in the biomechanics of the hand and wrist.* (pp. 197-223). NATO ASI Series. Series A: Life Sciences Vol. 256. New York: Plenum Press.

Cavanagh, P.R. & Wu, G. (1992). The ISB recommendation for standardization in the reporting of kinematic data—the next step. *ISB Newsletter, 47,* 2-5.

Chao, E.Y.S. (1980). Justification of triaxial goniometer for the measurement of joint rotation. *Journal of Biomechanics, 13,* 989-1006.

Chao, E.Y.S. (1990). Goniometry, accelerometry, and other methods. In N. Berme & A. Cappozzo (Eds.), *Biomechanics of human movement: Applications in rehabilitation, sports and ergonomics* (pp. 130-139). Worthington, OH: Bertec Corporation.

Chao, E.Y. & Morrey, B.F. (1978). Three-dimensional rotation of the elbow. *Journal of Biomechanics, 11,* 57-73.

Chao, E.Y.S., Rim, K., Smidt, G.L., & Johnston, R.C. (1970). The application of a 4 × 4 matrix method to the correction of the measurements of hip joint rotations. *Journal of Biomechanics, 3,* 459-471.

Chèze, L. & Dimnes, J. (1995). Modeling human body motions by the techniques known to robotics. In P. Allard, I.A.F. Stokes, & J.-P. Blanchi (Eds.), *Three-dimensional analysis of human movement* (pp. 177-200). Champaign, IL: Human Kinetics.

Cole, G.K., Nigg, B.M., Ronsky, J.L., & Yeadon, M.R. (1993). Application of the joint coordinate system to 3-D joint attitude and movement representation: A standardization proposal. *Journal of Biomechanical Engineering, 115,* 344-349.

Denavit, J. & Hartenberg, R.S. (1955). A kinematic description for lower pair mechanisms based on matrices. *Journal of Applied Mechanics, 22, Transactions of the ASME, 77,* 215-221.

Fujie, H. (1993). The use of robotics technology to study human joint kinematics: A new methodology. *Journal of Biomechanical Engineering, 115,* 211.

Georgopulos, A.P., Taira, M., & Lukashin, A.V. (1993). Cognitive neurophysiology of the motor cortex. *Science, 260,* 47-52.

Grood, E.S. & Suntay, W.J. (1983). A joint coordinate system for the clinical description of three-dimensional motion: Applications to the knee. *Journal of Biomechanical Engineering, 105,* 136-144.

Kalaska, J.F., Gaminiti, R., & Georgopulos, A.P. (1983). Cortical mechanisms related to the direction of two-dimensional arm movements: Relations in parietal area 5 and comparison with motor cortex. *Experimental Brain Research, 51,* 247-260.

Kay, B.A. (1988). The dimensionality of movement trajectories and the degree of freedom problem: A tutorial. *Human Movement Science, 7,* 858-869.

Kinzel, G.L. & Gutkovski, L.J. (1983). Joint models, degrees of freedom, and anatomical motion measurements. *Journal of Biomechanical Engineering, 105,* 55-62.

Kinzel, G.L., Hall, A.S., & Hilberry, B.M. (1972). Measurement of the total motion between two body segments. 1. Analytical development. *Journal of Biomechanics, 5,* 93-105.

Kosslyn, S.M., Chabris, C.F., Marsolek, C.J., & Koenig, O. (1992). Categorical versus coordinate spatial relations: Computational analysis and computer simulation. *Journal of Experimental Psychology: Human Perception and Performance, 18,* 562-577.

Lafortune, M.A. (1984). *The use of intra-cortical pins to measure the motion of the knee joint during walking.* Unpublished doctoral dissertation, Pennsylvania State University.

Lafortune, M.A., Cavanagh, P.R., Sommer, H.J. III, & Kalenak, A. (1992). Three-dimensional kinematics of the human knee during walking. *Journal of Biomechanics, 25,* 347-357.

Morecki, A., Ekiel, J., & Fidelus, K. (1971). *Bionika ruchu,* Warsaw, Poland: Panstwowe Wydawnictwo Naukowe.

Mussa-Ivaldi, F.A. & Giszter, S.F. (1992). Vector field approximation: A computational paradigm for motor control and learning. *Biological Cybernetics, 67,* 491-500.

Owen, B.M. & Lee, D.N. (1987). Establishing a frame of reference for action. In M.G. Wade & H.T.A. Whiting (Eds.), *Motor developments: Aspects of coordination and control.* Dordrecht, Holland: Martinus Nijhoff.

Paillard, J. (1991). Knowing where and how to get there. In J. Paillard (Ed.), *Brain and space* (pp. 461-481). Oxford, United Kingdom: Oxford University Press.

Panjabi, M.M., White, A.A., & Brand, R.A. (1971). A note on defining body parts configurations. *Journal of Biomechanics, 7,* 385-387.

Pennock, G.R. & Clark, K.J. (1990). An anatomy-based coordinate system for the description of the kinematic displacements in the human knee. *Journal of Biomechanics, 23,* 1209-1218.

Raikova, R. (1992). A general approach for modeling and mathematical investigation of the human upper arm limb. *Journal of Biomechanics, 25,* 857-865.

Reynolds, H.M. & Hubbard, R.P. (1980). Anatomical frames of reference and biomechanics. *Human Factors, 22,* 171-176.

Saltzman, E. (1979). Levels of sensorimotor representation. *Journal of Mathematical Psychology, 20,* 91-163.

Schwartz, A.B., Kettner, R.E., & Georgopulos, A.P. (1988). Primate motor cortex and free arm movements to visual targets in three-dimensional space. 1. Relations between single cell discharge and direction of movement. *Journal of Neuroscience, 8,* 2913-2927.

Shiavi, R., Limbird, T., Frazer, M., Stivers, K., Strauss, A., & Abramovitz, J. (1987). Helical motion analysis of the knee. 1. Methodology for studying kinematics during locomotion. *Journal of Biomechanics, 20,* 459-469.

Smidt, G.L., Day, J.W., & Gerleman, D.G. (1984). Iowa anatomical position system. A method of assessing posture. *European Journal of Applied Physiology, 52,* 407-413.

Söderkvist, I. & Wedin, P.-A. (1993). Determining the movements of the skeleton using well-conditioned markers. *Journal of Biomechanics, 26,* 1473-1477.

Soechting, J.F. & Flanders, M. (1989). Errors in pointing are due to approximation in sensorimotor transformations. *Journal of Neurophysiology, 62,* 595-608.

Soechting, J.F. & Flanders, M. (1992). Moving in three-dimensional space: Frames of reference, vectors, and coordinate systems. *Annual Review of Neuroscience, 15,* 167-191.

Sommer, H.J. & Miller, N.R. (1980). A technique for kinematic modeling of anatomical joints. *Journal of Biomechanical Engineering, 102,* 311-317.

Soutas-Little, R.W. & Verstrate, M.C. (1990). Use of a joint coordinate system between the foot and shank for gait analysis. In N. Berme & A. Cappozzo (Eds.), *Biomechanics of human movement: Applications in rehabilitation, sports and ergonomics,* (pp. 274-278). Worthington, Ohio: Bertec Corporation.

Spoor, C.W. & Veldpaus, F.E. (1980). Rigid body motion calculated from spatial coordinates of markers. *Journal of Biomechanics, 13,* 391-393.

van der Helm, F.C.T. & Pronk, G.M. (1995). Three-dimensional recording and description of motions of the shoulder mechanism. *Journal of Biomechanical Engineering, 117,* 27-40.

Vaughan, C.L., Hay, J.G., & Andrews, J.G. (1982). Closed loop problems in biomechanics. Part 1. A classification system. *Journal of Biomechanics, 15,* 197-200.

Veeger, H.E.J., van der Helm, F.C.T., & Rozendal, R.H. (1993). Orientation of the scapula in a simulated wheelchair push. *Clinical Biomechanics, 8,* 81-90.

Woltring, H.J. (1991). Representation and calculation of 3D joint movement. *Human Movement Science, 10,* 603-616.

Woltring, H.J. (1994). 3-D attitude representation of human joints: A standardization proposal. *Journal of Biomechanics, 27,* 1399-1414.

Woltring, H.J., Huiskes, R., De Lange, A., & Veldpaus, F.E. (1985). Finite centroids and helical axis estimation from noisy landmark measurements in the study of human joint kinematics. *Journal of Biomechanics, 18,* 379-389.

Zajac, F.E. & Gordon, M.E. (1989). Determining muscle force and action in multiarticular movement. *Exercise and Sports Science Reviews, 17,* 187-230.

3
CHAPTER

DIFFERENTIAL KINEMATICS OF HUMAN MOVEMENT

In Chapters **1** and **2**, the position of the human body and its parts have been discussed. This chapter is about movement of the body, specifically the movement of biokinematic chains. Throughout this chapter, joint motion is considered pure rotation. Translational motion within a joint is ignored. Unless stated otherwise, the reader should assume that one end of the chain does not move (for clarity, let it be a proximal terminal point). Only open kinematic chains are analyzed.

The chapter starts with discussing the velocity of kinematic chains (Section **3.1**). By reason of didactics, the text advances from simple to complex models, from planar movement to movement in three dimensions and from two-link chains to multilink chains. Movement of planar two-link chains is discussed in detail. It serves as the simplest example of various phenomena of human limb kinematics. The following concepts are addressed in this subsection: joint versus segment velocity, joint velocity versus the end-point velocity (the important notion of a Jacobian is introduced here), direct and inverse kinematic problems, singularity of the kinematic chain, degeneracy in human motion, inverse kinematics of the two-link planar chain, kinematics of pushing motion, maximizing end-point velocity, and—last but not least—the concept of instantaneous center of rotation and how to locate it. The chapter then progresses to planar three-link chains. Three-link chains, as well as multilink chains, are redundant; the position of the end point does not dictate angular positions at the joints. The issue of the chain redundancy is considered and instant centers of rotation are discussed. The main part of the discussion on instant centers of rotation is Kennedy's theorem, which defines the location of the instant centers of rotation in multilink chains. It is typical for higher animals to have a zigzag arrangement of the three-link extremities. Subsection **3.1.1.3** is devoted to multilink chains. The use of the Moore-Penrose pseudoinverse matrices for analysis of these chains is briefly highlighted here.

Section **3.1.2** concentrates on the problems associated with motion in three dimensions; it deals with angular velocity of a body. Initially, infinitesimal rotation is considered and angular velocity as a vector is discussed. Then, Poisson's formula relating derivatives of the rotation matrix and angular velocity is examined. The end of this section contains an explanation of Codman's paradox; the explanation is based on the concepts of pseudoaxial and real axial rotation.

Section **3.2** is intended to familiarize the students with acceleration of kinematic chains. The description is limited to planar movement and includes the two-link and multilink chains. The concept of acceleration difference is discussed along with Coriolis's acceleration and the relationship between the end-point acceleration and joint. The concepts of jerk and snap are also highlighted.

In Section **3.3** several applications of the differential kinematics are described and investigated. Three-dimensional representation of the vestibulo-ocular reflex is an example of using traditional methods of kinematics. The tau hypothesis, which is aimed at an explanation of the control of approach, is discussed.

It is presumed that the reader is familiar with main notions of calculus and basic mechanics of plane rotatory movement.

3.1 VELOCITY OF A KINEMATIC CHAIN

Consider first planar movement and then spatial movement of biokinematic chains.

••• *MATHEMATICS REFRESHER* •••

Calculus: Derivatives

The infinitesimal is a variable approaching zero. The derivative of y with respect to x is the limit of the ratio $\Delta y/\Delta x$ when Δx approaches zero. The symbol for the derivative is dy/dx. When the derivative is calculated with respect to time (t), the symbols are dx/dt and $\dot{x}$. The second derivative is the derivative of the first derivative. The second derivative, with respect to time, is designated as d^2x/dt^2 or $\ddot{x}$.

The partial derivative is the derivative with respect to one of several independent variables. If $y = f(x_1, x_2, \ldots, x_n)$, and $x_2, \ldots, x_n$ are kept constant, y is the function of x_1 alone. The derivative of y with respect to x_1 is called the partial derivative and is denoted by $\partial y/\partial x_1$.

••• *Classical Mechanics Refresher* •••

Velocity

Displacement is the difference between the coordinates of the body in its final and initial positions. Displacement is a vector; it does not depend on the route traveled. Distance is the magnitude of traveled path; it is a scalar, not a vector, quantity. For a round trip, the displacement equals zero (the object is in same place after as before the trip) and the distance equals the length of the track along which the body has moved.

Velocity is the time rate of change of position; it is a vector quantity. The velocity of a point is tangent to its path of motion. Speed is the time it takes to cover a distance; it is a scalar quantity (a number). Instantaneous velocity is the velocity when the time interval is infinitesimally small (approaches zero). This is the first derivative of displacement and the first integral of the acceleration with respect to time. Instantaneous speed is the magnitude of the instantaneous velocity.

Angular displacement is the change of angular position. For planar movement, this is a positive or negative number. Angular displacement does not exceed one revolution, but angular distance can be any number. Angular velocity is the rate of movement in rotation; it is the first time derivative of angular displacement. Angular velocity can be represented by a vector having a direction along the axis of rotation and a sense according to the right-hand thumb rule. When a solid body rotates about axis O, the linear velocity of a point P of the body is $v = r\omega$, where v is linear velocity of the point, r is the shortest distance between O and P (a radius), and ω is the angular velocity (in radians per second).

When a body both translates and rotates, the velocity of a point in the body can be represented as a vector sum of the velocity caused by the translation and the velocity caused by the rotation (Figure 3.1). An axis or point for which the velocity equals zero is an *instantaneous axis* or *center of velocity*. The rotational velocity at this point cancels the translational velocity. In general, the instantaneous axis or center of rotation remains fixed only for very small displacements or for a small time increment.

A rotation around any axis may be replaced by an equal rotation around any parallel axis and a motion of translation.

See **VELOCITY**, *p. 150*

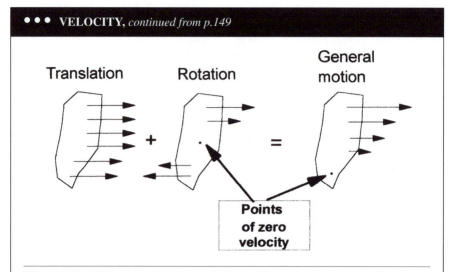

● ● ● **VELOCITY,** *continued from p.149*

Figure 3.1 Velocities due to translation and rotation are summed up (or subtracted) in a general motion.

If the location of point P is given by its radius-vector, the velocity of the point at a given instant can be represented by two orthogonal components: a radial component directed along the position vector and a transverse component directed perpendicular to the position vector.

3.1.1 Planar Movement

Fortunately, the main properties of a kinematic chain's planar motion can be illustrated with simple examples. To begin, we consider a two-link chain similar to the chain analyzed previously in Section **2.2.5.1** (see Figure 2.19).

3.1.1.1 Velocity of a Planar Two-Link Chain

The chain consists of links L_1 (proximal) and L_2 (distal) connected by a hinge joint J_2 (Figure 3.2). According to the convention adopted in Section **2.2.5**, the external reference frame fixed with the immovable environment is counted as an additional link of the chain, L_0, and the "joint" J_1 is included in the model. The angular velocities of segments L_1 and L_2 with regard to an immovable reference are $\dot{\theta}_1$ and $\dot{\theta}_2$. The joint angular velocity of joint 2 is $\dot{\alpha}_2$. The point P is the terminal tip (end point) of link 2.

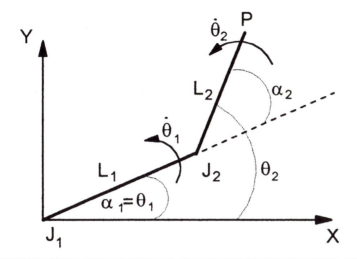

Figure 3.2 Movement of a two-link kinematic chain.

3.1.1.1.1 Joint Versus Segment Velocity

Angular velocities can be represented either in an absolute system of coordinates (in "segment space") or relative to a neighboring segment (in "joint space"). The transformation is simple:

$$\dot{\alpha}_2 = \dot{\theta}_2 - \dot{\theta}_1 \tag{3.1}$$

Thus, the joint velocity can be viewed as the difference between the corresponding segment velocities. In general, planar angular velocities are additive quantities. They can be added and subtracted from each other. In particular, equation 3.1 can also be written as

$$\dot{\theta}_2 = \dot{\theta}_1 + \dot{\alpha}_2 \tag{3.2}$$

If the second (distal) link does not change its orientation in space, its angular velocity, $\dot{\theta}_2$, equals zero. This situation occurs when both $\dot{\theta}_1$ and $\dot{\alpha}_2$ are zero (none of the links is rotating) or when $\dot{\theta}_1 = -\dot{\alpha}_2$. In the second case, the distal link rotates about the proximal link with an angular velocity of $\dot{\alpha}_2$, which is equal in magnitude and opposite in direction to the angular velocity of the proximal link in its rotation relative to the environment, $\dot{\theta}_1$. For example, when a pedal of a bicycle is rotating relative to the crank with an angular velocity $-\dot{\alpha}_2$, which is opposite in direction and equal in magnitude to the crank angular velocity $\dot{\theta}_1$, the pedal orientation in external space does not change, $\dot{\theta}_2 = 0$. As a result of the two rotations, the crank about the axle and the pedal about a

■ ■ ■ *From the Scientific Literature* ■ ■ ■

Gaze Control: Vestibulo-Ocular Reflex

Source: Tweed, D. (1994). Rotational kinematics of the human vestibuloocular reflex. II. Velocity steps. *Journal of Neurophysiology, 72,* 2480-2489.

When a person turns his or her head while keeping the eyes on a stationary target, two rotations take place at the same time. First, the head is rotating with respect to the environment. Second, the eyes are counterrotating to keep the gaze on the object of the interest. To keep the eyes fixed on the target, the eyeball turning velocity, ω_e (in orbital coordinates), must be equal in magnitude and opposite in direction to the head angular velocity, $\dot{\theta}_h$, taken with respect to the external space. Therefore, $\omega_e = -\dot{\theta}_h$. This is an essential condition for an angular velocity couple. The eyes' counterrotation in response to the head rotation is very precisely controlled. It has been reported that the control is based on vestibular feedback. This conclusion has been based on two main facts. First, in normal people the gaze direction is not changed when the head is passively turned in the absolute dark. The eyes' counterrotation matches precisely the head rotation, despite the absence of visual support. Second, in monkeys with a disrupted vestibular system, the eyes do not counterrotate during the head turning. The eyes' counterrotation is called the vestibuloocular reflex (VOR).

When a visual target is presented to the side, the eyes travel to the target more quickly than the head. The typical sequence of events is as follows:

1. The eyes and the head begin rotating toward the target; the eyes' angular velocity is much higher then the head's angular velocity, $\omega_e > \dot{\theta}_h$.
2. The eyes reach the target.
3. The eyes begin to counterrotate (with a latent period of about 15 ms from the start of the head turning).
4. The head continues to rotate toward the target while the eyes are counterrotating, $\omega_e = -\dot{\theta}_h$. During this period, the head and the eyes form the angular velocity couple.

crank rod, the pedal is moving translatory along the circular trajectory. The attitude of the pedal in space is, however, being kept constant. This is called the *angular velocity couple (pair of rotation)*. Pairs of rotation are customarily seen in human movement when a body segment is moving in a translatory manner, but not changing its orientation in space. For example, during walking the trunk orientation is kept constant, $\dot{\theta}_2 = 0$. To achieve that, the hip angular velocity $-\dot{\alpha}_2$ must be equal in magnitude and opposite in direction to the thigh angular velocity, $\dot{\theta}_1$.

3.1.1.1.2 Joint Velocity Versus the End-Point Velocity (Jacobian)

First, consider rotation of one link about a fixed axis (Figure 3.3). The instantaneous position of a point P of the link is given by

$$P_X = \ell \cdot \cos \alpha$$
$$P_Y = \ell \cdot \sin \alpha \qquad (3.3)$$

where ℓ is the length of the radius to P from the axis of rotation. The linear velocity of the point is $\mathbf{v} = \boldsymbol{\ell} \times \dot{\boldsymbol{\alpha}}$. The direction of $\mathbf{v}$ is established by the right-hand thumb rule. The projections of the linear velocity vector on the coordinate axes are

$$v_X = -\ell \sin \alpha \cdot \dot{\alpha}$$
$$v_Y = \ell \cos \alpha \cdot \dot{\alpha} \qquad (3.4)$$

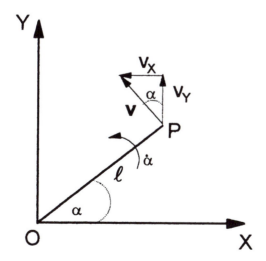

Figure 3.3 Velocity of a point P located in a rigid body that is rotating about a fixed axis.

The products $-\ell \sin \alpha$ and $\ell \cos \alpha$ relate the angular velocity of the link and the linear velocity of the point P. They are derivatives of the X and Y components of vector **l** with regard to α and, thus, they depend on the angular position of the link. The products can be written as a vector:

$$\begin{bmatrix} -\ell \sin \alpha \\ \ell \cos \alpha \end{bmatrix} \tag{3.5}$$

For the two-link chain under consideration, the relationship between the joint configuration and the end-point location is given by equation 3.6:

$$\begin{aligned} X_p &= \ell_1 \cos \alpha_1 + \ell_2 \cos(\alpha_1 + \alpha_2) \\ Y_p &= \ell_1 \sin \alpha_1 + \ell_2 \sin(\alpha_1 + \alpha_2) \end{aligned} \tag{3.6}$$

A relationship between the infinitesimal displacements of joint angles and the end-point location can be obtained by simply differentiating equation 3.6 with regard to α_1 and α_2.

$$\begin{aligned} dX &= \frac{\partial X(\alpha_1, \alpha_2)}{\partial \alpha_1} d\alpha_1 + \frac{\partial X(\alpha_1, \alpha_2)}{\partial \alpha_2} d\alpha_2 \\ dY &= \frac{\partial Y(\alpha_1, \alpha_2)}{\partial \alpha_1} d\alpha_1 + \frac{\partial Y(\alpha_1, \alpha_2)}{\partial \alpha_2} d\alpha_2 \end{aligned} \tag{3.7}$$

Equation 3.7 can also be written in a matrix form:

$$\begin{bmatrix} dX \\ dY \end{bmatrix} = \begin{bmatrix} \dfrac{\partial X(\alpha_1, \alpha_2)}{\partial \alpha_1} & \dfrac{\partial X(\alpha_1, \alpha_2)}{\partial \alpha_2} \\ \dfrac{\partial Y(\alpha_1, \alpha_2)}{\partial \alpha_1} & \dfrac{\partial Y(\alpha_1, \alpha_2)}{\partial \alpha_2} \end{bmatrix} \begin{bmatrix} d\alpha_1 \\ d\alpha_2 \end{bmatrix} \tag{3.8}$$

or
$$d\mathbf{P} = \mathbf{J} d\boldsymbol{\alpha} \tag{3.9}$$

where d**P** stands for the infinitesimal displacement vector of point P; d$\boldsymbol{\alpha}$ is the vector of infinitesimal joint displacement; and **J** is a Jacobian, which can also be written as

$$\mathbf{J} = \begin{bmatrix} \dfrac{\partial X}{\partial \alpha_1} & \dfrac{\partial X}{\partial \alpha_2} \\ \dfrac{\partial Y}{\partial \alpha_1} & \dfrac{\partial Y}{\partial \alpha_2} \end{bmatrix} \tag{3.10}$$

Elements of the Jacobian are partial derivatives that relate the infinitesimal displacement of the end point to the infinitesimal joint displacement at the present joint configuration. Note that the elements of the Jacobian depend on the joint configuration. Thus, to determine a Jacobian, the joint angles must be known.

Differentiating **P** with regard to time and taking into account equation 3.10 we obtain

$$\frac{d\mathbf{P}}{dt} = \mathbf{J}\frac{d\boldsymbol{\alpha}}{dt} \qquad \text{or} \qquad \mathbf{v} = \mathbf{J}\dot{\boldsymbol{\alpha}} \qquad (3.11)$$

where $\mathbf{P} = [X, Y]^T$, $\boldsymbol{\alpha} = [\alpha_1, \alpha_2]^T$, $\mathbf{v} \equiv \dot{\mathbf{P}}$ is the velocity vector $[\dot{X}, \dot{Y}]^T$, and $\dot{\boldsymbol{\alpha}} = [\dot{\alpha}_1, \dot{\alpha}_2]^T$ is the vector of joint angular velocities.

Because the Jacobian changes as a function of angular position of the kinematic chain, the same joint angular velocity produces various end-point velocities depending on the workspace in which a movement is performed. For example, assume that the task is to draw several straight lines by the terminal point of the distal segment of a two-link chain (e.g., a child is trying to draw several parallel lines in a notebook). Each line begins from a different starting position. Note that a point moves along a straight line if, and only if, a ratio of the horizontal and vertical components of its velocity is maintained. To produce a similar trajectory, the angular velocity should change—because of the Jacobian dependence on the joint configuration—when the starting position changes. Hence, when the child learns how to draw straight lines it is not enough for her to merely memorize the joint angular velocities and their ratios. Something else should be learned. This is true for other skills as well.

3.1.1.1.3 Direct and Inverse Kinematic Problems

For the two-joint chain under discussion, the Jacobian is

$$\mathbf{J} = \begin{bmatrix} -\ell_1 \sin \alpha_1 - \ell_2 \sin(\alpha_1 + \alpha_2) & -\ell_2 \sin(\alpha_1 + \alpha_2) \\ \ell_1 \cos \alpha_1 + \ell_2 \cos(\alpha_1 + \alpha_2) & \ell_2 \cos(\alpha_1 + \alpha_2) \end{bmatrix} \qquad (3.12)$$

When both the joint configuration and joint velocities are known, the Jacobian allows determination of the end-point linear velocity in terms of the joint angular velocities. Thus, the direct kinematic problem can be easily solved. If the joint space coordinates and velocities are given, the velocity of the end point in the external reference frame can be found.

The Jacobian given by equation 3.10 can be represented by two vector columns:

$$\mathbf{J}_1 = \begin{bmatrix} \dfrac{\partial X}{\partial \alpha_1} \\ \dfrac{\partial Y}{\partial \alpha_1} \end{bmatrix} \qquad \text{and} \qquad \mathbf{J}_2 = \begin{bmatrix} \dfrac{\partial X}{\partial \alpha_2} \\ \dfrac{\partial Y}{\partial \alpha_2} \end{bmatrix} \qquad (3.13)$$

J_1 and J_2 consist, correspondingly, of the first and the second columns of the Jacobian J. They represent the end-effector velocity induced by one unit of the appropriate joint velocity when the second joint is frozen. Thus, equation 3.11 can be written as

$$v = J_1\dot{\alpha}_1 + J_2\dot{\alpha}_2 \qquad (3.14)$$

The terms $J_1\dot{\alpha}_1$ and $J_2\dot{\alpha}_2$ represent the end-effector velocity resulting solely from the first and second joints. The resultant velocity is a vector sum of the two terms $J_1\dot{\alpha}_1$ and $J_2\dot{\alpha}_2$. For the elementary chain under consideration, J_1 and J_2 have simple mechanical meanings. If we draw radius-vectors from the joint centers 1 and 2 to the end point, J_1 and J_2 are derivatives of these vectors with regard to the joint angle. They are analogous to the vector column introduced for a single rotating link (equation 3.5).

To solve for joint velocities in terms of known end-point velocity (in the external frame), the Jacobian inverse must be known. If J^{-1} exists (in other words, if J is nonsingular), the vector of joint velocities can be obtained by simply premultiplying both sides of equation 3.11 by J^{-1}. The joint angular velocity vector $\dot{\alpha}$ is given by the inverse velocity kinematic equation

$$\dot{\alpha} = J^{-1}v \qquad (3.15)$$

Equation 3.15 is valid only when J is nonsingular. Thus, the singularity or nonsingularity of J is important.

3.1.1.1.4 Singularity Problem

It is well known that a matrix has an inverse if and only if (a) it is a square matrix and (b) its determinant does not equal zero. Consider what this means for a two-link planar chain. Because the number of coordinates or, in other words, the number of DOF in the segment space (total two, X and Y) equals the number of coordinates in the joint space (α_1 and α_2), the Jacobian J is a square matrix. Thus, for two-link chains the inverse solution exists in principle (except in the case of some joint configurations). However, because the Jacobian is a function of joint configuration it may become singular. For such a joint configuration, which is called a singular joint configuration, the inverse Jacobian does not exist and equation 3.15 is not valid. In a singular joint position, restrictions are imposed on the motion of the end effector. In Section **1.2.5.2**, a singular joint position was defined as a joint configuration at which the values of the Euler's angles cannot be determined. Here, the same term is used to designate a specific joint position at which the end effector cannot move freely in all directions. Let us explore an example of this phenomenon.

Again, consider the two-link chain presented in Figure 3.2 and the corresponding equation 3.6. For simplicity assume that $\ell_1 = \ell_2 = \ell$. Equation 3.10 can now be written as

$$\mathbf{J} = \begin{bmatrix} -\sin\alpha_1 - \sin(\alpha_1 + \alpha_2) & -\sin(\alpha_1 + \alpha_2) \\ \cos\alpha_1 - \cos(\alpha_1 + \alpha_2) & \cos(\alpha_1 + \alpha_2) \end{bmatrix} \qquad (3.16)$$

The determinant of this Jacobian is

$$\det |\mathbf{J}| = \sin\alpha_2 \qquad (3.17)$$

The sine function equals zero when the angle is either $0°$ or $180°$. Therefore, the chain (i.e., the arm or the leg) is in a singular position when it is fully extended, $\alpha_2 = 0°$. The singular position also occurs when the extremity is completely flexed, $\alpha_2 = \pi$; however, this configuration is not attainable in human motion. At the singular joint configuration, the extremity can only move perpendicularly to the chain links, not in other directions. It hampers free manipulation of objects.

In general, a joint position at which freedom of motion is constrained is denoted as a *degenerate configuration*. Singular joint positions, that is, the joint configurations at which the Jacobian determinant is zero, are always degenerate configurations. The inverse is not true; some degenerate positions are not singular. For example, the joint positions at the boundary of range of motion are degenerate but not compulsory singular.

3.1.1.1.5 Joint Degeneracy (Singularity) in Human Motion: Manipulability

At first glance, a human should avoid the singular joint configuration because of restrictions imposed on the motion of the end effector. However, singular joint configurations are used regularly. This is because the joint configurations in which minimal velocity is attainable are exactly the same positions at which maximal forces can be generated or a maximal load supported (this issue will be described in detail later). For example, when a leg or an arm is fully stretched, the velocity cannot be further provided along the extremity. In this position, however, the extremity can bear an infinite load. Therefore, when high forces are required, the singular joint positions are preferred. For example, weight lifters performing a "clean-and-jerk" lift first pull the bar from the floor with the arms extended (the arms are in a singular position), second, they catch the bar with the arms fully flexed (the position is not singular but still degenerate), and finally, they fix the bar overhead with the arms stretched (a second singularity).

The mobility of biokinematic chains in a given joint configuration can be described by the *manipulability ellipses*. If the maximal joint velocities in joints 1 and 2 are equal, the chain mobility can be represented by a circle in the joint space. When the two-link chain moves at equal joint velocities, the velocity of the end effector depends on the joint configuration. In the task space, the end-effector velocity capability is represented by an ellipse rather than by a circle (except for some joint configurations). The axis length of the ellipse is proportional to the attainable velocity in the associated direction (Figure 3.4).

Near a singular configuration, the manipulability ellipses collapse in one direction. The attainable velocity in this direction is minimal. According to the results of the simulation (see Figure 3.4) people should select different wrist trajectories when the requirement is to minimize joint velocity or joint torque.

3.1.1.1.6 Inverse Kinematics of the Two-Link Planar System

Suppose the end-point velocity **v** is known. Suppose also that the length of each link equals 1. The problem is to find the joint angular velocities that produced the end-point velocity. To arrive at the joint angular velocities, $\dot{\alpha}_1$ and $\dot{\alpha}_2$, equation 3.9 can be used. If the current chain configuration is not singular (det $|\mathbf{J}| = \sin \alpha_2 \neq 0$), the inverse of the Jacobian matrix is

$$\mathbf{J}^{-1} = \begin{bmatrix} \dfrac{\cos(\alpha_1 + \alpha_2)}{\sin \alpha_2} & \dfrac{\sin(\alpha_1 + \alpha_2)}{\sin \alpha_2} \\ -\dfrac{\cos \alpha_1 + \cos(\alpha_1 + \alpha_2)}{\sin \alpha_2} & -\dfrac{\sin \alpha_1 + \sin(\alpha_1 + \alpha_2)}{\sin \alpha_2} \end{bmatrix} \quad (3.18)$$

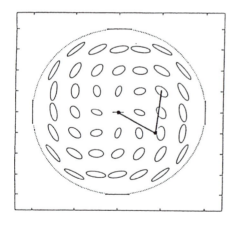
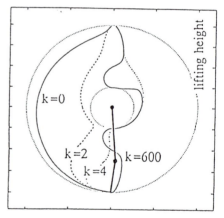

Figure 3.4   Manipulability ellipses for a two-link planar model of the arm with equal links (left figure) and the wrist trajectories during a simulation of weight lifting (right figure). During the simulation the requirements to the joint velocities and joint torques were varied in a systemic manner: k = 0 represents the minimal joint velocity; k = 600 represents the minimal joint torque.

From: Kieffer, J. & Lenarcic, J. (1993). Kinematic singularities in human arm motion. In *XIVth Congress of the International Society of Biomechanics, Abstracts* (pp. 680-681). Paris, July 4-8, 1993. Reprinted by permission.

Substitution of equation 3.18 into equation 3.15 yields the joint angular velocities:

$$\dot{\alpha}_1 = \frac{v_X \cos(\alpha_1 + \alpha_2) + v_Y \sin(\alpha_1 + \alpha_2)}{\sin \alpha_2}$$

$$\dot{\alpha}_1 = \frac{v_X [\cos \alpha_1 + \cos(\alpha_1 + \alpha_2)] + v_Y [\sin \alpha_1 + \sin(\alpha_1 + \alpha_2)]}{\sin \alpha_2}$$

(3.19)

Note, to apply equation 3.15, the inverse Jacobian $\mathbf{J}^{-1}$ must be known for all the values of joint angles α_1 and α_2.

In many applications, the velocity of the end effector (e.g., the hand) is known or desired. The joint velocities producing the end-effector velocity are then given by equation 3.19. Even for the simplest possible model considered here (planar movement of a two-link chain with links of equal length), the dependence between the end-effector velocity and joint angular velocities does not appear simple or evident. Notwithstanding the complexity of this relationship, when the hand is tracking a given trajectory with a certain velocity profile, the joint velocities must satisfy equation 3.15. This does not imply, however, that the central nervous system (CNS) calculates the Jacobian and its inverse. The CNS somehow solves this problem in its own way.

3.1.1.1.7 Pushing Motion (Leg and Arm Extension)

Representative paper: Ingen Schenau, 1989.

In this section, we take advantage of the simplicity of the two-link chain and analyze the relationship between the angular joint velocity and the linear velocity of the terminal tip without invoking the notion of the Jacobian. This is hard to do for multilink spatial chains.

The extremity (arm or leg) stretch, that is, the distance from the joint 1 to the terminal tip of the distal segment, is given by equation 2.17, which is repeated here for convenience (see Section **2.2.5.1.1** and Figure 3.2).

$$S = [\ell_1^2 + \ell_2^2 - 2\ell_1\ell_2 \cos \alpha_2']^{0.5}$$

(3.20a)

where S is the stretch, ℓ_1 and ℓ_2 are the lengths of the first and the second segments, respectively, α_2' is the included joint angle between the segments, $\alpha_2' = \pi - \alpha_2$. Note that $\dot{\alpha}_2' = -\dot{\alpha}_2$. When the two-link chain is extending without rotation (i.e., the distance between the proximal joint and the end point increases but orientation of the line drawn between these two points does not change), the links are rotating in opposite directions. The velocity of the extremity extension is then given by a time derivative of the stretch:

$$\dot{S} = \left[\frac{\ell_1 \ell_2 \cdot \sin \alpha_2}{\sqrt{\ell_1^2 + \ell_2^2 - 2\ell_1 \ell_2 \, \cos \alpha_2}} \right] \cdot \dot{\alpha}_2 \qquad (3.20b)$$

The expression in brackets, similar to the Jacobian, enables us to determine the stretch velocity, $\dot{S}$, in terms of the joint angular velocity, $\dot{\alpha}_2$. In other words, it determines the transfer of joint angular velocity into translational velocity along the radius. The transfer gradually decreases when the extremity approaches full extension. At full extension (when $\alpha_2 = 0°$ and $\alpha_2' = \pi$), sin α_2' is zero and, thus, $\dot{S}$ is zero too (a singular joint configuration). Hence, in a pushing movement, the same joint angular velocity is less effective the more the joint (the knee or elbow) is extended. This restriction is termed a *geometric constraint*.

In addition, human extremities are constrained on one side; joint motion of the knee or elbow cannot exceed full extension, roughly 180° of an included joint angle. Hyperextension of the knee or elbow is not possible without serious injury. Because of that, the joint angular velocity in the vicinity of full extension should be low to prevent damaging hyperextension. Imagine what would happen to the elbow joint if the forearm approached the fully extended arm position with the maximal attainable velocity and had to be stopped by the passive structures of the elbow joint. In this case, an injury is almost unavoidable.

In real human movements, injury to the knee or elbow joint is avoided by having the velocity decrease before the extremity assumes a fully extended position. This is called an *anatomic constraint*. As a consequence of both the geometric and anatomic constraints, the stretch velocity reaches its maximum far before full extension. This constraint can be illustrated with several examples. In shot putting, the shot loses contact with the arm before the arm is fully extended. In its most evident form, this phenomenon is seen during pushoff in speed skating. In this activity, coaches recommend the foot not be plantar flexed, to prevent abrading the ice surface with the blade. Because the ankle joint is frozen during the pushoff, the leg is realistically modeled as a two-link chain. In an analysis of elite skaters, researchers found that ice contact during pushoff ended while the athlete's hip and knee were still flexed (Figure 3.5). Full leg extension takes place after the skate is lifted from the ice. The peak leg extension velocity is reached at a knee joint angle of about 145°. Note that at this moment the knee angular velocity is still increasing. The knee velocity, $\dot{\alpha}_2$, is maximal at about 155° of the included joint angle. Then, $\dot{\alpha}_2$ is decelerated to 0° before full extension.

In biomechanical research, equation 3.20b has been used not only for determining the velocity of the end effector but also for calculating the velocity of the muscle length change—its shortening or lengthening. To perform such a calculation, ℓ_1 and ℓ_2 should be set equal to the distances from the joint center

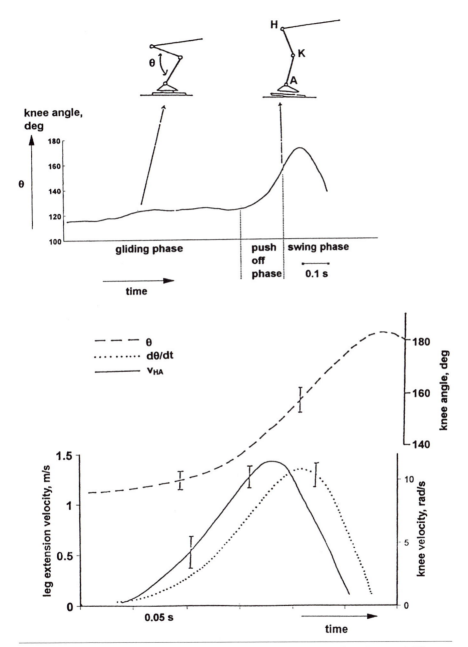

Figure 3.5   (Top) Knee angle during the speed skating stroke. (Bottom) The stretch velocity, knee angle velocity, and knee angle during the pushoff phase. Adapted from *Human Movement Science, 8,* Ingen Schenau, G.J. van, From rotation to translation: Constraints on multi-joint movements and the unique action of bi-articular muscles, 301-337, 1989, with kind permission from Elsevier Science - NL, Sara Burgerhartstraat 25, 1055 KV Amsterdam, The Netherlands.

to the points of the muscle origin and insertion, respectively. From equation 3.20b it follows that a constant joint angular velocity results in a nonconstant velocity of the muscle shortening or lengthening. The instant linear velocity of the muscle depends not only on the joint angular velocity but also on the instantaneous value of the joint angle. Hence, during *isokinetic* strength exercises, when joint angular velocity is being fixed, the muscle shortening or lengthening velocity is varying.

3.1.1.1.8 Maximizing End-Point Velocity

In many athletic events, the goal is to attain the maximal velocity of the most distal segment, specifically its terminal point. Depending on the movement, the aim is to maximize (a) the velocity of arm or leg extension (examples are push-type movements, such as shot put, takeoff in standing jumps, and so on); (b) the resultant velocity, disregarding the movement direction; or (c) the tangential velocity of the wrist or foot (as in pitching, javelin throwing, discus throwing, and kicking).

Maximizing Velocity of a Pushing Movement. The translational velocity of the terminal point of the end effector during a pushing motion depends on two factors (see equation 3.20b): the magnitude of the joint angular velocity $\dot{\alpha}_2$ and the transfer of the joint angular velocity into translational velocity along the radius, which is described by the following expression taken from equation 3.20b.

$$\left[\frac{\ell_1 \ell_2 \sin \alpha_2'}{\sqrt{\ell_1^2 + \ell_2^2 - 2\ell_1 \ell_2 \cos \alpha_2'}} \right] \tag{3.21}$$

The transfer depends on the included joint angle α_2' and on the relative length of links 1 and 2. To determine the angle at which the transfer is maximal, consider a triangle with links 1 and 2 as sides (Figure 3.6). Because the area of the triangle is $(\ell_1 \ell_2 \sin \alpha_2')/2$, the numerator of formula 3.21 is twice the area of the triangle. The denominator is simply the third side of the triangle (the stretch). Therefore, the whole expression is equal to the height of the triangle, h, multiplied by 2. If $\ell_1 \neq \ell_2$ (for clarity let ℓ_1 be greater than ℓ_2), $h = \ell_2 \cos \beta$, where β is the angle between the height and the smaller side, l_2. The height, h, is maximal when $\beta = 0$; at this instant, $\cos \alpha_2' = \ell_2/\ell_1$. Therefore, the maximal transfer is attained at an angle of $\alpha_2' = \arccos \ell_2/\ell_1$. If $\ell_1 = \ell_2$, $\ell_2/\ell_1 = 1$ and arccos $\ell_2/\ell_1 = 0$. If the proximal and distal links are of equal length, the height, h, is maximal at an angle of $\alpha_2' = 0°$ (the "extremity" is completely flexed). At this angular position, the transfer of the joint angular velocity into the stretch velocity is maximal. Note that this advantage is never used in natural human movements, evidently because of the low joint extension velocity, $\dot{\alpha}_2$, that can be produced at this body position.

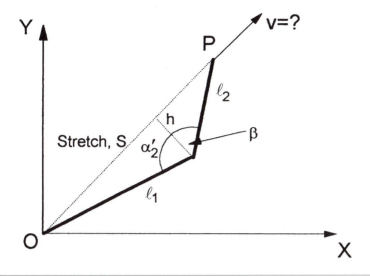

Figure 3.6 A model under consideration.

Maximizing Resultant Velocity. Assume that joint angular velocities are known. The problem is to find a joint configuration at which maximal linear velocity of the terminal tip of the distal segment is attained. To simplify the problem we assume that the two segments are of equal length, $\ell_1 = \ell_2 = 1$. The Jacobian of such a chain is given by equation 3.16. The projections of the end-point velocity on the X and Y axes are

$$v_X[-\sin \alpha_1 - \sin(\alpha_1 + \alpha_2)] \cdot \dot\alpha_1 + [-\sin(\alpha_1 + \alpha_2)] \cdot \dot\alpha_2$$
$$v_Y[\cos \alpha_1 + \cos(\alpha_1 + \alpha_2)] \cdot \dot\alpha_1 + \cos(\alpha_1 + \alpha_2) \cdot \dot\alpha_2 \qquad (3.22)$$

It is evident that v_X is maximal when both $\sin \alpha_1$ and $\sin (\alpha_1 + \alpha_2)$ equal 1, that is, when $\alpha_1 = \pi/2$ and $(\alpha_1 + \alpha_2) = \pi/2$. In such a configuration, the extremity is fully extended and oriented vertically. Similarly, the maximal vertical velocity, v_Y, is attained when an extended extremity is oriented horizontally, $\cos \alpha_1 = \cos (\alpha_1 + \alpha_2) = 1$. In general, if the joint angular velocities do not depend on the joint angle, the maximal end-point velocity is reached (a) when the extremity is completely extended and (b) at any value of the angle α_1. Because such a joint configuration is singular, the velocity vector is directed perpendicular to the longitudinal axis of the segment; that is, it is a tangential velocity.

In human movements, contrary to the assumption made above, the joint velocity depends on the joint angle. The maximal joint velocity cannot be attained at the beginning of motion because it cannot be reached instantaneously, and it cannot be maintained to the end of motion because of the anatomic constraint. As a result, the joint "velocity-angle" curve is usually bell-like, with the maximal value located somewhere in the middle part of the joint range of motion. For such a situation, when both $\dot{\alpha}_1$ and $\dot{\alpha}_2$ vary, to generate the maximal end-point velocity, a performer must find a compromise between the most efficient body configuration and the joint configuration at which maximal joint velocity can be reached. The body configurations at which the transfer of joint angular velocity into linear velocity is maximal are absolutely different for a pushing motion (a fully flexed extremity) and for a general motion not restricted by the requirement to produce high velocity along the stretch (a fully extended extremity oriented perpendicular to the velocity vector).

The human solution to this problem will be described in subsequent volumes of this textbook. Preliminarily, I will mention that the contribution of maximal angular velocity from individual joints to the end-point velocity is seemingly not the most important mechanism used to maximize the velocity of an end effector in human movements. In throwing, for example, the proximal joints decelerate rather than accelerate immediately before the release of an implement. A temporal proximal-to-distal ($P \rightarrow D$) sequence has been reported in many human movements requiring maximal end-point velocity (a so-called whip-like movement). Thus, other factors, not simply transfer of maximal angular velocity into translatory velocity of the end point, play a decisive role.

3.1.1.1.9 Instantaneous Centers of Rotation

In the diagram of the two-link chain shown in Figure 3.7, the global frame O-XY has its origin at the first joint and the local frame o-xy is anchored at the second joint with the x axis along the second link. The chain is rotating in both joints.

The instantaneous center of rotation (ICR) for the second link is sought. By definition (see Section **1.2.5.3** and Figure 1.30), the ICR is a point that is momentarily at rest in both the O-XY and o-xy frames. Because the point is stationary for an instant in both frames, the relative motion of the frames at this instant is rotation about the ICR. Hence, the ICR is the instantaneous location of a point on one plane or frame about which another plane or frame is instantaneously rotating. The velocity of any point fixed in the local frame is equal to the product $\dot{\theta} \times \mathbf{r}$, where $\dot{\theta}$ is the angular velocity of the local frame and $\mathbf{r}$ is the radius-vector of the point in the local frame. The velocity vector is evidently perpendicular to $\mathbf{r}$. The ICR is located somewhere in point C. To find C, we use the velocity equation from "Velocity in the Global and Local Frames" (see page 165). Because for point C, $\mathbf{V}_P = 0$ and $\mathbf{v}_P = 0$, the velocity equation for this point is

$$0 = \mathbf{V}_L + (\dot{\theta} \times \mathbf{r}_C) + 0 \tag{3.23}$$

where $\mathbf{r}_C$ is the radius-vector of point C in the local frame. For the equation to be true, the vectors $\mathbf{V}_L$ and $\dot{\theta} \times \mathbf{r}_C$ must be equal in magnitude but in opposite

••• CLASSICAL MECHANICS REFRESHER •••

Velocity in the Global and Local Frames

The following relationship exists between the velocity of point P in the global and local reference frames:

$$\mathbf{V}_P = \mathbf{V}_L + \bar{\omega} \times \mathbf{r}_P + \mathbf{v}_P$$

where $\mathbf{V}_P$ is the velocity of P in the global frame, $\mathbf{V}_L$ is the velocity of the origin of the local frame relative to the global frame, $\bar{\omega}$ is the angular velocity of the local frame relative to the global, $\mathbf{r}_P$ is the radius-vector of point P in the local frame, and $\mathbf{v}_P$ is the velocity of P in the local frame. The product $\bar{\omega} \times \mathbf{r}_P$ is the velocity of a point fixed in the local frame that momentarily coincides with P relative to a reference frame that has the same origin as the local system but does not rotate. Thus, the absolute velocity of a point P, $\mathbf{V}_P$, may be found if both the velocity of P relative to a human body segment and the motion of that segment are known.

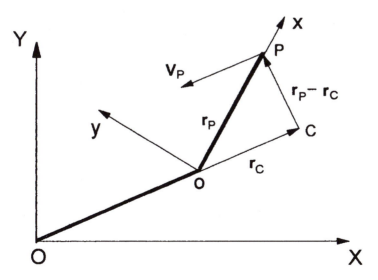

Figure 3.7   The point C is the instantaneous center of rotation of the second link of the chain. The center is at the interception of the line along the first link and the normal from P to $\mathbf{V}_P$.

directions. Note that $\mathbf{V}_L$ is perpendicular to the first link and $\dot{\theta} \times \mathbf{r}_C$ is perpendicular to $\mathbf{r}_C$. Hence, $\mathbf{r}_C$ is collinear to the first link. In other words, the ICR lies on the line along the first link, Oo. To locate C, we first write the velocity equation for the point P

$$\mathbf{V}_P = \mathbf{V}_L + (\dot{\theta} \times \mathbf{r}_P) \tag{3.24}$$

From equation 3.23 it follows that $\mathbf{V}_L = -(\dot{\theta} \times \mathbf{r}_C)$. Replacing $\mathbf{V}_L$ in equation 3.24, we find

$$\mathbf{V}_P = (\dot{\theta} \times \mathbf{r}_P) - (\dot{\theta} \times \mathbf{r}_C) = \dot{\theta} \times (\mathbf{r}_P - \mathbf{r}_C) \tag{3.25}$$

Consequently, the vector $\mathbf{r}_P - \mathbf{r}_C$ is perpendicular to $\mathbf{V}_P$. Hence, C lies at the intersection of the line along the first link and the normal to $\mathbf{V}_P$ through P. Equation 3.25 shows that the velocity of point P in the global frame may be considered as due to the rotation of the second link around the point C with an angular velocity of $\dot{\theta}$. This is true only for the moment shown. An instant later, the position of C will have changed.

The position of C can be found from a pure geometric consideration. The origin of the local system, o, which is at joint J_2, is rotating about the origin of the global system, O. Its velocity is normal to the first link. The same velocity might have been obtained by a suitable rotation around any point on the line along the first link, including point C. In a similar manner, the velocity of point P could have been obtained by a rotation about any point on the line CP. However P and o are anchored to the same body, the second link, which cannot rotate about two different centers at one time. For that reason, the point at the intersection of the two lines, C, is the only point that allows both P and o their motions.

The ICR moves with regard to both the global and the local reference frames. The ICR trajectory is called *centrode*, or *polode*. The locus of the ICR relative to the global frame (the fixed plane) is the *fixed centrode*, and the locus in the local frame (the moving plane) is the *moving centrode*. The fixed and the moving centrodes always touch each other at the pole, the ICR. At this point, they are tangent to each other. The two curves roll about one another; their relative velocities at the point of contact is zero. To visualize this movement, imagine a circle (or wheel) rolling without slipping on a surface. The ICR at the point of contact is moving along the circle (the moving centrode) as well as along the surface (the fixed centrode).

Representation of a body motion as an instant rotation around the ICR is only one of innumerable other representations. The motion of a solid body can be perceived alternatively as a combination of translation of a point within the body and rotation around this point. For any given ICR, an infinite number of combinations of translation and rotation can be found—each combination depends on which point is selected. However, because a rotation about an axis can be replaced by an equal rotation about any parallel axis combined with a translation (see "Velocity" on page 149), the magnitude of rotation does not depend on the selected center. Hence, a rotation about the ICR is equal to the rotation about any parallel axis.

3.1.1.2 Velocity of a Planar Three-Link Chain

Compared with two-link chains, three-link chains possess some specific properties.

3.1.1.2.1 Chain Redundancy

For a three-link chain, the Jacobian is not a square matrix and therefore is singular. The Jacobian has two rows, one for each coordinate of the terminal point of the distal link, and three columns, one column per joint. Therefore, the number of DOF of the chain (3) exceeds the number of the imposed constraints (2). Thus, the same end-point position can be assumed by various joint configurations. The kinematic chain is redundant. However, if both the location of the end-point tip and the attitude of the distal segment are prescribed, the chain has a square Jacobian matrix that may be used to solve both direct and inverse kinematic problems (except for singular joint configurations).

We now proceed to clarify the meaning of the Jacobian from a geometric standpoint. Consider the kinematic chain presented in Figure 3.8. The radii connecting point P with joint 1 (the shoulder) and joint 2 (the elbow) are shown by the dotted lines. A similar radius is assumed from joint 3 (the wrist) to point P. Rotations of these radii move point P along the vectors **d1**, **d2**, and **d3**. Note that these vectors are perpendicular to the radii. Displacement of point P, represented by the vector **dP**, is the vector sum of the elementary displacements in all three joints. Column-vectors J_1, J_2, and J_3 of the 2×3 Jacobian matrix represent the projections of these elementary displacements **d1**, **d2**, and **d3** on coordinate axes X and Y.

When constraints are imposed on one of the links, the number of DOF decreases and the joint velocities are interrelated. For example, the closed kinematic three-link chain, shown in Figure 3.9, has only one DOF. The angular velocity or position of any of the three joints specifies the velocity or position of the remaining two joints.

In many movements, intentional constraints are imposed on the orientation of a body segment. For instance, during walking, the trunk should be oriented vertically. To satisfy this requirement during the flat foot phase of the support period, the hip, knee, and ankle joint velocities must be coupled (Figure 3.10). In particular,

$$\dot{\alpha}_{knee} = \dot{\alpha}_{ankle} + \dot{\alpha}_{hip} \qquad (3.26)$$

The proof of this statement is left to the reader.

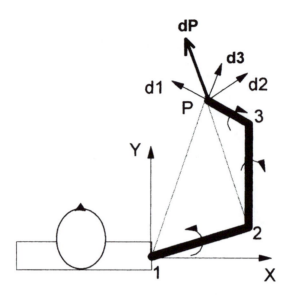

Figure 3.8 A three-link kinematic chain. Because of the angular motion at joint 1 (or 2 or 3), point P moves along the vector **d1** (or **d2** or **d3**). The movement of point P due to the combined effect of all of three joint motions is shown by the thick arrow, **dP = d1 + d2 + d3**. The vector-columns of the pertinent 2 × 3 Jacobian represent projections of the elementary vectors **d1**, **d2**, **d3** on coordinate axes.

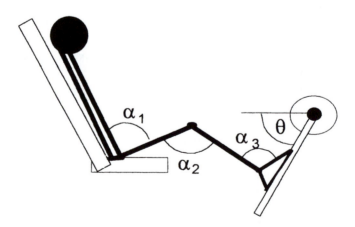

Figure 3.9 A multijoint motor task with one DOF. Any joint angle specifies the body posture. Although the angle θ is not a joint angle, it identifies the joint configuration.

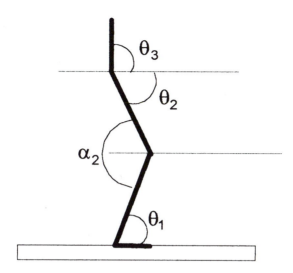

Figure 3.10   The leg positions during a flat foot phase in walking (a schematic). The shank has an angle θ_1 and the thigh has an angle θ_2 with the horizontal line. Note that the included knee angle equals $\theta_1 + \theta_2$. Use this equality to prove equation 3.26.

3.1.1.2.2 Instant Centers (Kennedy's Theorem)

In a three-link chain, there is a relative motion between the links. When systems of coordinates are imbedded in joints 1, 2, and 3, the frames move with regard to one another. Hence, three instantaneous centers exist, C_{12}, C_{13}, and C_{23}, where C_{ij} is the ICR of the relative motion of frame i relative to frame j. At C_{12}, for example, the second frame has zero velocity with regard to the first, and vice versa. Self-evidently $C_{ij} = C_{ji}$.

The Kennedy theorem—named after A.B.W. Kennedy, an American scientist of the 19th century—states: *When three bodies, 1, 2, and 3, have plane motion, their instant centers, C_{12}, C_{13}, and C_{23}, lie on a straight line.* The line is called the *line of centers*. To prove the theorem, consider a point A attached to the second reference frame and coinciding with C_{12}. The point does not move with regard to the first frame, but it has a velocity relative to the third frame, $\mathbf{v}_{23}$. The velocity is perpendicular to the radius drawn from C_{23} to C_{12}. Because point A does not move relative to the first frame, it can be attached to the frame. Its velocity with regard to the third frame is then $\mathbf{v}_{13}$. The velocity is perpendicular to the radius from C_{13} to C_{12}. Thus, at C_{12}, $\mathbf{v}_{23} = \mathbf{v}_{13}$. Hence, the radii from C_{23} to C_{12} and from C_{13} to C_{12} are collinear. The three poles are on the same straight line.

The fixed centers of rotation are also instant centers. Therefore, in a kinematic chain with revolute joints, C_{12} and C_{23} are at the joint centers. For a chain

••• COMPLEX CALCULUS REFRESHER •••

Geometric Representation of Complex Numbers

Any point in a plane can be represented as a vector, **c** = (a,b), as a *complex number*, c = a + ib (where i = $\sqrt{-1}$), or both. The complex number consists of a *real part* (a) and an *imaginary part* (ib) (both a and b are real numbers).

Geometrically, a real number, say b, can be represented as a point, or coordinate, along an axis (Figure 3.11). The product of 1 × b leaves the number at the same place along the axis, whereas the product −1 × b puts the number 180° from its initial position. The expression ib means that the number b should be turned 90° counterclockwise (the counterclockwise direction is chosen by agreement). The product i × i × b = −1 × b rotates b 180° from its original position.

The complex number can be represented in several forms:

$$c = a + ib = r(\cos \alpha + i \sin \alpha) = re^{i\alpha}$$

where e is the base of the systems of natural logarithms ($\approx$2.718) and r = $\sqrt{a^2 + b^2}$. When the equation is differentiated with respect to time, the velocity results. If the movement is a pure rotation, r is constant and the velocity is

$$\frac{d}{dt} re^{i\alpha} = re^{i\alpha} \times \frac{d(i\alpha)}{dt} = re^{i\alpha} \times i\dot{\alpha}$$

Hence, the velocity of a point or vector expressed as a complex number is the product of the vector times the angular velocity and rotated 90° counterclockwise.

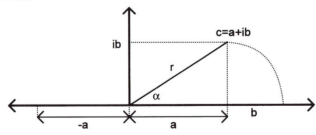

Figure 3.11 Geometric representation of the complex number c = a + ib on a complex plane (the Causs-Argand plane). The following equations are valid: (1) r² = a² + b², (2) a = r cos α, and (3) b = r sin α.

consisting of the thigh, shank, and foot, C_{12} is at the knee joint center and C_{23} at the ankle joint center, for example. During leg movement, the foot is also rotating with regard to the thigh. The instant center of this rotation is on the line of centers of the knee and ankle joints. If a three-link chain consists of the trunk, thigh, and shank, the ICR of the shank with regard to the trunk lies on the line through the hip and knee joint centers, and so on. To find the location of the ICR on the line of centers, complex calculus is customarily used.

For the system shown in Figure 3.12, and with the distance between the joints assumed to be equal to 1, the ICR of link 3 with regard to link 1 (global frame) is given by

$$\mathbf{C} = \mathbf{A} + i\,\frac{\dot{\mathbf{A}}}{\dot{\theta}_2} = \left(\mathbf{A} + i\,\frac{\mathbf{A} \times i\dot{\theta}_1}{\dot{\theta}_2}\right) = \mathbf{A}\left(1 - \frac{\dot{\theta}_1}{\dot{\theta}_2}\right) \qquad (3.27)$$

where $\mathbf{C}$ and $\mathbf{A}$ are radius-vectors of the ICR and of the second, distal, joint, and $\dot{\theta}_1$ and $\dot{\theta}_2$ are the angular velocities of the joints. Note that both angular velocities are given in the absolute system of coordinates. The expression $i \times \dot{\mathbf{A}}/\dot{\theta}_2$ gives a position of the ICR along the line of centers starting from point A. It can be thought of as a radius-vector from A to C in the global frame. The ratio $\dot{\mathbf{A}}/\dot{\theta}_2$ determines the magnitude of the vector and multiplication by i turns the velocity vector 90° counterclockwise. Hence, link 3 momentarily rotates around C with instantaneous angular velocity $\dot{\theta}_2$. Evidently, the position of the ICR of link 3 with regard to link 1 depends on the ratio of the two angular velocities. When there is no rotation in the first joint, $\dot{\theta}_1 = 0$, C coincides with A. When link 3 does not rotate with regard to link 1, $\dot{\theta}_2 = 0$ and $\dot{\alpha}_2 = \dot{\theta}_1$ (*pair of*

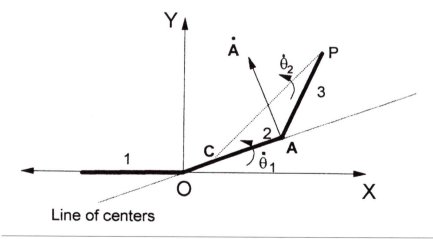

Figure 3.12 Locating the ICR for a three-link chain, 1, 2, and 3. The ICR of link 3 with respect to link 1 (global frame) is at point C.

■ ■ ■ *FROM THE SCIENTIFIC LITERATURE* ■ ■ ■

Instantaneous Center of Rotation of the Head

Source: Woltring, H.J., Kong, K., Osterbauer, P.J., & Fuhr, A.W. (1994). Instantaneous helical axis estimation from 3-D video data in neck kinematics for whiplash diagnostics. *Journal of Biomechanics, 27,* 1415-1432.

Rear-end car collisions often result in whiplash trauma. If the car seat has no head support, the head is jerked in extension. With the support, the head "rebounds" from the support in flexion, and the chin can reach the sternum. Both movements—especially the first one—may provoke a serious injury to the cervical spine.

To analyze the whiplash head motion, the head-neck system was modeled as a three-link two-joint system (a dual-pivot model) (Figure 3.13). In the model, the lower pivot is at the level of C7-T1, which is also selected as an origin of the trunk-based coordinate system. The second pivot is at the occipital-atlanto-axial complex. The radius-vector from the origin to the complex is **p**. The magnitude of **p** is assumed to be equal to 1. The angle

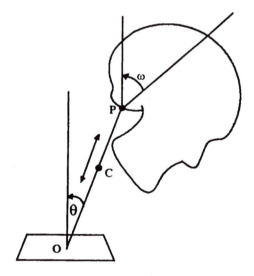

Figure 3.13 Dual-pivot, planar neck model for locating the ICR.
Reprinted from *Journal of Biomechanics, 27,* Woltring, H.J., et al., Instantaneous helical axis estimation from 3-D video data in neck kinematics for whiplash diagnostics, 1415-1432, 1994, with kind permission from Elsevier Science Ltd, The Boulevard, Langford Lane, Kidlington 0X5 1GB, UK.

that the neck is making with the vertical is θ. The angular velocities of the neck and head rotation are $\dot{\theta}$ and ω, respectively. The position of the instant center of rotation was found from the equation

$$\mathbf{c} = \left(1 - \frac{\dot{\theta}}{\omega}\right)\mathbf{p}$$

which is essentially equation 3.27.

---

rotation case, see Section **3.1.1.1.1**), the ICR is undefined. When $\dot{\theta}_1 = \dot{\alpha}_2 = \dot{\theta}_2/$ 2, the ICR is positioned on the midpoint between the two joints. When the ratio $\dot{\theta}_1/\dot{\theta}_2$ is negative (links 2 and 3 are moving in opposite directions), the ICR is located outside of the link connecting the two joints, but still on the line of centers. This case occurs when people flex or extend the arm or leg, the segments of which are arranged in a zigzag manner.

3.1.1.2.3 Zigzag Arrangement of the Leg Segments

Many primitive animals have legs with only two segments (stylopodium and zeogopodium). In higher animals, including birds and mammals, the legs have gained a third segment. According to the results of mathematical modeling, the addition of the third segment to the leg (a foot) decreases energy expenditure during ambulation up to 50% compared with the energy requirement for the two-segment leg model (Figure 3.14).

In three-segment legs, the segments are always arranged in such a way that during support and takeoff the adjacent segments rotate in opposite directions: the so-called zigzag arrangement. For example, during a takeoff in the human gait, when the thigh is rotating counterclockwise the shank is rotating clockwise and the foot, similar to the thigh, is moving in a counterclockwise direction.

From a purely kinematic standpoint, when the maximal velocity of the end effector is being sought, all links must be rotating in the same direction at the maximal joint velocity, and all the joints must be fully extended. This conclusion follows from equation 3.22 (this equation is for a two-link chain, but a similar equation can be written for a three-link chain). Theoretically, this motor pattern could be used in throwing and striking activities, although it never happens. Throwing and striking skills are characterized by the sequential motion of body segments progressing from the proximal to the distal segment. This proximal-to-distal sequential pattern cannot be explained in the framework of kinematics only. Because of the limb's zigzag arrangement, during a pushing movement when the extremity stretch increases, the links of the chain

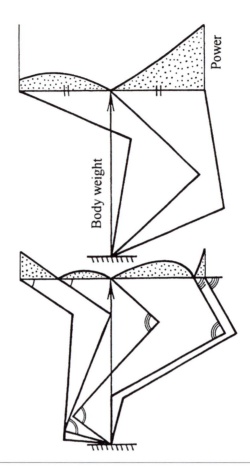

Figure 3.14 Mechanical energy spent for ambulation for two-link and three-link leg models. By assumption, the apex of the leg is moving at a constant height and constant velocity. Also, the apex travels equal distances behind and in front of the point of support. The length of the traveled path is the same for both models.
From Kuznetsov, A.N. (1993). Third segment saves energy in the flat walking leg. In: *XIVth Congress of the International Society of Biomechanics, Abstracts* (pp. 736-737). Paris, July 4-8, 1993. Adapted by permission.

are rotating in opposite directions. Because of that, the angular velocity of the extremity decreases. Therefore, it is impossible to attain in the same movement maximal velocity of both rotation and extension. A certain compromise should be found. Good runners, both animals and people, choose a motor pattern at takeoff with a minimal contribution from leg extension, stressing mainly the rotational movement of the leg, which originates mainly in the most proximal joint. It has been reported that elite sprinters, such as the World Championship gold and silver medalists Carl Lewis and Leroy Burrell (both from the

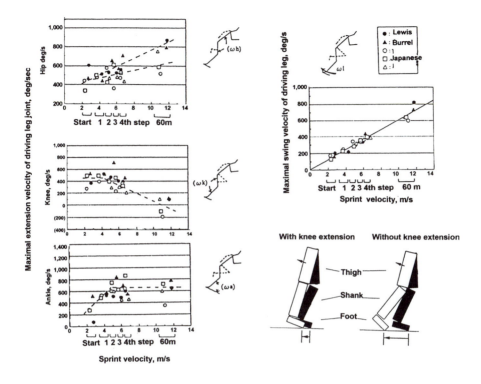

Figure 3.15 Leg movement analysis in sprinting in elite runners. The data were collected in the men's 100-meter final during the III World Championship in Athletics (Tokyo, 1991). The new world record was set in this run. The time of Carl Lewis was 9.86 seconds and the time of Leroy Burrel was 9.88 seconds. For comparison, the data from three Japanese sprinters (with running times of 10.47 to 10.58 seconds) are presented. The left-hand panel shows the correlation of the maximal joint angular velocities of the support leg with running velocity during first steps after the start and near the 60-meter point from the start line. The hip angular velocity increases, the knee extension velocity decreases, and the ankle joint velocity is almost constant when running velocity increases. The right-hand upper panel shows the relationship between the leg rotation velocity (represented by a line from the hip to the ankle) and the running velocity. A very high correlation is evident. The right-hand bottom panel shows the model of a support leg with and without leg extension. When the leg is extending, the shank, as compared with both the thigh and the ankle, is rotating in an opposite direction, decreasing the transmission of the hip extension velocity into the leg rotation velocity. The distance traveled is shown with the horizontal arrows, which are not of the relative magnitudes.

From Ito, A., Saito, M., Sagawa, K., Kato, K., Ae, M., & Kobayashi, K. (1993). Leg movement of gold and silver medalists in men's 100m final in III Word Championship in Athletics. In *XIVth Congress of the International Society of Biomechanics, Abstracts* (pp. 624-625). Paris, July 4-8, 1993. Reprinted by permission.

USA), demonstrate the highest leg rotation velocity values (Figure 3.15). Their leg extension velocities, however, were practically the same as in athletes of average skill.

3.1.1.3 Velocity of a Planar Multilink Chain (Moore-Penrose Pseudoinverse)

Representative papers: Hogan & Mussa-Ivaldi, 1992; Mussa-Ivaldi et al., 1988; Pellionisz, 1984.

The main features of simple two- and three-link chains highlighted in the two preceding sections are retained in multilink chains and will not be repeated here. For example, the joint and segment velocities are interrelated by the described transform (see equation 3.1), the maximal velocity of the end effector is attained when all the joints are fully extended and move with the maximal joint velocity (note again that this motor pattern is very seldom used in human movement), and the relationship between the joint velocities and the end-effector velocity can be conveniently described with the Jacobian matrix. For an open planar kinematic chain, the Jacobian matrix has either two rows (one for each coordinate of the terminal point of the end effector) or three rows (one for each DOF of the end effector, orientation included) and as many columns as there are joints in the chain.

$$\mathbf{J} = \begin{bmatrix} \partial X/\partial\alpha_1 & \partial X/\partial\alpha_2 & \dots & \partial X/\partial\alpha_n \\ \partial Y/\partial\alpha_1 & \partial Y/\partial\alpha_2 & \dots & \partial Y/\partial\alpha_n \\ \partial\theta_n/\partial\alpha_1 & \partial\theta_n/\partial\alpha_2 & \dots & \partial\theta_n/\partial\alpha_n \end{bmatrix} \tag{3.28}$$

where X and Y are linear coordinates of the end effector, θ_n is the angular attitude, n is the number of links in the chain (n > 3), and α_1, α_2, . . . ,α_n are the joint angles. Because the Jacobian is not a square matrix, the inverse matrix $\mathbf{J}^{-1}$ does not exist, and the inverse kinematic problem cannot be solved. Knowledge of the end-effector velocity does not allow for solving joint velocities in a unique way.

For redundant kinematic chains, a unique solution to the inverse kinematic does not exist. An optimal solution can, however, be sought. One such solution is based on the hypothesis that the brain resolves redundancy of kinematic chains, minimizing a certain cost-function. Such a cost-function is supposed to be a dot product of the joint velocity vector with itself, that is, the sum of the squared values of the joint angular velocity

$$\dot{\boldsymbol{\alpha}}^T \cdot \dot{\boldsymbol{\alpha}} = \dot{\alpha}_1^2 + \dot{\alpha}_2^2 + \dots + \dot{\alpha}_n^2 \tag{3.29}$$

where n is the number of the joints in the chain. Equation 3.29 represents a so-called *norm* of vector $\dot{\boldsymbol{\alpha}}$. A mechanical interpretation of equation 3.29 is evident. Because kinetic energy of a body is proportional to its velocity squared, the quadratic form (equation 3.29) corresponds to a minimal energy solution of equation 3.15. Note, however, that differences in the mass-inertial characteristics of the body segments are neglected. Hence, the kinetic energy is estimated not for the real kinematic chain but for an imaginary chain with links of equal mass and the same moments of inertia. To arrive at this solution, the *Moore-Penrose pseudoinverse matrix* is used. The Moore-Penrose matrix provides a minimal norm solution to the inverse kinematics problem. For a nonsingular Jacobian matrix $\mathbf{J}$, the Moore-Penrose pseudoinverse, $\mathbf{J}^+$, is defined as the matrix product

$$\mathbf{J}^+ = \mathbf{J}^T (\mathbf{J}\mathbf{J}^T)^{-1} \tag{3.30}$$

The optimal solution to the inverse kinematic problem is

$$\dot{\boldsymbol{\alpha}} = \mathbf{J}^+ \mathbf{v} = \mathbf{J}^T (\mathbf{J}\mathbf{J}^T)^{-1} \mathbf{v} \tag{3.31}$$

Note that by premultiplying equation 3.31 by the Jacobian matrix $\mathbf{J}$ we obtain equation 3.15. Equation 3.31 yields the solution to the inverse kinematic problem, which minimizes the quadratic cost function of the joint velocity vector.

In quadratic form (equation 3.29), all the joints have equal weight, or importance. If it is expected that movement in one joint is more important than that in another, for example, because of the larger mass of moving body segments, different weighting coefficients can be assigned to individual joints. If $\mathbf{W}$ is an $n \times n$ symmetric positive weighting matrix, the optimal solution to equation 3.15 is given by

$$\dot{\boldsymbol{\alpha}} = \mathbf{W}^{-1} \mathbf{J}^T (\mathbf{J}\mathbf{W}^{-1}\mathbf{J}^T)^{-1} \mathbf{v} \tag{3.32}$$

The matrix

$$\mathbf{J}^g = \mathbf{J}^T (\mathbf{J}\mathbf{W}^{-1}\mathbf{J}^T)^{-1} \tag{3.33}$$

is called the *generalized inverse*. The Moore-Penrose pseudoinverse and the generalized inverse are used to study human movement when there is reason to believe that a quantity analogous to energy expenditure is minimized. The application of these matrices to study three-dimensional movements is bound, however, with some additional problems, which will be discussed later in the text.

3.1.2 Motion in Three Dimensions

Any rigid-body motion involves both rotation and translation (Chasles' theorem). However, in human motion, either one point of the body is constrained to remain fixed, as in joint motion, or the motion can be represented as the motion of the center of mass and a rotation about the center of mass, as during airborne whole-body motion. In what follows, we neglect the translation and confine our attention to the rotation alone.

In Chapter **1**, we dealt with the spatial orientation of a rigid body. The finite rotation can be represented by a 3×3 rotation matrix, Euler's angles, or the screw method. We saw that (a) it is impossible to specify the orientation of the rigid body by three mutually independent and nonordered rotations about the X, Y, and Z axes of a global reference system and (b) finite rotations cannot be treated as vectors because they do not commute (the resultant rotation depends on the sequence of elementary rotations). In this section, I analyze the infinitesimal rotation and angular velocity of a rigid body and then of a biokinematic chain.

3.1.2.1 Angular Velocity of a Body

Consider again the example of an athlete performing a somersault with a twist (see Figure 1.23). By definition, the body configuration during rotation does not change. Therefore, we may treat the human body in this example as a solid body. The athlete is rotating about the axes belonging to the different coordinate systems:

- the Z axis of the global system (the somersaulting is performed around this axis),
- the floating axis (the tilting, i.e., the lateral inclination of the body is performed about this axis), and
- the longitudinal axis of the body (the twisting is performed about this axis).

I intend to prove that, in contrast to finite rotations, infinitesimal rotations do not depend on the order of rotation and that the complex spatial rotation can be represented by only one angular rotation. This means that if the body has three angular velocities about the three different axes at the same time, this complex movement can be described by only one instantaneous angular velocity about an axis. I will also prove that—again, opposite to finite rotations—infinitesimal rotations, as well as angular velocities, can be considered vectors.

3.1.2.1.1 Infinitesimal Rotation

A body has an angular velocity $\dot{\theta}$ about an axis when every point of the body, after rotating around this axis angle $\dot{\theta}dt$, assumes at the time $t + dt$ a new position. Suppose that the body rotated around three axes of rotation in a sequence. These axes can be X, Y, and Z of the global reference system. The duration of each rotation equals dt. The resultant orientation of the body is described by the three angles $\dot{\theta}_1 dt$, $\dot{\theta}_2 dt$, and $\dot{\theta}_3 dt$. Let P be any point in the body. The distances from P to the axes of rotation are r_1, r_2, and r_3. The displacement of P caused by the rotation around axis 1 is $\dot{\theta}_1 r_1 dt$. As a result of this rotation, let r_2

be increased to $r_2 + dr_2$. The displacement of P caused by the rotation about axis 2 equals $\dot\theta_2(r_2 + dr_2)dt$. The product of the two infinitesimal values (dr_2dt) may be ignored—this is usually done in calculus when the limits are calculated. Therefore, the displacement becomes $\dot\theta_2 r_2 dt$. Note that the displacement is independent of dr_2, the change of magnitude of r_2 due to the rotation around axis 1. In a similar way, the displacement due to the remaining rotation is $\dot\theta_3 r_3 dt$. Each of the three angular displacements do not depend on the other two infinitesimal displacements and, correspondingly, they do not depend on the order of rotation. In other words, the infinitesimal rotations are commutative. The final displacement is the same regardless of the sequence of the rotations that take place. Therefore, the three elementary rotations may be said to take place at the same time and may be characterized by one resultant rotation (and, thus, by one instantaneous angular velocity). Graphically, the displacement resulting from the three elementary rotations can be represented as the diagonal of the parallelepiped with the elementary displacements as the sides.

Consider infinitesimal rotations about the axes of a global reference system, X, Y, and Z. The rotation matrix representing infinitesimal rotation around the Z axis, $d\theta_Z = \dot\theta_Z dt$, is

$$[R_Z(d\theta_Z)] = \begin{bmatrix} \cos(d\theta_Z) & -\sin(d\theta_Z) & 0 \\ \sin(d\theta_Z) & \cos(d\theta_Z) & 0 \\ 0 & 0 & 1 \end{bmatrix} \tag{3.34}$$

Because $d\theta_Z$ is infinitesimal, cos $(d\theta_Z)$ equals 1, and sin $(d\theta_Z) = d\theta_Z$. Therefore, we can rewrite equation 3.34 as

$$[R_Z(d\theta_Z)] = \begin{bmatrix} 1 & -d\theta_Z & 0 \\ d\theta_Z & 1 & 0 \\ 0 & 0 & 1 \end{bmatrix} \tag{3.35}$$

In general, any infinitesimal rotation can be represented by the rotation matrix

$$[R(d\theta)] = \begin{bmatrix} 1 & -d\theta_Z & d\theta_Y \\ d\theta_Z & 1 & -d\theta_X \\ -d\theta_Y & d\theta_X & 1 \end{bmatrix} \tag{3.36}$$

It is possible to prove that infinitesimal rotations are not only commutative but also additive. The elements of the rotation matrix resulting from infinitesimal rotations are given by the algebraic sum of the elements of the corresponding rotation matrices.

••• **MATHEMATICS REFRESHER** •••

Derivative of a Matrix

The *derivative of matrix* [A] is the matrix of which all elements are derivatives of the corresponding elements of [A].

Because infinitesimal rotations are commutative and additive they can be treated as vectors. In particular, they can be added and subtracted vectorially.

3.1.2.1.2 Angular Velocity as a Vector

To work with angular velocities as vectors they must be explicitly defined. By agreement, the definition is done as follows. In vector form, a rotation about an axis is represented by the vector directed along the axis of rotation. The magnitude of the angular velocity is represented by a length of the vector measured along the axis, and the direction of the angular velocity, clockwise or counterclockwise, is represented by the vector direction according to the right-hand thumb rule (see "Cross Product of Vectors" on page 9). Angular velocity vectors may be added, subtracted, and resolved by the same rules as other vectors, according to the parallelogram rule. When a body rotates at an angular velocity $\dot{\theta}$ around an axis that makes angles α and $\pi/2 - \alpha$ with axes of a Cartesian coordinate frame, $\dot{\theta}$ can be resolved into two angular velocity projections, $\dot{\theta} \cos \alpha$ and $\dot{\theta} \sin \alpha$. Also, when the body is subjected to two angular motions, $\dot{\theta}_1$ and $\dot{\theta}_2$, the resultant angular velocity is $\dot{\theta} = \dot{\theta}_1 + \dot{\theta}_2$. Therefore, if a body has angular velocities $\dot{\theta}_X$, $\dot{\theta}_Y$, and $\dot{\theta}_Z$ about three axes X, Y, and Z at right angles, the magnitude of the resultant angular velocity is

$$\dot{\theta} = \sqrt{\dot{\theta}_X^2 + \dot{\theta}_Y^2 + \dot{\theta}_Z^2} \qquad (3.37)$$

The magnitude of resultant angular velocity $\dot{\theta}$ does not depend on the reference frame chosen. Therefore, this quantity is called the *invariant of the rotation*. The angular velocity vector makes angles with the coordinate axes whose cosines are $\dot{\theta}_X/\dot{\theta}$, $\dot{\theta}_Y/\dot{\theta}$, and $\dot{\theta}_Z/\dot{\theta}$. The relationship between the values of the angular velocity vector measured in the global and local frame is

$$\dot{\theta}_G = [R]\dot{\theta}_L \qquad (3.38)$$

where [R] is the rotation matrix.

It is convenient to represent the angular velocity vector in matrix form in terms of the components of $\dot{\theta}$, $[\dot{\theta}]$. To do that, both sides of equation 3.36 should be differentiated with regard to time:

$$[\dot{\theta}] = \begin{bmatrix} 0 & -\dot{\theta}_Z & \dot{\theta}_Y \\ \dot{\theta}_Z & 0 & -\dot{\theta}_X \\ -\dot{\theta}_Y & \dot{\theta}_X & 0 \end{bmatrix} \tag{3.39}$$

Note that the *angular velocity matrix* $[\dot{\theta}]$ is skew-symmetrical, therefore $[\dot{\theta}] = -[\dot{\theta}]^T$. The elements on the main diagonal equal zero because rotation around an axis does not change the position of a point along this axis. Once the position of any point P on a body, $[P_X, P_Y, P_Z]^T$, and the angular velocity $\dot{\theta}$ are specified, the velocity of P, $\mathbf{v} \equiv \dot{\mathbf{P}}_G$, can be determined:

$$\mathbf{v} = [\dot{\theta}]\mathbf{P}_G = \begin{bmatrix} 0 & -\dot{\theta}_Z & \dot{\theta}_Y \\ \dot{\theta}_Z & 0 & -\dot{\theta}_X \\ -\dot{\theta}_Y & \dot{\theta}_X & 0 \end{bmatrix} \begin{bmatrix} P_X \\ P_Y \\ P_Z \end{bmatrix} = \begin{bmatrix} -\dot{\theta}_Z P_Y + \dot{\theta}_Y P_Z \\ \dot{\theta}_Z P_X - \dot{\theta}_X P_Z \\ -\dot{\theta}_Y P_X + \dot{\theta}_X P_Y \end{bmatrix} = \begin{bmatrix} v_X \\ v_Y \\ v_Z \end{bmatrix} \tag{3.40}$$

Equation 3.40 enables us to solve for the linear velocity of point P in the global system when both the location of P in the global system, $\mathbf{P}_G$, and the angular velocity matrix, $[\dot{\theta}]$, are given. The equation can also be written in vector form

$$\mathbf{v} = \dot{\theta} \times \mathbf{P}_G \tag{3.41}$$

where $\dot{\theta} \times \mathbf{P}_G$ represents the vector product of two vectors $\dot{\theta}$ and $\mathbf{P}_G$. Because the cross product is not commutative, i.e., $\dot{\theta} \times \mathbf{P} \neq \mathbf{P} \times \dot{\theta}$, the order of the vectors in this formulation is important. The direction of $\mathbf{v}$ is established by the right-hand thumb rule—the fingers should be curled from the angular velocity vector to the position vector, $\dot{\theta}$ cross $\mathbf{P}_G$.

As compared with linear velocity, the introduction of the angular velocity has been much more complicated. The reason for this is evident. Unlike linear displacements, angular displacements are not representable by vectors. Consequently, the angular velocity vector cannot be obtained by simple differentiation of angular displacement. Also, the angular velocity vectors are said to be nonintegrable. Although the integral of angular velocity over time can be calculated—it represents the traveled angular distance—the final position of the body cannot be found in a unique way as an integral of an angular velocity vector. Consider, for example, a body that rotates first around the Z axis at the angular velocity $(0, 0, \pi/2)$ rad/sec and then around the X axis at the velocity $(\pi/2, 0, 0)$ rad/sec (Figure 3.16A). After 1 second of rotation, the body assumes the position A1, and after the next second the position A2. If the order of rotation is changed (Figure 3.16B) so that the rotation around the X axis is now the first and that around the Z axis is the second, the body assumes the position B2, which is different from A2.

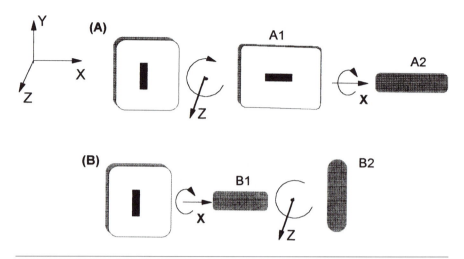

Figure 3.16 Illustration of the fact that in three-dimensional space the final body position cannot be determined as an integral of the angular velocity in a unique way. See explanation in the text.

3.1.2.1.3 Derivative of the Rotation Matrix and Angular Velocity

The relative attitude of two reference frames is characterized by the rotation matrix [R], whose elements are the direction cosines. When the coordinates of point P in the local coordinate system L are given by the radius-vector $\mathbf{P}_L$ and the rotation matrix [R] is known, the location of P in the global system G, in other words $\mathbf{P}_G$, can be found using equation 1.11, $\mathbf{P}_G = [R]\mathbf{P}_L$. To describe the angular motion, the matrix [R] should be given as a function of the time. Differentiating equation 1.11 with regard to time yields

$$\dot{\mathbf{P}}_G = \left[\dot{R}\right]\mathbf{P}_L + [R]\dot{\mathbf{P}}_L \tag{3.42}$$

Because $\mathbf{P}_L$, expressed in the local coordinate system, does not change during rotation and is time invariant, $\dot{\mathbf{P}}_L = 0$, equation 3.42 can be written simply as

$$\dot{\mathbf{P}}_G = \left[\dot{R}\right]\mathbf{P}_L \tag{3.43}$$

Equation 3.43 enables us to solve for the linear velocity of point P in the global system, or $\dot{\mathbf{P}}_G$, when both the location of P in the local system, $\mathbf{P}_L$, and the matrix, $[\dot{R}]$, are given. The elements of the matrix $[\dot{R}]$ are time derivatives of the direction cosines. They are abstract quantities that are difficult to grasp and work with. However, both the rotation matrix [R] and its derivative $[\dot{R}]$ are indispensable tools for computing. Our immediate goal is to find the relationship between the angular velocity $\dot{\boldsymbol{\theta}}$ on the one hand, and the matrices [R] and [$\dot{R}$] on the other hand. Combining equations 3.41 and 3.43, we have

$$\dot{\mathbf{P}}_G = [\dot{\theta}]\mathbf{P}_G = [\dot{R}]\mathbf{P}_L \qquad (3.44)$$

Because the rotation matrix [R] is orthogonal, that is, its inverse equals its transpose, and according to equation 1.12, $\mathbf{P}_L = [R]^{-1}\mathbf{P}_G = [R]^T\mathbf{P}_G$,

$$\dot{\mathbf{P}}_G = [\dot{\theta}]\mathbf{P}_G = [\dot{R}][R]^T\mathbf{P}_G \qquad (3.45)$$

or, after postmultiplying equation 3.45 by $\mathbf{P}_G^{-1}$,

$$[\dot{\theta}] = [\dot{R}][R]^{-1} \text{ or } [\dot{\theta}] = [\dot{R}][R]^T \qquad (3.46)$$

Thus, angular velocity depends on both the rate of change of direction cosines, [$\dot{R}$], and the body attitude, which is represented by $[R]^T$.[*]
[*] Equation 3.46 is credited to S.D. Poisson (1781-1840).

It is common knowledge that linear velocity, as well as planar angular velocity, are time derivatives of positional data (location and orientation angle). In the three-dimensional case, because of the nonvectorial nature of the finite angular displacement, angular velocity is not equal to a time derivative of the orientation angle and cannot be obtained by immediate differentiation of any set of attitude angles. The Poisson equation (equation 3.46) should be used to arrive at the angular velocity.

3.1.2.1.4 Pseudoaxial and Real Axial Rotation: Clarification of Codman's Paradox

Representative paper: Miazaki & Ishida, 1991.

In some human body motions, the so-called induced twist phenomenon is observed. Although the rotation of a body segment around its long axis is not performed, the segment is found in a twisted position. For example, when a person standing in a dependent posture (military posture at attention) performs the following movements at the shoulder joint, (1) flexion of 90°, (2) horizontal extension (abduction) of 90°, and (3) adduction, the initial and final orientation of the arm differs. The thumb, which was initially pointing anteriorly, is now pointing laterally. Hence, the arm was supinated without axial rotation. This result, when seen in shoulder movement, is called Codman's paradox (see Section **2.1.2**). There is nothing paradoxical about it, however, nor is it specific to human movements. It is a general phenomenon of three-dimensional movement that can be seen anywhere (see Figure 1.25).

■ ■ ■ *From the Scientific Literature* ■ ■ ■

Three-Dimensional Behavior of the Vestibulo-Ocular Reflex (VOR)

Source: Robinson, D.A. (1982). The use of matrices in analyzing the three-dimensional behavior of the vestibulo-ocular reflex. *Biological Cybernetics, 46,* 53-66.

Because of the vestibulo-ocular reflex (VOR), the visual axes in space maintain the same direction, without regard to head motion. Hence, when the head is rotating, the eye is rotating in the opposite direction about the same axis and at the same speed. Mathematically, this observation is described by a negative identity matrix:

$$
\begin{bmatrix} \dot{\alpha}_{E/x} \\ \dot{\alpha}_{E/y} \\ \dot{\alpha}_{E/z} \end{bmatrix} = \begin{bmatrix} -1 & 0 & 0 \\ 0 & -1 & 0 \\ 0 & 0 & -1 \end{bmatrix} \cdot \begin{bmatrix} \dot{\theta}_{H/X} \\ \dot{\theta}_{H/Y} \\ \dot{\theta}_{H/Z} \end{bmatrix} = [\text{VOR}] \cdot \begin{bmatrix} \dot{\theta}_{H/X} \\ \dot{\theta}_{H/Y} \\ \dot{\theta}_{H/Z} \end{bmatrix}
$$

where $\dot{\alpha}_{E/x}$, $\dot{\alpha}_{E/y}$, and $\dot{\alpha}_{E/z}$ are the components of angular velocity of the eye about the local, head-centered axes x, y, and z; $\dot{\theta}_{H/X}$, $\dot{\theta}_{H/Y}$, and $\dot{\theta}_{H/Z}$ are the components of the angular velocity of the head about the axes of the global reference frame, and [VOR] is the negative identity matrix of the vestibulo-ocular reflex. The matrix [VOR] does not, however, immediately represent the transformation of the input afferent signal into the motor command by the CNS. The reason is that neither the vestibular semicircular canals nor the muscles rotating the eye are perfectly aligned with the main planes of the human body. Therefore, [VOR] can be thought of as a product of three matrices: sensory, [S]; central, [C]; and motor, [M].

The sensory matrix, [S], represents a misalignment of the plane of the canals with the main planes of the human body. The elements of [S] are cosines of the projection angles between the axes of the head-fixed reference frame and the so-called sensitivity vectors in the plane of the canal. These angles are measured experimentally. The sensory matrix, [S], and its relation with the head angular velocity vector [$\dot{\theta}_H$] is as follows:

$$
[A] = [S] \cdot [\dot{\theta}_H] = \begin{bmatrix} -0.673 & 0.723 & 0.156 \\ 0.673 & 0.723 & 0.156 \\ 0 & -0.374 & 0.927 \end{bmatrix} \cdot \begin{bmatrix} \dot{\theta}_{H/X} \\ \dot{\theta}_{H/Y} \\ \dot{\theta}_{H/Z} \end{bmatrix}
$$

where [A] is the vector of the afferent signal from the vestibular organ. When the head is rotating about the vertical axis Z, $\dot{\theta}_{H/X}$ and $\dot{\theta}_{H/Y}$ are zero. The horizontal canals are activated by 92.7% of $\dot{\theta}_H$, and the anterior and posterior canals, which are not perfectly vertical, are stimulated by 15.6%. During pitch movement of the head performed in the vertical plane about the X axis, the anterior and posterior canals are excited equally and oppositely by 67.3%. The horizontal canal does not respond to the pitch.

The motor matrix, [M], relates the level of muscle activation with eye movement. The six muscles serving the eye are combined in three pairs: lateral and medial recti (M_{lmr}), superior and inferior recti (M_{sir}), and superior and inferior oblique (M_{sio}). The planes of the individual muscles do not coincide with the cardinal planes of the human body. As a result, the eye movement is expressed as a matrix product:

$$[\dot{\alpha}_E] = [M] \cdot \begin{bmatrix} M_{lmr} \\ M_{sir} \\ M_{sio} \end{bmatrix}$$

Finally, the overall vestibulo-ocular reflex is equal to the product of the three matrices:

$$[\dot{\alpha}_E] = [M][B][S][\dot{\theta}_H] = [VOR][\dot{\theta}_H]$$

where [B] is the so-called brain stem matrix representing the relative strength of the neural pathways from the canals to the muscles. Because [VOR] is a negative unit matrix, [B] can be easily found:

$$[B] = [M]^{-1}[-1][S]^{-1}$$

It is useful to distinguish between pseudoaxial rotation and real axial rotation. The first is defined as the difference in the initial and the final attitude of a human body segment. The attitude is measured with regard to the external axis, which initially aligns with the longitudinal axis of the segment. In our example, this is the vertical axis. The pseudoaxial rotation of the arm equals 90°. The real axial rotation is the total amount of rotation of a segment about its own long axis occurring during the whole motion. In the example, the real axial rotation is 0°. Clinicians are usually interested in the real rather than in the pseudoaxial rotation. Therefore, the measurement of the real axial rotation is practicable.

To calculate the real axial rotation, the angular velocity vector is projected onto the long axis of the body segment and then integrated. Because in this case the total magnitude of rotation about an axis, rather than the body orientation, is the object of interest, integration is possible. After such an integration we do not know anything about the final orientation of the segment in the space. Only the traveled angular distance, i.e., the accumulated infinitesimal rotation, is determined. The method is further explained through an example of the shoulder movement (Figure 3.17).

The movement starts from the abducted position $(90° - θ)$ in the frontal plane. The palm is facing anteriorly. The arm rotates around the vertical axis, Z, until it reaches the sagittal plane. Now, the palm is facing medially. The initial and the final attitudes of the arm are described by the following nautical angles:

| | Arm attitude (in degrees) | |
Angle	Initial	Final
Yaw, the rotation about the vertical axis, Z, of the global frame, $φ$	0	90
Pitch, the rotation about the horizontal (nodes) axis, $θ$	$θ$	$θ$
Roll, the rotation about the x axis, $ψ$	0	0

Consider two extreme starting postures. In the first, the arm is hanging vertically along the trunk, $θ = 90°$. In the second, the arm is abducted, $θ = 0°$. The first joint configuration is singular. In this posture, the Z axis of the global frame and the x axis are collinear, and the rotation about the vertical axis can be considered real axial rotation. When the arm rotates around the vertical Z axis from the abducted position, the real axial rotation is $0°$. Hence, the real axial rotation decreases monotonically when $θ$ decreases and becomes $0°$. The projection of the angular velocity vector on the x axis, $ω_x$, is

$$ω_x = -\dot{φ} \sin θ + \dot{ψ} \tag{3.47}$$

where $\dot{φ}$ is the yaw angular velocity, $θ$ is the pitch angle, and $\dot{ψ}$ is the roll angular velocity. The integral of $ω_x$ over time gives the angular distance, that is, the total amount of rotation about the x axis measured from the global frame. When $θ = 0°$, $\sin θ = 0°$, and real axial rotation is possible only because of the

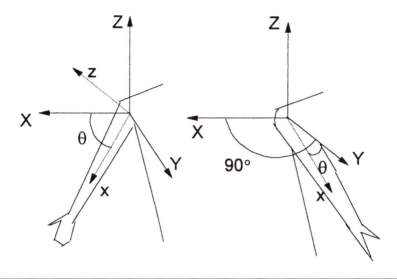

Figure 3.17   Initial (left panel) and final (right panel) attitude of the arm. The x axis of the local frame is along the longitudinal axis of the arm.

roll. When $\theta = 0°$ and $\dot{\psi} = 0$, the real angular rotation equals $0°$. When θ is between $0°$ and $90°$, the arm twists because of both the yaw and roll movement. The contribution of these two separate rotations into the real axial rotation is given by equation 3.47. When the arm is in a singular position ($\theta = 90°$), the contribution of the yaw and roll is indistinguishable.

3.1.2.2 Angular Velocity of Kinematic Chains

First, let us discuss the movement of a two-link chain and then the movement of a chain with N links.

3.1.2.2.1 Movement of a Two-Link Chain

Consider a two-link chain similar to that analyzed previously but not restricted to planar movement (Figure 3.18). The joints J_1 and J_2 are ball-and-socket joints with three DOF. The joint velocities are three-dimensional vectors, $\dot{\boldsymbol{\alpha}}_1$ and $\dot{\boldsymbol{\alpha}}_2$. Because the link ℓ_1 is rotating, the center of joint J_2 is moving along a circular path.

The local system of coordinates, x, y, z, which is fixed with the link ℓ_2 at the center of joint J_2, is translating and rotating. The translation, however, does not influence angular velocity. Therefore, the angular velocity of the second link

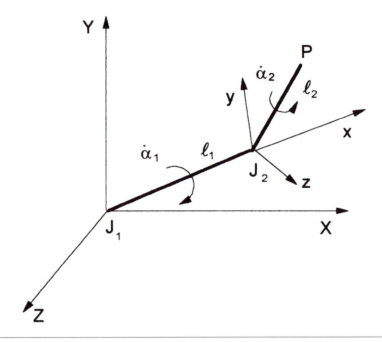

Figure 3.18   Movement of a two-link chain in three-dimensional space.

of the chain, ℓ_2, as seen from a fixed reference X, Y, Z, is simply the vector addition of the two joint motions. Thus,

$$\dot{\boldsymbol{\theta}}_2 = \dot{\boldsymbol{\alpha}}_1 + \dot{\boldsymbol{\alpha}}_2 \tag{3.48}$$

The form of equation 3.48 is the same as that developed in Section **3.1.1.1.1** for planar motion.

To determine the velocity of point P, located at the terminal tip of link l_2, it is first necessary to find the linear velocity of the joint center J_2 with regard to the fixed reference and the velocity of P with regard to the moving reference. They can be determined as vector products:

$$\begin{aligned}
\mathbf{v}(J_2)_G &= \dot{\boldsymbol{\alpha}}_1 \times \boldsymbol{\ell}_1 = \dot{\boldsymbol{\theta}}_1 \times \boldsymbol{\ell}_1 \\
\mathbf{v}(P)_L &= \dot{\boldsymbol{\alpha}}_2 \times \boldsymbol{\ell}_2
\end{aligned} \tag{3.49}$$

where $\mathbf{v}(J_2)_G$ is the velocity vector of the joint center J_2 in the global system; $\mathbf{v}(P)_L$ is the velocity of point P in the local system, which is fixed at the joint center J_2; $\dot{\boldsymbol{\alpha}}_1$ is the joint angular velocity at joint i (i = 1,2); $\dot{\boldsymbol{\theta}}_1$ is the velocity of rotation of link 1 measured from the fixed X, Y, Z axes ($\dot{\boldsymbol{\alpha}}_1 = \dot{\boldsymbol{\theta}}_1$); and ℓ_1 and ℓ_2 are the position vectors along the limbs 1 and 2. Here, ℓ_1 defines the position of the joint center J_2 with regard to the joint center J_1, and ℓ_2 defines the position of P measured from the joint center J_2. To calculate the velocity of P with regard to the global frame, $\dot{\mathbf{P}}_G \equiv \mathbf{v}(P)_G$, we also need to know the effect caused by rotation of the local reference frame. This effect is equal to the vector product of the angular velocity of rotation of the local frame with regard to the global, $\dot{\boldsymbol{\alpha}}_1$ and vector $\ell_2, \dot{\boldsymbol{\alpha}}_1 \times \ell_2$. Note that because the length ℓ_2 is constant, vector ℓ_2 changes only when the angular position of joint 2 changes. Summing these quantities yields

$$\mathbf{P}_G = \dot{\boldsymbol{\alpha}}_1 \times \ell_1 + \dot{\boldsymbol{\alpha}}_2 \times \ell_2 + \dot{\boldsymbol{\alpha}}_1 \times \ell_2 \tag{3.50}$$

Because the distributive law is valid for cross products,

$$\dot{\mathbf{P}}_G = \dot{\boldsymbol{\alpha}}_1 \times (\ell_1 + \ell_2) + \dot{\boldsymbol{\alpha}}_2 \times \ell_2 \tag{3.51}$$

or simply

$$\dot{\mathbf{P}}_G = \dot{\boldsymbol{\alpha}}_1 \times \mathbf{P}_G + \dot{\boldsymbol{\alpha}}_2 \times \ell_2 \tag{3.52}$$

where $\mathbf{P}_G$ is the position vector of P in the global frame. Vector $\mathbf{P}_G$ is a function of the link parameters and the chain configuration. The cross product $\dot{\boldsymbol{\alpha}}_1 \times \mathbf{P}_G$ represents the velocity of P generated by the proximal joint J_1 and the cross product $\dot{\boldsymbol{\alpha}}_2 \times \ell_2$ characterizes the velocity of P caused by the distal joint J_2.

In applications, the contributions of individual joints to the end-point velocity are often considered. For example, in throwing tasks, the relative proportion of the implement release velocity that comes from the shoulder or elbow joint movement is sought. Because the vectors $\dot{\boldsymbol{\alpha}}_1 \times \mathbf{P}_G$ and $\dot{\boldsymbol{\alpha}}_2 \times \ell_2$ are, as a rule, not collinear and added vectorially, their magnitudes do not immediately represent the contributions of each part. To find the percentage of the release velocity generated by individual joints, the projections of the vectors $\dot{\boldsymbol{\alpha}}_1 \times \mathbf{P}_G$ and $\dot{\boldsymbol{\alpha}}_2 \times \ell_2$ on the release velocity vector $\dot{\mathbf{P}}_G \equiv \mathbf{v}(P)_G$ should be determined first. In summary, to determine the end-point velocity from joint movement, the following information is needed: length of links, joint angular velocities, and the angle at the distal joint.

3.1.2.2.2 Movement of a Multilink Chain

The methods described in the previous section are not convenient for investigating complex chains. In the following analysis, a method based on the Jacobian matrices is developed. Because of its generality, this method is conventionally used to analyze complex multilink chains. Consider first a general, spatial n-link chain with hinge joints (Figure 3.19). The chain moves in three-dimensional space but each of the joints has only one DOF. Evidently, it is possible only when the joint offset angle, as it is defined in the Denavit-

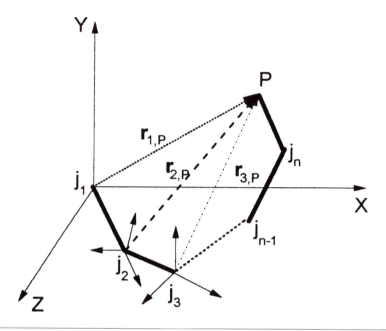

Figure 3.19 A multilink open chain with n revolute joints of the 5th class, j_i $(i = 1, \ldots, n)$. Position vectors drawn from the joint center i to the terminal point of the end effector are denoted $\mathbf{r}_{iP}$. Only three such vectors of n are shown.

Hartenberg convention (see Section **2.2.5.3**), differs from zero in at least one joint.

When not only the position of P but also the orientation of the end effector is important, the linear and angular velocity of the end effector is represented by a six-dimensional vector $\dot{\mathbf{P}}$.

$$\dot{\mathbf{P}} = \begin{bmatrix} \mathbf{v}_n \\ \dot{\boldsymbol{\theta}}_n \end{bmatrix} \tag{3.53}$$

where $\mathbf{v}_n$ and $\dot{\boldsymbol{\theta}}_n$ are three-dimensional vectors of linear and angular velocity of the *n*th link measured in the global frame X, Y, Z. When we represent the end-effector velocity and angular velocity as linear functions of joint velocity and use the Jacobian matrix, we find

$$\dot{\mathbf{P}} = \mathbf{J}\dot{\boldsymbol{\alpha}} \tag{3.54}$$

where $\dot{\boldsymbol{\alpha}}$ is an n-dimensional joint velocity vector. The Jacobian matrix has six rows and as many columns as there are joints in the chain. Each column-vector characterizes the linear and angular velocity of the end effector generated by the particular joint. Partitioning the Jacobian matrix, we have

$$
\mathbf{J} = \begin{bmatrix} \mathbf{J}_{L1} & \mathbf{J}_{L2} & \cdots & \mathbf{J}_{Ln} \\ \mathbf{J}_{A1} & \mathbf{J}_{A2} & \cdots & \mathbf{J}_{An} \end{bmatrix} = \begin{bmatrix} \mathbf{J}_{L} \\ \mathbf{J}_{A} \end{bmatrix} \tag{3.55}
$$

where $\mathbf{J}_{Li}$ and $\mathbf{J}_{Ai}$ are column-vectors associated with the linear and angular velocities, respectively. The matrix $\mathbf{J}_{L}$, which includes the first three rows of $\mathbf{J}$ is called the *positional Jacobian*. The matrix $\mathbf{J}_{A}$, which contains the remaining three rows is the *orientation Jacobian*.

Representing Linear Velocity. The linear velocity of the end effector can be expressed as

$$
\mathbf{v}_n = \mathbf{J}_L \dot{\boldsymbol{\alpha}} = \mathbf{J}_{L1} \dot{\alpha}_1 + \mathbf{J}_{L2} \dot{\alpha}_2 + \ldots + \mathbf{J}_{Ln} \dot{\alpha} \tag{3.56}
$$

Each term of this equation, $\mathbf{J}_{Li} \dot{\alpha}_i$, represents the linear velocity at the end effector produced by the ith joint. This is the velocity of the terminal point of the position vector $\mathbf{r}_{iP}$ from joint i to P. For example, vector $\mathbf{r}_{1P}$ goes from the most proximal joint to P. Because of the motion at joint i, the vector rotates at the same angular velocity as the composite of all the distal links from links i to n. Denoting the angular velocity of the position vector $\boldsymbol{\theta}_{r_i}$ and using equation 3.54, we have

$$
\mathbf{J}_{Li} \dot{\alpha}_i = \dot{\boldsymbol{\theta}}_{r_i} \times \mathbf{r}_{iP} = \left[\dot{\boldsymbol{\theta}}_{r_i} \right] \mathbf{r}_{iP} \tag{3.57}
$$

where both $\boldsymbol{\theta}_{r_i}$ and $\mathbf{r}_{iP}$ are measured with regard to the global frame, and the joint angular velocity $\dot{\alpha}_i$ is taken with regard to the local reference system at the joint.

Because the axes in individual joints are not collinear, the joint angular velocities are measured with regard to the local frames, which are at a certain angle to each other. To convert the angular velocity values measured with regard to one reference frame into another frame, equation 3.38 should be used, $\dot{\boldsymbol{\theta}}_G = [R] \dot{\boldsymbol{\theta}}_L$. This equation can be simplified, however. According to the Denavit-Hartenberg convention, rotation in the joint i is performed about the z_i axis. Therefore orientation of this axis, which is given by the third column of $[R]$, is the sole object of interest. Hence, instead of equation 3.38, the following vector equation can be employed:

$$
\dot{\boldsymbol{\theta}}_{r_i} = \mathbf{z}_i \dot{\alpha}_i \tag{3.58}
$$

where $\boldsymbol{\theta}_{r_i}$ is the angular velocity of the vector $\mathbf{r}_{iP}$ observed in the global reference; $\mathbf{z}_i$ is the unit vector pointing along the direction of the axis of rotation at joint i, measured likewise in the global frame; and $\dot{\alpha}_i$ is the joint angular velocity, a scalar. After substituting equation 3.58 into equation 3.57 and evident transformations, we have

$$\mathbf{J}_{Li}\dot{\alpha}_i = \dot{\boldsymbol{\theta}}_{r_i} \times \mathbf{r}_{iP} = \mathbf{z}_i\dot{\alpha}_i \times \mathbf{r}_{iP} = (\mathbf{z}_i \times \mathbf{r}_{iP})\dot{\alpha}_i \quad\quad (3.59)$$

or

$$\mathbf{J}_{Li} = (\mathbf{z}_i \times \mathbf{r}_{iP}) \quad\quad (3.60)$$

In words, equation 3.60 states that the element of the positional Jacobian for joint i equals the vector product of the unit vector along the axis of rotation in the joint, $\mathbf{z}_i$, and the position vector from i to P, $\mathbf{r}_{iP}$. The position vector $\mathbf{r}_{iP}$ can be obtained as a vector difference: the position vector from joint 1 to P, $\mathbf{r}_{1P}$, minus the position vector from joint 1 to joint i, $\mathbf{r}_{1i}$. The vectors $\mathbf{r}_{1P}$ and $\mathbf{r}_{1i}$ can be computed using the structure equation of the chain (equation 2.20). Hence,

$$\mathbf{r}_{iP} = \mathbf{r}_{1P} - \mathbf{r}_{1i} = \left[T_{01}\right]\left[T_{12}\right],\ldots,\left[T_{(n-1)n}\right]\hat{\mathbf{X}} - \left[T_{01}\right]\left[T_{12}\right],\ldots,\left[T_{(i-1)i}\right]\hat{\mathbf{X}} \quad\quad (3.61)$$

where $[T_{i(i+1)}]$ is the 4 × 4 transformation matrix between link i and link (i + 1) introduced in Section **2.2.5.2**, and $\hat{\mathbf{X}} = [1,0,0,0]^T$ is the augmented position vector representing the origin of the reference frame at joint 1. Note that this joint is between link 0 (environment) and link 1. A transformation matrix $[T_{(n-1)n}]$ is taken with. regard to point P located on link n.

Representing Angular Velocity. The angular velocity of the end effector can be represented as

$$\dot{\boldsymbol{\theta}}_n = \mathbf{J}_A\dot{\boldsymbol{\alpha}} = \mathbf{J}_{A1}\dot{\alpha}_1 + \mathbf{J}_{A2}\dot{\alpha}_2 + \ldots + \mathbf{J}_{An}\dot{\alpha}_n \quad\quad (3.62)$$

From equations 3.61 and 3.57, it follows that

$$\mathbf{J}_{A1}\dot{\alpha}_1 = \dot{\boldsymbol{\theta}}_{r_1} = \mathbf{z}_i\dot{\alpha}_1 \qu\quad\quad (3.63)$$

or

$$\mathbf{J}_{Ai} = \mathbf{z}_i \qu\quad\quad (3.64)$$

In words, equation 3.64 states that the element of the orientation Jacobian for joint i equals the unit vector along the axis of rotation in the joint, $\mathbf{z}_i$. In summary, the column of the Jacobian that represents the linear and angular velocity of the end effector produced by joint i can be written as

$$\mathbf{J}_i = \begin{bmatrix} \mathbf{J}_{Li} \\ \mathbf{J}_{Ai} \end{bmatrix} = \begin{bmatrix} \mathbf{z}_i \times \mathbf{r}_{iP} \\ \mathbf{z}_i \end{bmatrix} \qu\quad\quad (3.65)$$

When the kinematic chain includes ball-and-socket joints, the angular velocity in an individual joint is not a scalar but a vector and $[\dot{\alpha}]$ is an n $\times$ 3 joint velocity matrix. The general equation 3.54 can still be applied, however, if instead of the joint angular velocities their projections on the local coordinate axes are used.

3.2 ACCELERATION OF A KINEMATIC CHAIN

To obtain the acceleration of body segments one must determine the forces produced during human movement. Compared with the notion of velocity,

---

■ ■ ■ *FROM THE SCIENTIFIC LITERATURE* ■ ■ ■

Effectiveness of Arm Segment Rotations in Producing Racquet-Head Speed

Source: Sprigings, E., Marshall, R., Elliott, B., & Jennings, L. (1994). A three-dimensional kinematic method for determining the effectiveness of arm segment rotations in producing racquet-head speed. *Journal of Biomechanics, 27,* 245-254.

An end-effector velocity results from a series of joint rotations. In this study of a tennis serve, the racquet head was considered the end effector. The employed procedure included, in short, three main steps:

1. The joint angular velocities for the shoulder, elbow, and wrist, $\dot{\alpha}_i$, were computed (i = 1, 2, 3).

2. The vectors from the individual joints to the center of the racquet head, $\dot{r}_{iP}$, were determined.

3. The vector products $\dot{\alpha}_i \times r_{iP}$ were calculated and added vectorially.

$$v_P = v_{sh} + \dot{\alpha}_1 \times r_{1P} + \dot{\alpha}_2 \times r_{2P} + \dot{\alpha}_3 \times r_{3P}$$

where v_{sh} is the shoulder velocity in translation, the contribution of the legs and torso to the racquet-head velocity. The projections of the vector products on the direction of the resultant end-point velocity vector characterize the contribution of the individual joints to the resultant velocity. In this study, the components of the joint angular velocity vectors , e.g., flexion-extension, were calculated and their contributions into the end-point velocity vector computed (Figure 3.20).

See **RACQUET-HEAD,** *p. 194*

■ ■ ■ **RACQUET-HEAD,** *continued from p. 193*

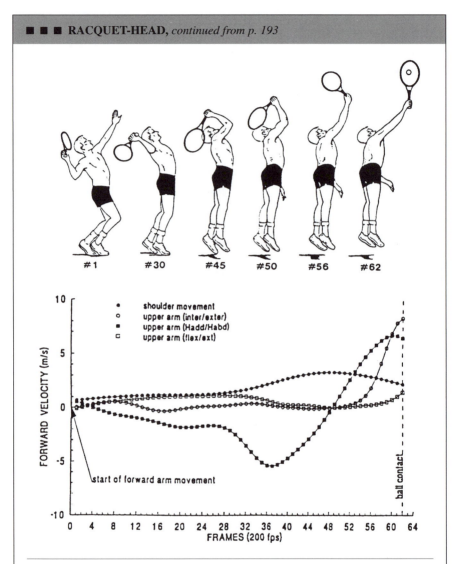

Figure 3.20 Sequential images of the tennis serve and the individual contributions provided by the joint rotations to the forward speed of the racquet. The racquet-head speed for the center of the racquet at the ball contact was 27 m/sec.

Adapted from *Journal of Biomechanics, 27,* Springings, E., et al., A three-dimensional kinematic method for determining the effectiveness of arm segment rotations in producing racquet-head speed, 245-254, 1994, with kind permission from Elsevier Science Ltd, The Boulevard, Langford Lane, Kidlington OX5 1GB, UK.

See **RACQUET-HEAD,** *p. 195*

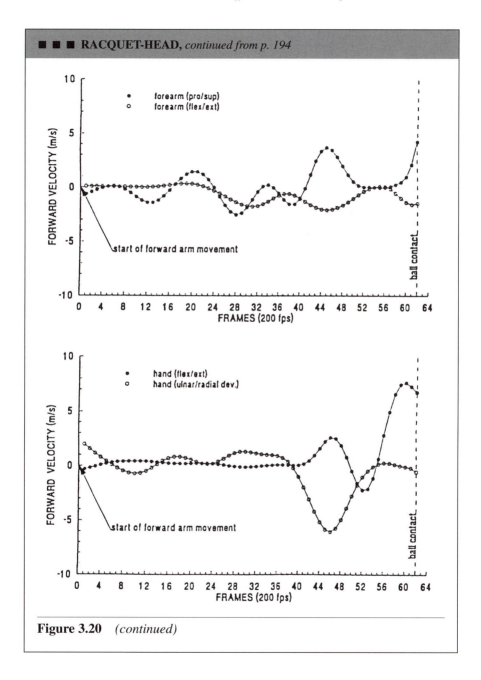

RACQUET-HEAD, *continued from p. 194*

Figure 3.20 *(continued)*

acceleration is much less obvious. Although both the joint angular and end-point velocity can be directly perceived, some sorts of acceleration cannot be observed directly. We detect the effect of acceleration, the change in veloc-

---

••• CLASSICAL MECHANICS REFRESHER •••

Acceleration

Acceleration is the rate of change of velocity. Acceleration results in a change of magnitude, direction, or both of a velocity vector. Acceleration is a vector quantity; this is the second time derivative of displacement and the first derivative of velocity with respect to time.

The acceleration of the point can be represented by two orthogonal vector components: centripetal, or normal, acceleration, pointing toward the center of curvature, and tangential acceleration, pointing along the path in the direction of increasing speed. The magnitude of centripetal acceleration of a material point P rotating around center O equals $v^2/r = \omega^2 \cdot r$, where v is the linear velocity, ω is the angular velocity, and r is the radius, the distance between P and O. The tangential acceleration is the product of the angular acceleration and the radius.

---

ity, rather than the acceleration itself. The velocity may change magnitude or direction. The rate of change of the direction of the velocity is *normal acceleration*; the rate of change of the magnitude of velocity is *tangential acceleration*.

3.2.1 Acceleration of a Planar Two-Link Chain

Once again, we analyze the two-link kinematic chain shown in Figure 3.2. The proximal link is rotating about the fixed point, joint 1, and the distal link is rotating about an axis at joint 2, which is itself rotating. The relationship between the joint and segment velocity for such a system is simple: joint angular velocity equals the difference between the relevant segment velocities, $\dot{\alpha}_2 = \dot{\theta}_2 - \dot{\theta}_1$ (see Section **3.1.1.1**). The relationship between the accelerations, however, is more complex. I will now discuss three approaches used for obtaining the acceleration of the end point when either the joint or segment angular positions, velocities, and accelerations are given (direct kinematics). The inverse kinematics will also be briefly discussed.

3.2.1.1 Acceleration Difference

To begin, consider two points affixed to a rigid body segment. One point, P, is located at the tip and the second point, J, is at the joint center (Figure 3.21).

Their absolute linear accelerations are $\mathbf{a}_P$ and $\mathbf{a}_J$ ($\mathbf{a}_P > \mathbf{a}_J$). By definition, the acceleration difference of these two points is

$$\mathbf{a}_{P/J} = \mathbf{a}_P - \mathbf{a}_J \tag{3.66}$$

which can be rewritten as

$$\mathbf{a}_P = \mathbf{a}_J + \mathbf{a}_{P/J} \tag{3.67}$$

Thus, the absolute linear acceleration of one point on a rigid body, P, equals the vector sum of the absolute linear acceleration of the second point on the same body, J, and the acceleration difference of the point P with regard to the point J. Not surprisingly, the distance between P and J is constant. In a global reference system, point P moves (rotates) with regard to point J with the acceleration difference $\mathbf{a}_{P/J}$. In the moving reference system, located in point J, the position of point P does not change.

Because the movement of point P relative to point J is a pure rotation, the acceleration difference, $\mathbf{a}_{P/J}$, may be decomposed into the normal and tangential components. Therefore Equation 3.67 can be rewritten as

$$\mathbf{a}_P = \mathbf{a}_J + \mathbf{a}_{P/J}^n + \mathbf{a}_{P/J}^t \tag{3.68}$$

where $\mathbf{a}_{P/J}^n$ is the normal acceleration of the point P relative to the point J; and $\mathbf{a}_{P/J}^t$ is the tangential acceleration of the point P with regard to point J. The

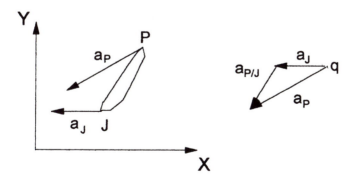

Figure 3.21   Acceleration of two points, P and J, on a body segment (left panel) and a corresponding acceleration polygon (right panel). To construct the acceleration polygon, the vectors representing absolute acceleration of the points J and P were drawn from the pole q. The acceleration difference of the point P relative to the point J was obtained by solving vectorially the equation $\mathbf{a}_{P/J} = \mathbf{a}_P - \mathbf{a}_J$. The vector of acceleration difference, $\mathbf{a}_{P/J}$, does not originate and terminate at the pole.

method of obtaining the acceleration difference based on equation 3.68 is termed the *relative acceleration method*. It can be applied when the length of the radius does not change, i.e., the distance between point P and point J is constant.

Let us apply the relative acceleration method to the two-link chain rotating uniformly at joints 1 and 2 (Figure 3.22). We are trying to find the magnitude of the end-point acceleration. Because both joint angular velocities, $\dot{\alpha}_1$ and $\dot{\alpha}_2$, are constant, the tangential acceleration is zero. The end-point velocity alters because of the combined action of the two centripetal accelerations; one directed toward joint 1 and the second toward joint 2. Using the angular velocities of the segment rotation, $\dot{\theta}_1$ and $\dot{\theta}_2$, as input variables and writing the equation in vector form, we have

$$\mathbf{a}_P = \dot{\boldsymbol{\theta}}_1^2 \times \mathbf{l}_1 + \dot{\boldsymbol{\theta}}_2^2 \times \mathbf{l}_2 \qquad (3.69)$$

Note that when applying the relative acceleration method, the absolute angular velocity and acceleration of the body segments must be used.

3.2.1.2 Coriolis' Acceleration

Coriolis' acceleration—named after the French scientist G. Coriolis (1792-1843)—acts when a body of interest (link 2, in our example) is moving with regard to a rotating frame (fixed with the proximal link). The Coriolis' accel-

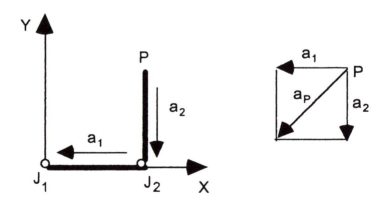

Figure 3.22 A two-link chain with both links rotating in the counterclockwise direction with uniform angular velocities. The centripetal acceleration of the joint center 2, $\mathbf{a}_1$, is directed to the left (from J_2 toward J_1). The centripetal acceleration of the point P relative to joint center 2, $\mathbf{a}_2$, is directed downward (from P toward J_2). The resultant acceleration of the point P in the segment space, $\mathbf{a}_P$, can be determined as the vector sum of the two mentioned accelerations (shown on the right panel).

eration is due to changes in magnitude and direction of the velocity vector when the point of interest changes its position in the rotating reference frame. For instance, when the end-point position changes from point P_1 to point P_2, the stretch, i.e., the length of the radius-vector drawn from the joint 1 to the end point, increases and, thus, the same joint velocity, $\dot{\alpha}_1$, induces a greater linear end-point velocity (Figure 3.23). Any variation in the stretch value changes the linear end-point speed. It also induces a change in the direction of the velocity vector.

Coriolis' theorem states: The absolute linear acceleration of a point P, which is moving with regard to a local reference system that is also in motion, is equal to the vector sum of

1. the acceleration the point would have if it were fixed to the moving system (in our example, this is a centripetal acceleration due to the rotation at joint 1), $\mathbf{a}_G$;

2. the acceleration of P with regard to the local moving system (the centripetal acceleration due to the rotation at joint 2), $\mathbf{a}_L$; and

3. a compound supplementary acceleration (Coriolis' acceleration), $\mathbf{a}_C$.

$$\mathbf{a} = \mathbf{a}_G + \mathbf{a}_L + \mathbf{a}_C \qquad (3.70)$$

For a complete proof of the theorem the reader is referred to textbooks of classical mechanics.

When the joint angular velocities are not constant, the terms $\mathbf{a}_G$ and $\mathbf{a}_L$ reflect not only centripetal but also tangential acceleration. The Coriolis' acceleration is twice the cross product of the angular velocity of the local reference frame and the relative velocity of P with regard to this frame. If an observer were

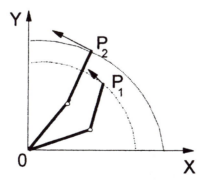

Figure 3.23 When position of the end point changes from P_1 to P_2, the same angular velocity at joint 1 induces greater linear end-point velocity.

positioned in the origin of the local rotating frame, she would measure this velocity of point P. The Coriolis' acceleration arises for two reasons: the changes in both magnitude and direction of the relative velocity vector (that is why the coefficient 2 appears). In a general planar case, the magnitude of $\mathbf{a}_C$ is

$$a_C = 2\omega v_L \tag{3.71}$$

where ω is the magnitude of the angular velocity of the local frame, and v_L is the speed of the point of interest with regard to the moving frame. For the end point of the two-link chain,

$$a_C = 2\ell\dot{\alpha}_1\dot{\alpha}_2 \tag{3.72}$$

The Coriolis' acceleration is zero when at least one joint does not move ($\dot{\alpha}_1$, $\dot{\alpha}_2$, or both equal 0). The direction of the Coriolis' acceleration is always perpendicular to the velocity vector $\mathbf{v}_L$. Its sense is in accordance with the right-hand thumb rule. The following simple rule is also implied: to find the sense of the Coriolis' acceleration, turn vector $\mathbf{v}_L$ 90° in the sense of the absolute angular velocity. The direction of Coriolis' acceleration for the second link of the two-link chain is shown in Figure 3.24.

Absolute linear acceleration of the end point of the planar two-link chain is

$$\mathbf{a} = \mathbf{a}_G + \mathbf{a}_L + \mathbf{a}_C = \overline{\dot{\alpha}_1^2\left(\ell_1^2 + \ell_2^2 - 2\ell_1\ell_2\cos\alpha_2\right)^{1/2}} + \overline{\dot{\alpha}_2^2\ell_2} + \overline{2\ell_2\dot{\alpha}_1\dot{\alpha}_2} \tag{3.73}$$

where α_2' is an included angle at joint 2. The horizontal lines above the terms in equation 3.73 indicate that the terms are vectors and should be added vectorially. Consider an example. In Figure 3.24 (upper left panel) the links are rotating counterclockwise. The angle at joint 2 is 90°. Link 1 is 0.4 m long and is rotating with a uniform absolute angular velocity $\dot{\alpha}_1$ of 2 rad/sec. Link 2 is 0.3 m long and is rotating, also uniformly, with a joint velocity $\dot{\alpha}_2$ of 3 rad/sec. The acceleration of the end point is sought.

Solution

1. Find $\mathbf{a}_G$. The acceleration of particle P due to rotation in joint 1 is $2^2 \times (0.4^2 + 0.3^2)^{1/2} = 4 \times 0.5 = 2.0$ m/sec^2. This is the absolute linear (centripetal) acceleration of a point P', which coincides with the point P. The acceleration is from P toward joint 1.

2. Find $\mathbf{a}_L$. The linear acceleration of P relative to joint center 2 equals $3^2 \cdot 0.3 = 2.7$ m/sec^2. The acceleration is from P toward joint 2. This is the relative (centripetal) acceleration due to rotation in that joint.

3. Find $\mathbf{a}_C$. The Coriolis' acceleration is $2 \times 2 \times (3 \times 0.3) = 3.6$ m/sec^2.

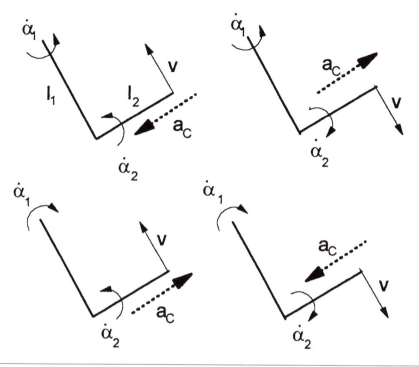

Figure 3.24 The direction of Coriolis' acceleration with different vectors of motion.

Because both links are rotating counterclockwise, the Coriolis' acceleration is directed from point P toward joint 2.

To find the resultant acceleration of the point P, the three accelerations must be summed vectorially. When Coriolis' acceleration is sought, the joint (relative) angular velocity and acceleration must be used. Compare equations 3.69 and 3.73. They define the linear acceleration of the end point in terms of either the segment angular velocity (equation 3.69) or the joint velocity (equation 3.73). Equation 3.69 is simpler, but the contribution of various kinds of acceleration, in particular Coriolis' acceleration, is not separated. When joint velocity is used as an input variable, the equation is more complex, but the acceleration due to the separate motion of individual segments is explicitly defined. Such a consideration is desired when the problem is to determine how movement is perceived. Evidently, the performer uses a somatic, not an external, reference system to perceive relative motion of his or her segments.

••• **ANATOMY AND PHYSIOLOGY REFRESHER** •••

The Vestibular Organ

The vestibular organ includes three semicircular canals containing a viscous fluid. The vestibular canals function mainly as angular velocity transducers. Each of the canals lies within a plane. The afferent signals from the canal are proportional to the projections of the head angular velocity vector on the plane of the canal. At the same magnitude of angular velocity, the afferent signal is maximal when the axis of rotation is perpendicular to the canal's plane.

The horizontal canals lie approximately in a horizontal plane. They serve the horizontal vestibulo-ocular reflex, which is induced by the yaw movement of the head (head rotation about the vertical axis). The vertical canals, anterior and posterior, lie in planes that make an angle approximately 45° with the sagittal (frontal) plane. The left and right canals work in three push-pull pairs: horizontal canals, left anterior and right posterior canals, and left posterior and right anterior canals. During any head rotation, the firing rate from one of the paired canals increases and from the other decreases. The direction that corresponds to the maximal activation of a pair is characterized by the sensitivity vector, which lies midway between the planes of the two canals.

3.2.1.3 End-Point Versus Joint Acceleration

The accelerations caused by the movement of individual body segments (tangential, centripetal, Coriolis' acceleration) have different directions. Because the individual accelerations must be summed vectorially, the resultant acceleration of the end effector depends not only on the segment or joint velocity and acceleration but also on the body posture at each instant of time. The use of the Jacobian matrix enables us to solve for the end-effector acceleration in terms of known joint positions, velocities, and accelerations. Differentiating equation 3.11 we obtain

$$\frac{d\mathbf{v}}{dt} = \frac{d(\mathbf{J}\dot{\boldsymbol{\alpha}})}{dt} \tag{3.74}$$

or

$$\dot{\mathbf{v}} = \frac{d\mathbf{J}}{dt}\dot{\boldsymbol{\alpha}} + \frac{d\dot{\boldsymbol{\alpha}}}{dt} = \dot{\mathbf{J}}\dot{\boldsymbol{\alpha}} + \mathbf{J}\ddot{\boldsymbol{\alpha}} \tag{3.75}$$

where $\mathbf{v}$ is the velocity vector $[\ \dot{X}, \dot{Y}\ ]^T$, $\dot{\mathbf{v}}$ is the acceleration vector $[\ \ddot{X}, \ddot{Y}\ ]^T$, $\dot{\boldsymbol{\alpha}} = [\ \dot{\alpha}_1, \dot{\alpha}_2]^T$ is the vector of joint angular velocities, $\ddot{\boldsymbol{\alpha}} = [\ddot{\alpha}_1, \ddot{\alpha}_2]^T$ is the vector of joint angular accelerations, and $\mathbf{J}$ is the Jacobian. $\dot{\mathbf{J}}$ can be directly obtained by differentiating $\mathbf{J}$ with regard to time. The implementation of equation 3.75 requires the knowledge of the joint configuration, joint velocity, and joint acceleration at all times. For the two-link planar chain,

$$\dot{\mathbf{J}} = \frac{d}{dt} \begin{bmatrix} -\ell_1 \sin\alpha_1 - \ell_2 \sin(\alpha_1 + \alpha_2) & -\ell_2 \sin(\alpha_1 + \alpha_2) \\ \ell_1 \cos\alpha_1 + \ell_2 \cos(\alpha_1 + \alpha_2) & \ell_2 \cos(\alpha_1 + \alpha_2) \end{bmatrix} \qquad (3.76)$$

$$= \begin{bmatrix} -\ell_1\dot{\alpha}_1 \cos\alpha_1 - \ell_2(\dot{\alpha}_1 + \dot{\alpha}_2) \cos(\alpha_1 + \alpha_2) & -\ell_2(\dot{\alpha}_1 + \dot{\alpha}_2) \cos(\alpha_1 + \alpha_2) \\ -\ell_1\dot{\alpha}_1 \sin\alpha_1 - \ell_2(\dot{\alpha}_1 + \dot{\alpha}_2) \sin(\alpha_1 + \alpha_2) & -\ell_2(\dot{\alpha}_1 + \dot{\alpha}_2) \sin(\alpha_1 + \alpha_2) \end{bmatrix}$$

■ ■ ■ *FROM THE SCIENTIFIC LITERATURE* ■ ■ ■

Coriolis' Effect Provokes Disorientation

Source: Lackner, J.R., & Di Zio, P. (1992). Gravitational, inertial, and Coriolis force influences on nystagmus, motion sickness, and perceived head trajectory. In A. Berthoz, W. Graf, & P.P. Vidal (Eds.) *The head and neck sensory motor system* (pp. 216-222). New York: Oxford University Press.

When a rotating subject tilts his or her head about an axis that is not parallel to the axis of rotation, the subject usually experiences disorientation and motion sickness symptoms. This is called the *Coriolis' effect*. The effect is thought to be caused by two mechanisms: cross-coupling stimulation of the semicircular canals and Coriolis' acceleration. We will explain the cross-coupling stimulation through an example. Consider a figure skater rotating counterclockwise at a constant angular velocity around the longitudinal axis of his or her body (Figure 3.25). The athlete tilts the head backward (some figure skaters really perform this maneuver while spinning on one toe). Now, the projection of the angular velocity vector on the yaw canal decreases and that on the roll canal increases. The pitch canal signals accurately the head movement. The cross-coupling prevents accurate sensation of the head and body movement. The second mechanism is Coriolis' acceleration, which is acting on the vestibular organ because the distance from the organ to the axis of rotation changes during the tilting motion. When subjects seated on a rotating chair try to perform the head tilt and the arm movement in the sagittal plane they cannot fulfill

See **CORIOLIS' EFFECT**, *p. 204*

■ ■ ■ **CORIOLIS' EFFECT,** *continued from p. 203*

this task precisely. Because of the misperception, both the head and the arm follow an "elliptical" trajectory (Figure 3.25).

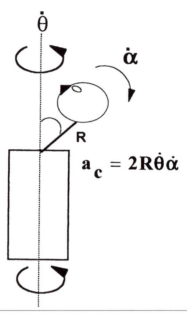

$$a_c = 2R\dot{\theta}\dot{\alpha}$$

Figure 3.25 Coriolis' effect during a backward head movement (a scheme). The body is rotating counterclockwise about the vertical axis. When the head tilts backward, the projection of the body angular velocity on the y axis decreases and is sensed as the angular deceleration (acceleration in clockwise direction). The projection on the roll axis increases and is sensed as the counterclockwise acceleration. Also, because the vestibular organ is moving with respect to the rotating body, the Coriolis' acceleration is acting on the organ. Due to these effects, the accuracy of sensation deteriorates.

In equation 3.75, the expression $\mathbf{J}\ddot{\alpha}$ represents the contribution of joint angular acceleration to the end-effector acceleration. If $\ddot{\alpha} = 0$, i.e., the joint angular velocities are constant, the acceleration of the end effector is $\dot{\mathbf{J}}\dot{\alpha}$. In such a case, the end-effector velocity changes only because of centripetal and Coriolis' accelerations. The centripetal acceleration is proportional to the joint angular velocity squared, $\dot{\alpha}_1^2$ and $\dot{\alpha}_2^2$, and the Coriolis' acceleration is proportional to the product $\dot{\alpha}_1\dot{\alpha}_2$. Hence, the elements in equation 3.75 depend on both the joint angles and the joint angular velocities. Sometimes it is more convenient to represent the contribution of the centripetal and Coriolis' acceleration as a product of a matrix that depends only on a joint angular configuration and a vector depending solely on the joint velocities. To do that, a vector of the joint velocity products, $\dot{\boldsymbol{\alpha}}_{pr} = [\dot{\alpha}_1^2, \dot{\alpha}_1\dot{\alpha}_2, \dot{\alpha}_2^2]^T$, is introduced. Then, the expression $\dot{\mathbf{J}}\dot{\boldsymbol{\alpha}}$ is written as $\mathbf{V}\dot{\boldsymbol{\alpha}}_{pr}$ ($\dot{\mathbf{J}}\dot{\boldsymbol{\alpha}} = \mathbf{V}\dot{\boldsymbol{\alpha}}_{pr}$), where V is a 2×3 matrix resulting from rearranging the terms of the expression $\dot{\mathbf{J}}\dot{\boldsymbol{\alpha}}$.

$$\mathbf{V} = \begin{bmatrix} -\ell_1\cos\alpha_1 - \ell_2\cos(\alpha_1 + \alpha_2) & -2\ell_2\cos(\alpha_1 + \alpha_2) & -\ell_2\cos(\alpha_1 + \alpha_2) \\ -\ell_1\sin\alpha_1 - \ell_2\sin(\alpha_1 + \alpha_2) & -2\ell_2\sin(\alpha_1 + \alpha_2) & -\ell_2\sin(\alpha_1 + \alpha_2) \end{bmatrix}$$

(3.77)

The inverse kinematic problem can be solved provided that **J** is nonsingular.

3.2.2 Acceleration of a Two-Link Chain in Three Dimensions

Consider again the two-link chain moving in three dimensions (see Figure 3.18). The links are connected by ball-and-socket joints. The position of the end point P in the global reference frame is described by the vector $\mathbf{P}_G$. Other symbols are as usual. We intend to find the acceleration of **P** in the global frame, $\ddot{\mathbf{P}}_G$, as a function of movement in the joints. To do that, the time derivative of equation 3.52, which states that $\dot{\mathbf{P}}_G = \dot{\boldsymbol{\alpha}}_1 \times \mathbf{P}_G + \dot{\boldsymbol{\alpha}}_2 \times \boldsymbol{\ell}_2$, should be taken. The velocity vector ($\dot{\boldsymbol{\alpha}}_1 \times \mathbf{P}_G$) is defined in the global reference and its time derivative can be computed using the conventional formula for derivative of a vector product. Because $\mathbf{l}_2$ is measured between two points in the rotating reference, the

---

••• CLASSICAL MECHANICS REFRESHER •••

Time Derivative of a Vector

When vector **P** is measured in a local system, which is itself translating and rotating, the time derivative of **P** in the global system is

$$(\dot{\mathbf{P}})_G = (\dot{\mathbf{P}})_L + \dot{\boldsymbol{\theta}} \times \mathbf{P}$$

where $(\dot{\mathbf{P}})_G$ and $(\dot{\mathbf{P}})_L$ are the time derivatives of vector **P** as seen from the fixed and the moving system, correspondingly, and $\dot{\boldsymbol{\theta}}$ is the angular velocity of the moving frame of reference measured from the fixed axes.

velocity vector ($\dot{\boldsymbol{\alpha}}_2 \times \boldsymbol{\ell}_2$) is defined in the local frame and its time derivative should be taken by using the equation from the mechanics refresher "Time Derivative of a Vector" on page 204. Expressing ($\dot{\boldsymbol{\alpha}}_1 \times \mathbf{P}_G$) as ($\dot{\boldsymbol{\alpha}}_1 \times \boldsymbol{\ell}_1 + \dot{\boldsymbol{\alpha}}_2 \times \boldsymbol{\ell}_2$) and then taking the time derivative of equation 3.52, after simplifying, yields

$$\ddot{\mathbf{P}}_G = \underbrace{\ddot{\boldsymbol{\alpha}}_1 \times \boldsymbol{\ell}_1 + \dot{\boldsymbol{\alpha}}_1 \times (\dot{\boldsymbol{\alpha}}_1 \times \boldsymbol{\ell}_1)}_{\text{(1) Acceleration of } J_2,} + \underbrace{\ddot{\boldsymbol{\alpha}}_1 \times \boldsymbol{\ell}_2}_{\text{(2)}} + \underbrace{\dot{\boldsymbol{\alpha}}_1 \times (\dot{\boldsymbol{\alpha}}_1 \times \boldsymbol{\ell}_2)}_{\text{(3)}} + \underbrace{2\dot{\boldsymbol{\alpha}}_1 \times (\dot{\boldsymbol{\alpha}}_2 \times \boldsymbol{\ell}_2)}_{\substack{\text{(4) Coriolis'} \\ \text{acceleration}}} + \underbrace{\ddot{\boldsymbol{\alpha}}_2 \times \boldsymbol{\ell}_2}_{\substack{\text{(5)} \\ \text{Tangential} \\ \text{acceleration} \\ \text{of P with} \\ \text{respect to } J_2}}$$

the origin of the local frame

$$(3.78)$$

Terms (2) and (3) resemble tangential and centripetal components of linear acceleration seen in rotation about a fixed axis. However, these acceleration terms are pointing, as a rule, neither perpendicularly to $\boldsymbol{\ell}_2$ nor along it and, consequently, cannot be called tangential or centripetal. Because in three-dimensional motion the vector of angular acceleration $\dot{\boldsymbol{\alpha}}_1$ depends on the change in both the magnitude and direction of the angular velocity $\dot{\boldsymbol{\alpha}}_1$, it should be computed by applying the rule from "Time Derivative of a Vector."

As the reader may see, the direct problem of kinematics for acceleration of the end effector is solvable, but it is not elementary, even for a chain with only two links.

3.2.3 Acceleration of a Multilink Chain

The methods described in the preceding paragraphs, such as Jacobian matrices, can be used for studying multilink chains. The computations, however, may become unmanageable without a computer. The acceleration of a multilink chain can be analyzed either in joint or in segment space. The representation of the movement in the joint space allows us to single out the physical sources of the segment acceleration and therefore enrich the movement analysis. The influence of every body segment on every other segment can be determined. The forces due to various accelerations can be separated. If velocity and acceleration for a given joint are zeroed, it is possible to find how much change occurs without movement in this particular joint. However, the use of joint space does not allow for easier calculation. The equations of motion are lengthy and complex. The writing of the equations is a cumbersome task. In summary, the use of the absolute reference system permits simple calculations and easy comprehension of what happens with each of the segments examined in isolation. However, the segments' interaction is more understandable when the movement is analyzed in joint space.

3.2.4 Jerk and Snap

Representative papers: Edelman & Flash, 1987; Flash & Hogan, 1985; Plamondon et al., 1993.

Jerk is the time rate of change of acceleration. In general, the change in acceleration is associated with the changes in the magnitude of the acting forces, and thus, the changes in the stresses within the moving body. In engineering applications, the jerk is registered when the propagation of the deformation waves is of interest. When movement is performed at only one joint, the jerk is associated with the rate of muscle force development. In biomechanics and motor control, there is one additional stimulus for studying jerk: it has been suggested that a skilled performance is characterized by a decrease in jerk magnitude. According to the minimum-jerk hypothesis, in skilled people the arm moves in a maximally smooth way. The square of the jerk integrated over the entire movement is minimized.

$$C_j = \frac{1}{2} \int_0^{t_f} \left\{ \left(\frac{d^3 X}{dt^3} \right)^2 + \left(\frac{d^3 Y}{dt^3} \right)^2 \right\} dt \xrightarrow{\quad min \quad} \tag{3.79}$$

where X and Y are Cartesian coordinates of the end effector and t_f is the movement duration. This criterion looks appealing for some movements, such as transporting a glass filled with liquid.

Kinematically, the jerk of the end effector depends on several variables. Differentiating equation 3.75 yields

$$\dddot{P} = \dddot{J} \dot{\alpha} + 2 \ddot{J} \ddot{\alpha} + J \dddot{\alpha} \tag{3.80}$$

where $\dddot{P}$ is the end-effector jerk and $\dddot{\alpha}$ is the vector of the joint angular jerks. From equation 3.80, it follows that the end-effector jerk depends on the system's Jacobian and its first and second derivatives, as well as on the angular velocities, accelerations, and jerks in individual joints. How the CNS manages to minimize $\dddot{P}$ in such a complex situation is still a matter of research.

Snap is the fourth time derivative of the displacement. In the human movement science, several models based on the idea of minimizing snap have been explored.

3.3 BIOLOGICAL SOLUTIONS TO THE PROBLEMS OF DIFFERENTIAL KINEMATICS: CONTROL OF MOVEMENT VELOCITY

This section describes several strategies supposedly used by animals and people to control velocity of their motor actions. The underlying concept is that the methods of differential kinematics developed by scientists, although powerful tools of analysis, are in no way used by the CNS for motor control. The brain of a frog evidently does not calculate a Jacobian to convert joint angular velocities into the velocity of the end effector. However, how exactly the CNS controls the movement is not known at this time. This is a field of very active research and many zestfully debated hypotheses. In Sections **3.3.1** and **3.3.2** several of these hypotheses are described and investigated. Some of the hypotheses, for instance Berkenblit's hypothesis, make use of the classic methods of differential kinematics. Others are based on new ideas, e.g., on the idea of nonlinear time, as in Gutman's hypothesis. My goal is to give a brief introduction to these hypotheses to prepare the reader for understanding the current scientific literature. The emphasis is on the main ideas and the formal methods rather than on validation of the hypotheses and experimental finding. I will restrain, however, from a discussion of the hypotheses in their entireties: the validation or rejection of the hypotheses is beyond the scope of this book.

3.3.1 Control of Approach: The Tau Hypothesis

Representative paper: Lee, Young, & Rewt, 1992.

In everyday life, as well as in athletics, people constantly adjust their movements to the external environment. Avoiding collision with unmovable and movable objects (walls, other people, cars), intercepting a moving object (catching, kicking), landing, and stopping a car at a required place are examples of such motor tasks. For animals, control of approach is one of the basic skills required for survival: successful hunting for some of them, avoiding the hunters for others. It is equally important for all animals, mammals, birds, fishes, and even for insects (a disoriented fly bumps a window glass repeatedly; it lands smoothly on an opaque wall).

To begin our investigation of the approach control, let us consider one example. An actor (i.e., a human or an animal) is moving rectilinearly toward a destination at velocity $\dot{x}$ (Figure 3.26). The actor should reach the target at a certain speed. In some cases, the speed at destination should be exactly zero (to avoid collision); in others, it should not exceed a given value (to avoid injury). For clarity, assume that the actor must stop exactly at the destination. The problem is how should the actor decelerate to fulfill the task.

One theoretically possible solution is to move toward the destination with a constant deceleration $\ddot{x}$. The stopping distance equals a product of the average velocity ($\dot{x}_0/2$) and the time to stop ($-\dot{x}_0/\ddot{x}$), where $\dot{x}_0$ is the velocity at the beginning of the approach. By convention, the time is negative and decreases during the movement. The stopping distance, D, is then

$$D = -\dot{x}_0^2 / 2\ddot{x} \tag{3.81}$$

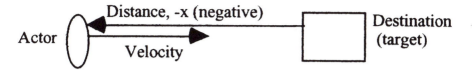

Figure 3.26 A scheme representing control of approach. The distance from the destination is $-x$. Note that the distance is negative. When the performer is moving toward the destination the distance decreases. More explanation is given in the text.

If the actor stops at the destination,

$$-x_0 = -\dot{x}_0^2 / 2\ddot{x} \tag{3.82}$$

and the deceleration is

$$\ddot{x} = \dot{x}_0^2 / 2x_0 \tag{3.83}$$

However, according to numerous observations, neither people nor animals control an approach, as a rule, with constant acceleration. They use another strategy to adjust braking. The hypothesis is that the time to the approach (called *tau*) is somehow estimated by the actor, and the rate of the time change (called *tau dot*) is controlled. The tau hypothesis is credited to D. Lee. Let us consider the kinematic aspects of this hypothesis only, i.e., the relationship between acceleration and deceleration and the tau dot values. Several examples supporting the tau hypothesis will also be given.

3.3.1.1 Tau Dot Versus Acceleration Kinematics

Let us begin with definitions. The time it would take the actor to reach the destination, if he or she were to continue at a constant approach velocity is called *tau,* or *tau margin:*

$$\tau(x) = \frac{x}{\dot{x}} \tag{3.84}$$

The rate of change of a tau margin is denoted as *tau dot.*

$$\dot{\tau}(x) = 1 - x\ddot{x}/\dot{x}^2 \tag{3.85}$$

According to equation 3.83, the tau dot and acceleration values are interrelated. Approaching the destination, the actor may maintain constant (1) acceleration or deceleration or (2) tau dot. These strategies are analyzed below.

3.3.1.1.1 Constant Acceleration Approach

The comparison of the constant acceleration or deceleration and the tau dot values for three different situations is presented in Table 3.1.

To avoid a collision, or to decrease the magnitude of the impact at collision, the actor should decelerate himself or herself ($\ddot{x} < 0$, or $\dot{\tau}(x) < 1$). From equation 3.83 it follows that the relationship between (constant) deceleration, tau dot values, and the outcome of the approach is determined by the magnitude of $x\ddot{x}/\dot{x}^2$. When the stopping distance equals x_0, the subject stops at the destination and

$$\frac{x_0\ddot{x}}{\dot{x}_0^2} = \frac{x_0}{\dot{x}_0^2} \cdot \frac{\dot{x}_0^2}{2x_0} = \frac{1}{2} \tag{3.86}$$

When $x\ddot{x}/\dot{x}^2$ exceeds 0.5, the subject stops short of the destination and when it is less than 0.5, the subjects collides with the target object. Hence, the requirement to stop exactly at the destination is $\ddot{x} = \dot{x}^2/2x$.

We introduce a new variable now, *tau prime*. This is the ratio

$$\tau' = \frac{\dot{x}}{\ddot{x}} \tag{3.87}$$

Tau prime is the time it would take the subject to stop if he or she were to continue at a constant deceleration. It is easy to note that the expression $x\ddot{x}/\dot{x}^2$ can be written as the quotient

$$x\ddot{x}/\dot{x}^2 = \tau/\tau' \tag{3.88}$$

Thus, the ratio of the tau and tau prime values determines the approach. A study of the approach can be based on either tau prime or tau dot data. The analysis based on the tau dot values dominates in the scientific literature, how-

Table 3.1   Acceleration or Deceleration Versus Tau Dot Values[1]

Acceleration/Deceleration	Tau dot	Time to the destination[2]	Outcome of the approach
$\ddot{x} > 0$ (acceleration)	$\dot{\tau}(x) > 1$	$< \tau_0$	Collision
$\ddot{x} = 0$ (constant velocity)	$\dot{\tau}(x) = 1$	$= \tau_0$	Collision
$\ddot{x} < 0$ (deceleration)	$\dot{\tau}(x) < 1$	$> \tau_0$	Uncertain

[1] A constant acceleration or deceleration is assumed.
[2] τ_0 is the value of the tau at the beginning of the approach.

ever the relationship between the tau dot values and the outcome of the approach is presented in Table 3.2.

3.3.1.1.2 Constant Tau Dot Approach

To stop at the destination the actor may vary deceleration so that the tau dot stays constant at a value of k, $0 < k < 0.5$. In this case,

$$-\tau(x) = -\tau_0 + kt \qquad (3.89)$$

Consider an example. The distance to a destination is 30 m, the initial velocity is 10 m/sec, and the initial tau is, not surprisingly, 3 sec. Assume $k = 0.5$. After 1 sec, $-\tau(x) = -2.5$ sec, that is, if the performer were to continue with the same velocity it would take him 2.5 sec to reach the destination. After 2 sec, $-\tau(x) = 2$ sec, and after 3 sec, $-\tau(x) = 1.5$ sec. Evidently, the performer will reach the destination in 6 sec. When this strategy is used, the deceleration is changed as

$$\ddot{x} = (1 - k)\frac{\dot{x}^2}{x} = (1 - k)\frac{\dot{x}}{\tau} \qquad (3.90)$$

This follows simply from equation 3.89. Note that the deceleration is not constant; it is the function of both the distance to the destination and velocity (or the velocity and tau). Constant tau dot value k can be conveniently represented through normalized time, t_n. The normalized time is the ratio of running time to initial time-to-contact with the destination under constant velocity, $t_n = -t/\tau_0$. Using equation 3.89 and assigning $x = 0$ yields

$$k = \tau_0 / t = 1 / t_n \qquad (3.91)$$

Thus, normalized time to reach the destination is the inverse of k.

Table 3.2   The Outcome of the Approach at Various Tau/Tau Prime and Tau Dot Values*

Approach	Tau/tau prime	Tau dot
	$\tau/\tau' = x\ddot{x}/\dot{x}^2$	$\dot{\tau}(x) = 1 - x\ddot{x}/\dot{x}^2$
Stop at destination	$= 0.5$	$= 0.5$
Short of destination	> 0.5	< 0.5
Collision	< 0.5	$0.5 < \dot{\tau}(x) < 1.0$

*A constant deceleration is assumed.

The equations of motion can be obtained by integrating equation 3.90:

$$x/x_0 = (1 + kt/\tau_0)^{1/k} \tag{3.92a}$$

$$\dot{x}/\dot{x}_0 = (1 + kt/\tau_0)^{(1/k)-1} \tag{3.93a}$$

The ratios $\dot{x}/\dot{x}_0$ and x/x_0 are the normalized velocity and normalized distance, respectively. That is, they show the instantaneous values of the velocity and distance to the destination expressed as a proportion of their initial value. Denoting the normalized distance x_n ($x_n = x/x_0$), the normalized velocity $\dot{x}_n$ ($\dot{x}_n = \dot{x}\dot{x}_0$), and using the normalized time, t_n ($t_n = -t/\tau_0$), we can rewrite equations 3.92a and 3.93a as

$$x_n = (1 - kt_n)^{(1/k)} \tag{3.92b}$$

$$\dot{x}_n = (1 - kt_n)^{(1/k)-1} \tag{3.93b}$$

Also, after introducing the normalized deceleration $\ddot{x}_n = \ddot{x}/(\dot{x}_0^2/x_0) = \ddot{x}(x_0/\dot{x}_0^2)$, where $\dot{x}_0^2/x_0$ is twice the constant deceleration needed to stop precisely at the destination, the deceleration is

$$\ddot{x}_n = (1 - k)(1 - kt_n)^{(1/k)-2} \tag{3.94}$$

Equations 3.92, 3.93, and 3.94 represent the distance, velocity, and deceleration, respectively, as functions of the time to the destination with k-values as a parameter. The different variants of braking behavior at various magnitudes of k are shown in Figure 3.27.

When tau dot remains constant at a value less than 0.5, the deceleration decreases monotonically. If this strategy is adopted, the actor brakes himself or herself very actively at the beginning of the approaching period; then the magnitude of the deceleration diminishes. When tau dot is 0.5, the deceleration remains constant until the destination is just reached. Finally, when $0.5 < \tau(x) < 1.0$ the deceleration increases monotonically. With such a strategy, the closer to the destination, the greater the deceleration and, correspondingly, the greater the braking force. Theoretically, the braking force at destination reaches infinity. However, in reality, the braking force produced by a subject cannot exceed a certain limit. Because of that, when tau dot is greater than 0.5 and less than 1.0, its value is maintained constant until maximal braking force is reached, and then the braking force is kept constant at the maximal level (the *deceleration ceiling*). This procedure is called the *controlled-collision strategy*. The strategy results in a controlled collision with the destination.

3.3.1.2 Experimental Studies of the Approach

The kinematic analysis made above is purely theoretical. It does, however, provide a background for experimental testing of the tau hypothesis. This

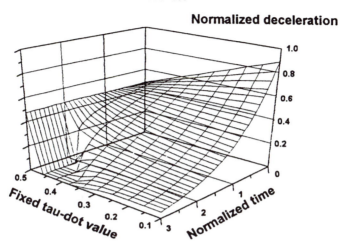

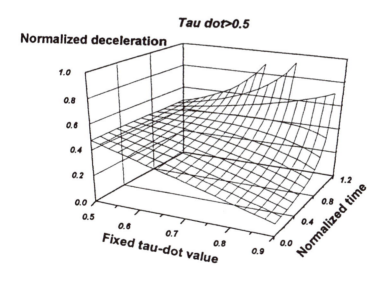

Figure 3.27   Acceleration profiles when approaching the destination at constant tau dot values, k<0.5 or k>0.5. Note the different direction of the time axes in the two graphs.

hypothesis has been tested actively from two perspectives: sensory and motor. When the sensory aspects are the objects of interest the problem is, do animals and people really determine tau (time-to-contact with an object in space) and the rate of change of this time (tau dot)? It seems they do. When an object, such as a ball, is moving toward the actor, or, conversely, the actor is approaching the object, the image of the object on the retina is becoming progressively larger. The rate of dilation of the image on the retina is triggering a specific motor response. The time-to-contact, it seems, is being perceived directly from the inverse of the rate of expansion of the image. Regarding the motor aspects of the approach, it has been shown that the optical variable tau is used to control both the timing of the movement and the pattern of braking. Examples of the timing are the consistent time of muscle preactivation in preparation for landing in dropping jumps from various heights (the average standard deviation was only 19 ms) takeoff time in ski jumping and wing folding in diving seabirds, which occurs 300 to 400 ms before entry into the water. The braking pattern at constant tau dot with k = 0.425 is used by experienced car drivers when they are asked to stop at a line. The controlled-collision strategy has also been observed. For example, a person performing a somersault prepares for landing by demonstrating an almost linear change of tau with k > 0.5 (Figure 3.28).

3.3.2 Control of Velocity in Reaching Movement

Reaching a target is one of the most common motor tasks. Even when the reaching movement is performed by the arm alone (the shoulder joint does not translate), the problem is kinematically redundant—the number of rotational DOF of the arm, seven, exceeds the number of DOF of the end effector, six. The Jacobian of the kinematic chain is singular. Hence, the inverse kinematic problem cannot be solved uniquely and the movement can be performed in a limitless number of ways. The formidable task is to find why the CNS prefers some movement patterns over others.

3.3.2.1 Berkenblit's Hypothesis

Representative paper: Berkenblit, Gelfand, & Feldman, 1986b.

This hypothesis was stimulated by the studies of the wiping reflex in the spinal frog. The reflex was initially described in the middle of the 19th century. If a piece of paper soaked in acid is placed on the frog's back, the frog wipes the paper from the body. It was observed that the limb—comprising three body segments—may come repeatedly to the same point of the skin with different values of joint angles. Hence, the frog does not "solve" the inverse kinematic problem in a unique way; the end-point location does not prescribe the particular joint configuration. The idea was to suggest a model that describes such a behavior.

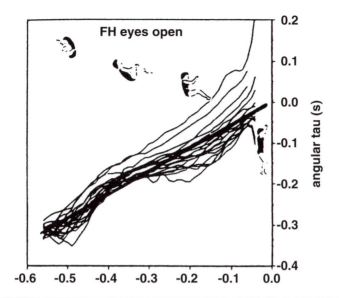

Figure 3.28 The change of the angular tau (angular distance/angular velocity) during preparation for landing after the forward somersault. The line joining the temple and the ankle was used to estimate the body attitude and angular velocity at rotation. Thin lines are plots for individual somersaults (n = 10). The thick line is the linear regression of angular tau on time-to-landing. For a period from -0.560 to -0.040 sec, the relationship is close to linear, $r^2 = 0.963$. The angular tau dot (the slope of the regression line) is k = 0.552. Because k is greater than 0.5, the somersaulter reaches the ground with non-zero angular velocity (which is inevitable because of the conservation of the angular momentum).

From Lee, D.N., Young, D.S., & Rewt, D. (1992). How do somersaulters land on their feet? *Journal of Experimental Psychology, 18,* 1195-1202. Copyright © 1992 by the American Psychological Association. Reprinted with permission.

For each joint, two vectors originating from the joint were introduced. One vector, P_j, goes from the joint to the end effector—it is similar to the vector introduced in Section **3.1.2.2.1**. The second vector is from the joint to the target, T_j. The vectors form an angle α_j. The model is based on the assumption that the joint angular velocity $\dot{\alpha}_j$ is proportional to (1) sin α_j ($0 \le \alpha_j \le \pi/2$; if $\alpha_j > \pi/2$, a correction is required) and (2) the length of the vectors P_j and T_j. Hence, for any joint

$$\dot{\boldsymbol{\alpha}}_j = a_j(\mathbf{P}_j \times \mathbf{T}_j) \tag{3.95}$$

where a_j is a positive number, $(\mathbf{P}_j \times \mathbf{T}_j)$ is the vector product, and $\dot{\boldsymbol{\alpha}}_j$ is the joint velocity vector. The vector is along the axis of rotation in the joint. It is normal to the plane containing (a) the joint center, (b) the point at which the end effector is initially located, and (c) the target point. According to the hypothesis, the joints act simultaneously but are controlled independently—the controlling signals to the joints are formed individually, ignoring what the other joints are doing. The task of a joint j is to turn the vector $\mathbf{P}_j$ in the direction to the target without regard to the joint configuration of the whole limb and the command signals arriving at this time to the other joints.

The suggested algorithm is able to solve the inverse kinematic problem without inverting the limb Jacobian. The mathematical operations required by the algorithm—summation, multiplication, and calculating sine and cosine functions—do not place unrealistic demands on the nervous system, the operations can be carried out by neural networks. In particular, it has been shown—recall Section **2.3.2.2** and Figures 2.28 and 2.29—that activity of some cortex neurons changes as a sinusoidal function of the direction of movement. For a given initial joint configuration and target location, the algorithm provides a unique movement pattern that resembles the natural limb trajectories.

3.3.2.2 Gutman's Hypothesis

Representative paper: Gutman, Gottlieb, & Corcos, 1990.

When a subject moves an arm to a target, the working point trajectory is a nearly straight line and the velocity-versus-time curve is bell-shaped. The bell-shaped angular velocity profile has also been observed in single-joint movements (Figure 3.29). Gutman's hypothesis is aimed at an explanation of the observed velocity patterns.

This hypothesis is based on the idea of the nonlinearity of subjective time—the perceived duration of temporal intervals differs from their real duration. Judgment of brief temporal intervals can be described by the *psychophysical law for time*, a power function $t' = bt^{\alpha}$, where t' and t are the subjective and real time, correspondingly, and b and a are parameters. (The existence of such a psychophysical law for time is still an object of discussion in the literature.) According to the hypothesis, people are using subjective time rather than real time for movement planning. If the hypothesis holds, the distance to the target during the movement execution as a function of time may be represented as

$$D(t) = D \cdot \exp\left(-\frac{t^{\alpha}}{\tau} \right) \tag{3.96}$$

and the velocity as its time derivative

$$V(t) = -\alpha \frac{D}{\tau} t^{\alpha-1} \exp\left(\frac{t^{\alpha}}{\tau} \right) \tag{3.97}$$

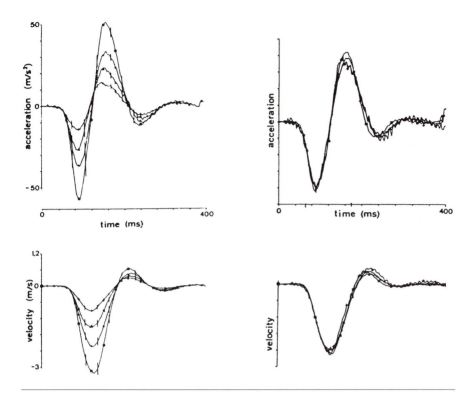

Figure 3.29 Traces for acceleration and velocity before scaling (left panels) and after scaling (right panels). The subjects performed movements of the same duration, 125 ms, and different amplitudes, d = 7.5, 15, 22.5, and 30 cm. For the scaling, the ratio $d:d_0$ was used where $d_0 = 22.5$ cm.
From Gielen, C.C.A.M., van der Oosten, K., & Pull ter Gunne, F. (1985). Relation between EMG activation patterns and kinematic properties of aimed arm movements. *Journal of Motor Behavior, 17,* 421-442. Reprinted with permission of the Helen Dwight Reid Educational Foundation. Published by Heldref Publications, 1319 Eighteenth St., N.W., Washington, D.C. 20036-1802. Copyright © 1985.

where D is the movement distance, α and τ are parameters representing the time nonlinearity and movement speed, correspondingly. Equation 3.97 describes the experimental velocity-time curves quite well.

3.3.2.3 Scaling Velocity Profiles: The Generalized Motor Programs

Representative papers: Carter & Shapiro, 1984; Gielen et al., 1985; Schmidt, 1975.

In these experiments, the subjects performed aimed arm movements of either the same duration and various amplitudes or the same amplitude and various duration. Then, the velocity-time and acceleration-time curves were scaled with regard to each other. In the experiments of the first group (same duration, various amplitude), the amplitude of the signal was scaled. In the second group of experiments (same amplitude, various duration), the time was scaled. It was observed that after the appropriate normalization, the traces of the velocity and acceleration coincide almost ideally (Figure 3.29).

It has been suggested that these movements are governed by a common underlying motor program that, depending on the motor task, is adjusted for various movement amplitudes, durations, or both. This hypothesis is called the theory of *generalized motor program*. The idea is that the CNS stores just a plain outline of the motor program. During the movement execution, the program is specified according to the parameters of the motor task—the required movement duration, amplitude, and so on.

The scaling method, however, is limited in that it can only be applied for studying simple movements. Two movements, A and B, are called simple if they can be transformed into each other by simple multiplication of the time and space scales—the transform A↔B is an *affine transformation* of the movement time and space. The method cannot be immediately applied for studying complex multiphase movements in which consecutive phases are adjusted independently. To do that the *local proportional scaling method*, which is not described here, can be employed.

3.4 SUMMARY

Movement of a kinematic chain is described with the joint angular coordinates or external coordinates. These ways of representation are discussed first for planar movement and then for movement in three dimensions.

A joint angular velocity can be computed as a difference between the segment angular velocities of the links forming the joint. When a joint angular velocity is equal in magnitude and opposite in direction to the segment angular velocity of one of the links the adjacent link maintains a fixed orientation in space (a pair of rotation).

The direct kinematic problem is defined as a search for an end-point velocity in terms of the joint angular velocities. An opposite problem—a quest for joint angular velocities in terms of known end-point velocity—is called the inverse kinematic problem. For a simple two-link chain these problems can be solved by using simple geometric considerations; however, analysis of multilink chains requires more complex methods, in particular using Jacobian matrices. If a Jacobian is known, the direct kinematic problem can be easily solved—the velocity of the end effector in the external reference frame is computed as a product of the Jacobian and the joint angular velocity vector. The column-vectors of the Jacobian represent the contribution of individual joints to the end-point velocity. These column-vectors are derivatives of the radius-vectors from the joints to the end point with regard to the joint angles.

The inverse kinematic problem can be unequivocally solved if, and only if, the Jacobian is not singular. The singularity occurs when either the number of DOF in the joints exceeds the number of DOF of the end effector or when the chain assumes a singular joint position at which restrictions are imposed on the motion of the end effector. In human movement, the singular joint configurations are often used when a maximal load is supported.

In a pushing motion resembling leg and arm extension, the transfer of the joint angular velocity into linear end-point velocity is limited by geometric and anatomic constraints. As a consequence, in human movement the stretch velocity reaches its maximum before the full extension. To produce maximal end-point velocity, a performer should find a compromise between (a) the body position at which maximal joint velocity is reached and (b) the joint configuration at which maximal transfer of the joint angular velocity into linear velocity of the end effector occurs.

Three-link chains, as well as multilink chains, are redundant. Their Jacobians are singular. The inverse kinematic problem for these chains cannot be solved in a unique way. Additional restrictions can help, however. For instance, if the requirement is to minimize mechanical work for movement, optimal joint angular velocities can be computed by using the generalized Moore-Penrose matrices. Although each link of a kinematic chain undergoes a complex movement—a combination of translation and rotation—in each moment the link motion can be viewed as rotation around an instant center. According to Kennedy's theorem, the instant centers of three consecutive links in a chain lie on a straight line, the line of centers. In humans, as well as in higher animals, the extremities consist of three main links arranged in a zigzag manner. The zigzag arrangement makes ambulation more economical.

Infinitesimal rotations, in contrast to final rotations, are commutative and considered vectors. Angular velocity in three-dimensional movement can be represented either as a vector or as an angular velocity matrix. The angular velocity is related to the derivative of the rotation matrix by Poisson's equation.

The direct kinematic problem for chains moving in three-dimensional space can be solved using vector and matrix methods. In the vector approach, position vectors from the individual joints to the end point are computed. Then, the velocity of the end point generated by a given joint is found as a cross product of the corresponding angular velocity vector and the position vector. In the matrix approach Jacobians are used. They are customarily decomposed into positional and orientational Jacobians. The elements of a positional Jacobian equal the vector product of the unit vector along the axis of rotation by the position vector. The elements of an orientational Jacobian are the unit vectors along the axis of rotation in the joint.

The accelerations due to the movement of individual body segments can be analyzed either in a global or in a local reference frame. The relative acceleration method is used when the absolute velocity and acceleration of the body segments are considered. When the joint angular velocity and acceleration are analyzed, the Coriolis' acceleration must be taken into account. The Coriolis' acceleration is acting when a body link is moving with regard to a rotating reference frame, e.g., a frame fixed with the neighboring proximal link. Because the individual accelerations must be summed vectorially, the resultant acceleration of the end effector depends not only on the segment or joint velocity and acceleration but also on the body posture at each instant of time. The Jacobian matrix provides a tool for computing the end-point acceleration in terms of known joint positions, velocities, and accelerations.

Jerk is the time rate of change of acceleration, the third time derivative of the displacement. Snap is the fourth time derivative of the displacement.

Biological solutions to the problems of differential kinematics are addressed in the last section of the chapter. Several hypothetical strategies used to control velocity of human movement are discussed. Control of approach velocity is represented by the tau hypothesis, and control of velocity of reaching movement by Berkenblit's and Gutman's hypotheses.

3.5 QUESTIONS FOR REVIEW

1. The trunk orientation during walking is kept constant. How are the joint angular velocity at the hip joint of the support leg and the thigh angular velocity related? How are the hip, knee, and ankle joint angular velocities related to each other during the flat foot phase of the support period?

2. Write down a Jacobian for a two-link planar chain with hinge joints. Explain the meaning of the column-vectors of the Jacobian.

3. Describe the direct and the inverse kinematic problem. Write down corresponding equations.

4. Discuss the singularity problem. Compare Jacobians for two-, three-, and multilink planar chains. Describe singular joint positions of a two-link chain.

5. Discuss chain redundancy.

6. What are the geometric and anatomic constraints during leg and arm extension? At which joint angular positions is the transfer of

joint angular velocity into the linear velocity of the end effector maximal?

7. Explain the Kennedy theorem.

8. What is a cost function used in the Moore-Penrose method of matrix (pseudo)inversion?

9. Can angular finite displacements and angular velocities be represented as vectors? Explain in terms of kinematics an integral of angular velocity over time.

10. Write down and explain Poisson's equation.

11. Describe the methods used to characterize the relationship between the joint angular velocities and the end-point velocity of a spatial kinematic chain.

12. Motion of a planar two-joint chain is observed in an absolute coordinate system and in the joint angular coordinates. Write down the equations for the end-point acceleration in terms of the segment angular velocities and the joint angular velocities.

13. Define jerk.

14. Discuss the tau hypothesis.

BIBLIOGRAPHY

Alexander, M.J.L. & Colbourne, J. (1980). A method of determination of the angular velocity vector of a limb segment. *Journal of Biomechanics, 13,* 1089-1093.

Atkeson, C.G. (1989). Learning arm kinematics and dynamics. *Annual Review of Neuroscience, 12,* 157-183.

Atkeson, C.G. & Hollerbach, J.M. (1985). Kinematic features of unrestrained vertical arm movements. *Journal of Neuroscience, 5,* 2318-2330.

Baker, D.R. & Wampler, C.W. (1988). On the inverse kinematics of redundant manipulators. *International Journal of Robotics Research, 7,* 3-21.

Berkenblit, M.B., Gelfand, I.M., & Feldman, A.G. (1986a). A model for the aiming phase of the wiping reflex. In S. Grillner, P.S.G. Stein, D. Stuart, H. Forssberg, & R.M. Herman (Eds.), *Wenner-Gren International Symposium Series: Vol. 45. Neurobiology of vertebrate locomotion* (pp. 217-227). Dobbs Ferry, New York: Sheridan House.

Berkenblit, M.B., Gelfand, I.M., & Feldman, A.G. (1986b). A model for the control of multi-joint movements. *Biofizika, 31,* 128-138.

Carter, M.C. & Shapiro, D.C. (1984). Control of sequential movements: Evidence for generalized motor programs. *Journal of Neurophysiology, 52,* 787-796.

Cruse, H., Bruer, M., & Dean, J. (1993). Control of three- and four-arm movement: Strategy for a manipulator with redundant degrees of freedom. *Journal of Motor Behavior, 25,* 131-139.

Dimnet, J. (1980). The improvement in the results of kinematics of *in vivo* joints. *Journal of Biomechanics, 13,* 653-661.

Edelman, S. & Flash, T. (1987). A model of handwriting. *Biological Cybernetics, 57,* 25-36.

Featherstone, R. (1983). Position and velocity transformations between robot end-effector coordinates and joint angles. *International Journal of Robotics Research, 2,* 35-45.

Feltner, M.E. (1989). Three-dimensional interactions in a two-segment kinetic chain. Part II. Application to the throwing arm in baseball pitching. *International Journal of Sport Biomechanics, 5,* 420-450.

Feltner, M.E. & Dapena, J. (1989). Three-dimensional interactions in a two-segment kinetic chain. Part 1. General model. *International Journal of Sport Biomechanics, 5,* 403-419.

Flash, T. & Hogan, N. (1985). The coordination of arm movements: An experimentally confirmed mathematical model. *Journal of Neuroscience, 5,* 1688-1703.

Gielen, C.C.A.M., van der Oosten, K., & Pull ter Gunne, F. (1985). Relation between EMG activation patterns and kinematic properties of aimed arm movements. *Journal of Motor Behavior, 17,* 421-442.

Goodman, S.R. (Gutman, S.R.) (1995). Inverse kinematics problem: Solutions by pseudoinversion, inversion and no-inversion. *Behavioral and Brain Sciences, 18,* 756-758.

Goodman, S.R., (Gutman, S.R.), & Gottlieb, G.L. (1995). Analysis of kinematic invariences of multi-joint reaching movement. *Biological Cybernetics, 73,* 311-322.

Gutman, S.R., Gottlieb G.L., & Corcos, D.M. (1990). Exponential model of a reaching movement trajectory with nonlinear time. *Comments on Theoretical Biology, 2,* 357-383.

Gutman, S.R. & Gottlieb, G.L. (1990). Non-linear "inner" time in reaching movement trajectory formation. In: *First World Congress on Biomechanics. Abstracts*, v. 1, p. 190, San Diego, CA.

Hogan, N. & Mussa-Ivaldi, F. (1992). Muscle behavior may solve motor coordination problem. In A. Berthoz, W. Graf, & P.P. Vidal (Eds.), *The head and neck sensory motor system* (pp. 153-157). New York: Oxford University Press.

Hurmuzlu, Y. (1994). Presenting joint kinematics of human locomotion using phase plane portraits and Poincaré maps. *Journal of Biomechanics, 27,* 1495-1500.

Ingen Schenau, G.J. van (1989). From rotation to translation: Constraints on multi-joint movements and the unique action of bi-articular muscles. *Human Movement Science, 8,* 301-337.

Ito, A., Saito, M., Sagawa, K., Kato, K., Ae, M., & Kobayashi, K. (1993). Leg movement of gold and silver medalists in men's 100 m final in III World

Championship in Athletics. In *XIVth Congress of the International Society of Biomechanics, Abstracts* (pp. 624-625). Paris.

Kanatani-Fujimoto, K., Lazareva, B., & Zatsiorsky, V. (1997). Local proportional scaling of time-series data: Method and applications. *Motor Control, 1,* 20-43.

Kieffer, J. & Lenarcic, J. (1993). Kinematic singularities in human arm motion. In *XIVth Congress of the International Society of Biomechanics, Abstracts* (pp. 680-681). Paris.

Kuznetsov, A.N. (1993). Third segment saves energy in the flat walking leg. In *XIVth Congress of the International Society of Biomechanics* (pp. 736-737). Paris.

Lackner, J.R. & Di Zio, P. (1992). Gravitational, inertial, and Coriolis force influences on nystagmus, motion sickness, and perceived head trajectory. In A. Berthoz, W. Graf, & P.P. Vidal (Eds.), *The head and neck sensory motor system* (pp. 216-222). New York: Oxford University Press.

Lee, D.N., Young, D.S., & Rewt, D. (1992). How do somersaulters land on their feet? *Journal of Experimental Psychology, 18,* 1195-1202.

Miazaki, S. & Ishida, A. (1991). New mathematical definition and calculation of axial rotation of anatomical joints. *Journal of Biomechanical Engineering, 113,* 270-275.

Mussa-Ivaldi, F.A., Morasso, P., & Zaccaria, R. (1988). Kinematic networks. A distributed model for representing and regularizing motor redundancy. *Biological Cybernetics, 60,* 1-16.

Pellionisz, A. (1984). Coordination: A vector-matrix description of transformations of overcomplete CNS coordinates and a tensorial solution using the Moore-Penrose inverse. *Journal of Theoretical Biology, 110,* 353-375.

Plagenhoef, S. (1971). *Patterns of human motion. A cinematographic analysis.* Englewood Cliffs, NJ: Prentice Hall.

Plamondon, R., Alimi, A.M., Yergeau, P., & Leclere, F. (1993). Modeling velocity profiles of rapid movements: A comparative study. *Biological Cybernetics, 69,* 119-128.

Robinson, D.A. (1982). The use of matrices in analyzing the three-dimensional behavior of the vestibulo-ocular reflex. *Biological Cybernetics, 46,* 53-66.

Schmidt, R.A. (1975). A schema theory of discrete motor skill learning. *Psychological Review, 82,* 225-260.

Sprigings, E., Marshall, R., Elliott, B., & Jennings, L. (1994). A three-dimensional kinematic method for determining the effectiveness of arm segment rotations in producing racket-head speed. *Journal of Biomechanics, 27,* 245-254.

Tweed, D. (1994). Rotational kinematics of the human vestibuloocular reflex. II. Velocity steps. *Journal of Neurophysiology, 72,* 2480-2489.

Verstraete, M.C. & Soutas-Little, R.W. (1990). A method for computing the three-dimensional angular velocity and acceleration of a body segment from three-dimensional position data. *Journal of Biomechanical Engineering, 112,* 114-118.

Woltring, H.J., Kong, K., Osterbauer, P.J., & Fuhr, A.W. (1994). Instantaneous helical axis estimation from 3-D video data in neck kinematics for whiplash diagnostics. *Journal of Biomechanics, 27,* 1415-1432.

Zajac, F.E. & Gordon, M.E. (1989). Determining muscle force and action in multi-articular movement. *Exercise and Sports Science Reviews, 17,* 187-230.

Zelaznik, H.N., Schmidt, R.A., & Gielen, S.C.A.M. (1986). Kinematic properties of rapid-aimed head movements. *Journal of Motor Behavior, 18,* 353-372.

4

CHAPTER

JOINT GEOMETRY AND JOINT KINEMATICS

In the previous chapters, an analysis of joint motion was based on several simplifying assumptions. It was explicitly accepted that joint movement is pure rotation about a fixed joint axis. Translatory joint motion was completely disregarded. In joints with two and three DOF, the axes of rotation for flexion-extension, abduction-adduction, and internal-external rotation were usually assumed to be orthogonal to each other, intersect at one joint center, and coincide with the main anatomic axes. Therefore, by this assumption, the joint rotation planes lie in the anatomic planes, i.e., sagittal, frontal, and transverse. Such joints—with fixed orthogonal axes parallel to the main anatomic axes and intercepting at one center—are called *nominal joints* or *geometrically ideal joints*. No human joints are perfectly ideal, however.

In biomechanics, the joints are called geometrically ideal when details of the joint kinematics are not important. For example, gross motor activities such as walking and running have commonly been investigated under the assumption that joint motion occurs about fixed axes that lie in the anatomic planes. This approach is too superficial when specific details of the joint motion are the object of inquiry. In such a case, the joint motion should be studied in all its complexity:

- joint motion does not occur about a fixed axis of rotation,
- the axis of rotation is oblique to one or more anatomic planes,
- joint motion involves translation,
- the axes for rotation and translation may be different, and
- the axes are not orthogonal to each other and may not intersect.

In this chapter, human joint motion is discussed with due attention to its real intricacy. However, some simplifying assumptions are made. In particular, articular surfaces and ligaments are assumed to be nondeformable.

This chapter starts by discussing intrajoint kinematics. In Section **4.1.1**, geometric analysis of the joint surfaces is presented, including elementary

••• ANATOMY REFRESHER •••

Joints

Joints are classified either as *synarthroses* or as *diarthroses* (also known as *synovial joints*). In a synarthrosis, the bones are joined by a deformable tissue and transmit forces in any direction (compression, tension, and shear). In a diarthrosis, the articulating surfaces glide easily with respect to each other and transmit only compression forces. In this chapter, with the exception of the motion segments of the spine, we will address only the synovial joints.

Kinematically, diarthroses and synarthroses are different. The axes of rotation in synovial joints are defined by their morphology—articular surfaces and ligaments—and do not substantially depend on the acting forces. Contrary to that, movement in synarthroses depends on the direction and magnitude of the applied forces.

Diarthroidal joints fall into two groups: simple and compound. A *simple joint* includes two bony surfaces. A *compound joint*, for example the radiocarpal joint, contains more than two surfaces. In *bicondylar joints*, such as the knee and temporomandibular joint, one bone articulates with another by two distinct surfaces called *condyles*. The movement of two condyles is coupled. The *mortice joints,* for example the ankle joint (talocrural joint), have a bar of bone on each side preventing movement in this direction.

examination of the joint surfaces. The examination is used as a basis for joint classification. A more sophisticated analysis of the joint surfaces based on the ideas of differential geometry is also presented. Among other things, the notion of joint congruence is introduced here. Relative movement of articular surfaces is addressed in Section **4.1.2**. The following issues are highlighted: close-packed and loose-packed joint positions; rolling and skidding of the conarticular surfaces; the slip ratio, which is used for estimating relative contribution of the rolling and skidding; and spin, which is rotation around an axis perpendicular to the joint surfaces at the point of contact. Section **4.1.3** is concerned with the geometry and algebra of intra-articular motion developed by M. A. MacConaill. The effect of ligament placement on joint motion is highlighted in Section **4.1.4**. In particular, the theory of the four-bar mechanism realized in the construction of the knee joint is explained.

Section **4.2** begins with a general introduction to the concept of centers and axes of rotations in joints. Planar joint movement is considered and three-dimensional joint movement is described. The following issues are also addressed: locating a fixed center of rotation, instant rotation in joints, locating an instant center of rotation (the Reuleaux method is also described here), and finite and instantaneous centrodes. Section **4.2.2.1** concentrates on joints with one DOF. Joints with two and three DOF are described and investigated in Section **4.2.2.2**. Section **4.2.2.3** is devoted to envelopes of joint motion.

4.1 INTRAJOINT KINEMATICS

Any joint motion is determined by acting forces and imposed constraints. The forces, which are not considered in this book limited to kinematics, are muscle forces and external forces. The constraints are bony constraints (articular surfaces) and ligamentous constraints. Synovial joints are said to be force locked: the articular surfaces direct joint motion when compressive forces are acting and the ligaments guide articular motion in the presence of tensile forces. The constraints determine passive motion characteristics of a joint including (a) geometry (shape) of joint motion; (b) range of motion, also called *flexibility*, which is the difference between the two extremes of joint movement; and (c) passive stiffness and passive viscosity of the joint (these will be addressed later in subsequent volumes of this textbook).

4.1.1 Articular Surfaces and Types of Joints

Articular surfaces have a complex geometry and cannot be precisely described by elementary equations. The articular surfaces can, however, be *approximated*

••• *Calculus Refresher* •••

Curves and Surfaces

Any differentiable function of one variable may be represented graphically as a plane curve. Any differentiable function of two variables may be represented graphically as a surface in a three-dimensional space.

Arc is any part of a curve. *Arc* (s) = *radius* (r) $\times$ *angle* (α).

Curvature is the change in direction per unit of arc, $\Delta\alpha/\Delta s = 1/r$. The *curvature at a point* is a limit of the quotient $\Delta\alpha/\Delta s$ when $\Delta\alpha$ and Δs approach zero. The *radius of curvature* is the reciprocal of the curvature at the point where $r = ds/d\alpha$. For a circle, the radius of the circle is equal to the radius of the curve.

by simple geometric surfaces or algebraic functions. These surfaces or functions do not describe the articular surface accurately and cannot be used to study joint motion in detail. They are, however, useful in classifying the articulating surfaces and the corresponding joints. The type of surface determines the permissible joint motion as well as the number of DOF in the joint. More advanced examination of the joint surfaces is based on the methods of differential geometry, a mathematical discipline that deals with the description of the shape of complex geometric objects.

4.1.1.1 Elementary Examination of Joint Surfaces

Representative paper: MacConaill, 1966.

The simplest articular surface is a *plane surface,* which is essentially a flat surface. The notion of a plane articular surface is an idealization. Real articular surfaces are always more or less curved. In some joints, however, the curvature is small and is commonly disregarded. The plane surfaces in the human body are the intermetacarpal joints; the anterior surface of the scapula is also flat. In joints with flat articular surfaces, in addition to axial rotation, gliding is kinematically possible. The bone surfaces in these joints permit three DOF (rotation and two translations in a plane); however, based on the real movements performed in the joints they are usually classified as having two DOF (Figure 4.1A).

Articular surfaces that are either completely concave or completely convex are called *synclastic* surfaces. The joints in which a convex ball or head fits into a concave socket are the ball-and-socket joints (Figure 4.1B*)*. Imagine the joint surface is cut by planes perpendicular to the surface. The line that is in the plane and wholly on the surface is an *arc*. An ideal ball-and-socket joint would be spherical. In a spherical joint, the shape and size of the arc, as well as the radius of the curvature, does not depend on the direction of the slice. In other words, every part of the sphere has the same curvature and the arcs of such a joint coincide after rotation. Because of this property, a spinning motion is feasible in these joints. These joints have three rotational DOF. In the human body, real ball-and-socket joints are more or less *spheroidal* (approximately spherical) rather then ideally spherical. The curvature of a spheroid is slightly different at different points, i.e., the spheroids are egg shaped. Thus, synclastic articular surfaces are called *ovoids*. The radii of curvature of an ovoid vary on the same articular surface. Examples of the spheroidal joints with ovoid or synclastic articular surface are the shoulder and hip joints. Although they are not ideally spherical, the joints still permit axial rotation. The deviation from sphericity in these joints is small. For the hip joint, it is less than 1% of the radius of curvature.

When the curvature in one direction much exceeds the curvature in the perpendicular direction, the synclastic joint is *ellipsoidal* (Figure 4.1C). In the ellipsoidal joints, one of the axes is longer than another and an oval condyle fits into an elliptical cavity. Movement in two planes at right angles, i.e., flexion-extension and abduction-adduction, is permitted in these joints. The wrist (radiocarpal) and metacarpophalangeal joints are ellipsoidal.

When articular surfaces are concave or convex in one direction and are (almost) flat in the perpendicular direction, joint motion is performed exclusively in one plane. Those joints, with one DOF, are further classified as hinge joints and pivot joints (Figure 4.1, E and F). In the hinge joints, a cylinder-like surface fits into a concave surface. Two examples are the elbow and finger joints.

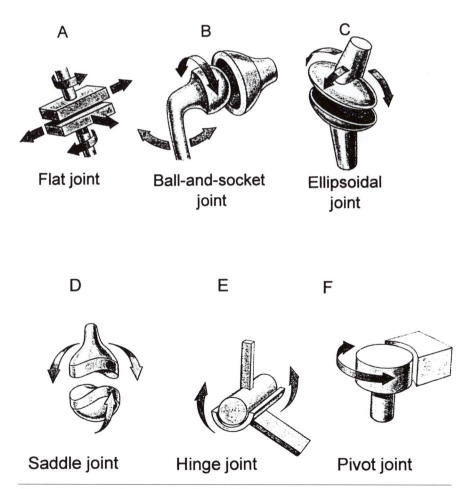

Figure 4.1 Models of joint surfaces and joint movement.

In the pivot joints, such as between the radius and ulna or in the atlantoaxial joint between the first and second cervical vertebrae, an arch-shaped surface rotates around a rounded pivot. The difference between the hinge and the pivot joints is not very strict, however.

The articular surfaces of saddle joints, called *anticlastic* or *sellars*, are convex in one direction and concave in the perpendicular direction (Figure 4.1D). The joint at the thumb, between the first metacarpal and the trapezium, is a saddle joint (Figure 4.2).

Saddle-shaped joints have two DOF. The joint motion includes flexion-extension and abduction-adduction. Kinematically, the saddle-shaped joints are different from ellipsoidal joints, which also have two DOF permitting flexion-extension and abduction-adduction. In the ellipsoidal joints, the two axes of rotation are on the same side of the joint, on the side of the bone with two convexities. In the saddle-shaped joints, the axes for flexion-extension and abduction-adduction are on the opposite sides of the joint. One axis is located on the proximal body segment and the second on the distal. In general, axes of joint rotation are positioned at the body segments with convex joint surfaces.

Convex surfaces are also called *male* and concave surfaces, *female*. The male articular surface always has more area than the female. This feature is used to classify male and female saddle joint surfaces, which are convex and

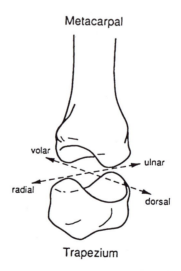

Figure 4.2   The carpometacarpal joint is a saddle joint with two axes of rotation. One axis, for thumb flexion combined with pronation, is at the trapezium. The second axis, for abduction-adduction, is at the base of the first metacarpal.

concave at the same time. The smaller of two mating sellar surfaces is called female and the larger is called male. Table 4.1 summarizes the described taxonomy of joint motion. The reference surface in this taxonomy is the male surface.

Consider now a point fixed to the distal end of a bone with the bone moving at the proximal joint. The locus of all the attainable positions of the point is a spheroidal convex surface. This surface is termed the *ovoid of motion* (Figure 4.3). It can be approximated by a sphere if the bone under deliberation is long enough. In this case, instead of the ovoid of motion the *sphere of motion* is considered and the joint movement is studied without opening the joint surgically. Generally, replacement of an ovoid of motion by the sphere of motion is used in motor control research and when biomechanical analysis is limited to gross motor patterns. The replacement cannot be done without an experimental proof of its validity when intrajoint kinematics is the object of interest.

The pattern of joint motion, e.g., whether it is sphere or ovoid, depends not only on the geometry of individual mating surfaces but also on their congruity. For the joint surfaces to be congruent, their radii of curvature should be equal. In other words, the geometric centers of the male and female surfaces should coincide.

Table 4.1   Taxonomy of Joints

	Type of Joint				
Characteristics	Ball-and-socket	Ellipsoidal	Saddle-shaped	Flat	Hinge/pivot
Number of DOF	3	2	2	2	1
Ideal articular surface	Ovoid (synclastic)	Ovoid (synclastic)	Sellar (anticlastic)	Flat	Cylinder
Curvature, X and Y directions	Convex/convex	Convex/convex	Convex/concave	Flat/flat	Convex/flat
Radii of curvature in X and Y directions	Equal	Not equal	Does not matter	Does not apply	Does not apply
Bone at which the joint rotation axes are anchored	Male	Male	Male and female	Does not apply	Male

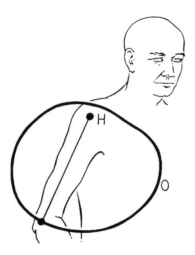

Figure 4.3   Ovoid of the joint motion.
After Benninghoff-Goerttler (1964). *Lehrbuch der Anatomie des Menschen.* (9th ed).
From MacConaill, M.A. & Basmajian, J.V. (1969). *Muscles and movements. A basis for human kinesiology.* Baltimore: Williams & Wilkins.

••• *Differential Geometry Refresher* •••

Curvature Characteristics

Consider a continuously differentiable surface S with a grid projected on it. The parametric coordinates (u,w) are assigned to the points on the surface at the intersections of the rows and columns of the grid. A point on the surface x (u,w) is characterized by a vector $\mathbf{x}(u,w) = x_k(u,w)$, where k = 1,2,3 indicates the three coordinate directions. At each point on the surface, a unique outward normal vector $\mathbf{n}(u,w)$ can be defined as follows:

$$\mathbf{n}(u,w) = \frac{\dfrac{\partial \mathbf{x}}{\partial u} \times \dfrac{\partial \mathbf{x}}{\partial w}}{\left| \dfrac{\partial \mathbf{x}}{\partial u} \times \dfrac{\partial \mathbf{x}}{\partial w} \right|}$$

where $\partial \mathbf{x}/\partial u$ and $\partial \mathbf{x}/\partial w$ are the partial derivatives of vector $\mathbf{x}(u,w)$ along coordinate directions u and w, $\times$ denotes a vector cross product, and the vertical lines stand for a magnitude of the vector. A planar curve C is at the intersection of the surface S and a plane containing vector $\mathbf{n}$ (Figure 4.4).

By definition, the curvature of C is negative, or *concave*, when the center of curvature is on the same side of the curve C as the vector **n**; otherwise, the curvature is called positive, or *convex*. Rotating the plane around the vector **n** produces a family of curves C. The curvature of the curves C assumes minimal and maximal values, k_{min} and k_{max}, called *principal curvatures* at a point x(u,w). The directions of the principal curvatures are called *principal directions*. The principal directions are mutually perpendicular.

As a measure of surface curvature the Gaussian curvature, $K=k_{min}k_{max}$, is commonly used. When K>0 the surface is ovoid; when K<0 the surface is sellar, saddle shaped; and when K=0 the surface is either cylindrical or flat.

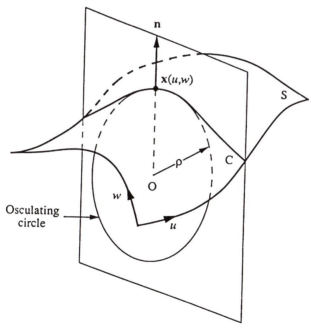

Figure 4.4 On a continuous surface S—think of a joint surface—a normal vector **n** can be defined at any point x (u,w). A planar curve C is along the intersection of the surface S and a plane containing **n**. The radius of curvature of C at point x (u,w) is the radius of the osculating circle.

Adapted from *Journal of Biomechanics, 25,* Ateshian, G.A., Rosenwasser, M.P., & Mow, V.C., Curvature characteristics and congruence of the thumb carpometacarpal joint: Differences between female and male joints, 591-607, 1992, with kind permission from Elsevier Science Ltd, The Boulevard, Langford Lane, Kidlington 0X5 1GB, UK.

4.1.1.2 Advanced Examination of Joint Surfaces: Surface Curvature, Joint Congruence

Representative paper: Ateshian et al., 1992.

The geometric form of joint surfaces influences the magnitude of the stress experienced by the different parts of the surfaces. The incongruity of the articular surfaces results in a reduction in contact areas and, thus, in an increase in contact pressure. Large local stresses cause focal lesions of the joint cartilage, osteoarthritis, and other joint diseases. To assess the size and location of the joint contact area, the geometry of the mating surfaces as well as the joint congruence should be known.

To characterize a joint surface's geometry, in addition to the principal curvatures, k_{min} and k_{max}, the root-mean-square curvature is used.

$$k_{rms} = \sqrt{\frac{k_{min}^2 + k_{max}^2}{2}} \tag{4.1}$$

The root-mean-square curvature accounts for both the minimal and maximal curvature at a given point. When the surface in the neighborhood of a point is flat, $k_{rms} = 0$. To visualize a joint surface, *curvature contour maps* are used (Figure 4.5). In these maps, a contour representing a curvature estimate, for example, the Gaussian curvature, K, or the root-mean-square curvature, k_{rms}, is superposed onto the topographic map of the surface.

Joint congruence is characterized by the difference in the profiles between two articular surfaces. To estimate the joint congruence, the original articular surfaces are replaced by a system consisting of an *equivalent surface*, calculated as the mentioned difference, and a plane (Figure 4.6).

The principal curvatures of the equivalent surface at the point of initial contact with the plane are called the *relative principal curvatures*. The relative principal curvatures are direct measures of the minimal and maximal joint congruence. An overall congruence of the joint at a contact point is defined by the root-mean-square curvature of the equivalent surface:

$$k_{rms}^e = \sqrt{\frac{\left(k_{min}^e\right)^2 + \left(k_{max}^e\right)^2}{2}} \tag{4.2}$$

Frequently, articular surface geometry is too complex to be described by a single equation. In such a case, a piecewise representation is used; a network of contiguous patches is constructed to provide a description of the entire surface. Each of the patches is described mathematically. The requirement for the patches is continuity of the second-order derivatives over the entire articular surface.

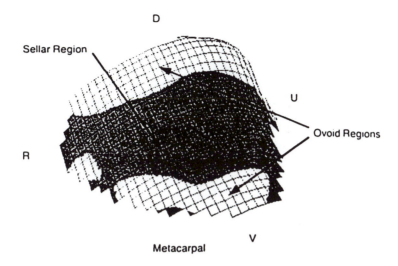

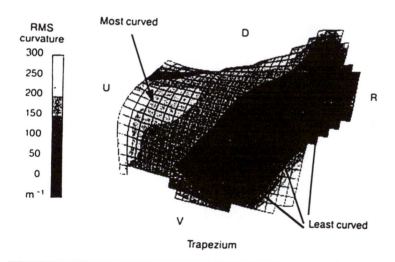

Figure 4.5 Curvature contour maps of the articular surface of the trapezius at the thumb metacarpal joint. The upper panel is a Gaussian curvature map. The dark areas correspond to the sellar regions of the surface (K < 0); the white areas represent the ovoid surface (K > 0). It follows from the map that the metacarpal surface is a sellar along a middle radioulnar band. Peripheral regions are ovoids. The bottom panel shows a root-mean-square curvature map. The map shows flat and curved regions of the articular surface, irrespective of concavity and convexity. D, dorsal; V, volar; U, ulnar; R, radial.

Adapted from *Journal of Biomechanics, 25,* Ateshian, G.A., Rosenwasser, M.P., & Mow V.C., Curvature characteristics and congruence of the thumb carpometacarpal joint: Differences between female and male joints, 591-607, 1992, with kind permission from Elsevier Science Ltd, The Boulevard, Langford Lane, Kidlington 0X5 1GB, UK.

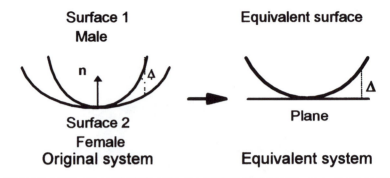

Figure 4.6 Representation of the joint surfaces by an equivalent system consisting of an equivalent surface coming into contact with a plane. To get the equivalent surface, surface 2, female, is subtracted from surface 1, male. The distances between the surfaces at the corresponding points in the original and equivalent system, Δ, measured along the direction of the normal at the point of initial contact, are equal.

■ ■ ■ *From the Scientific Literature* ■ ■ ■

Congruence of the Thumb Carpometacarpal Joint in Women and Men

Source: Ateshian, G.A., Rosenwasser, M.P., & Mao, V.C. (1992). Curvature characteristics and congruence of the thumb carpometacarpal joint: Differences between female and male joints. *Journal of Biomechanics, 25,* 591-607.

The prevalence of osteoarthritis at the thumb carpometacarpal joint is greater in women than in men. In this study, stereophotogrammetry was used to record the joint surfaces of the trapezium and the metacarpal and, then, the congruence of the joint was estimated (Table 4.2).

None of the relative principal curvatures were equal to zero; thus, the joint is not perfectly congruent. The two articular mating surfaces do not conform ideally to each other. The women's joints were less congruent than the men's; the relative principal curvatures were also larger. The difference between the men and women was statistically significant, $p \leq 0.01$. Consequently, under similar joint loading conditions, such as during pinch or grasp, the contact areas in the female carpometacarpal joint are smaller than in male joints. As a result, female carpometacarpal joints experience higher mechanical stresses, which predisposes women to degenerative disease of this joint.

Table 4.2 Congruence of the Carpometacarpal Joint in Men and Women

Group	Relative Principal Curvatures (means ± SD)		
	k^e_{min}, m^{-1}	k^e_{max}, m^{-1}	k^e_{rms}, m^{-1}
Women	109.5 ± 36.0	145.3 ± 36.2	129.3 ± 33.3
Men	43.9 ± 28.3	77.1 ± 25.8	63.5 ± 24.7
Comparison	$p \leq 0.01$	$p \leq 0.01$	$p \leq 0.01$

4.1.2 Movement of the Articular Surfaces

Representative papers: MacConaill, 1953; O'Connor & Zavatsky, 1990.

Joint motion can be thought of as the movement of one articulate surface with regard to another. The articular surfaces of a joint are termed *conarticular surfaces*. Articular surfaces form mating pairs; the surfaces of a mating pair are more or less congruent. In simple joints, there is only one mating pair and in compound joints there are two or more. A male surface from a compound joint articulates only with the female surface of that mating pair, not with a female surface of another pair.

4.1.2.1 Close-Packed and Loose-Packed Joint Positions

Mating pairs are fully congruent in only one position. The position of full congruency is called the *close-packed position*. In this position, the male surface is in contact with the female surface at each point on the latter. The positions other than the close-packed are called the *loose-packed positions* (Figure 4.7 and Table 4.3). The mobility of a joint in a close-packed position is limited.

4.1.2.2 Rolling and Skidding of the Conarticular Surfaces

There are two basic types of planar motion between surfaces, *rolling* and *skidding*. The majority of habitual joint movements are combinations of these two motions. Consider two surfaces, male and female, that are in direct contact (Figure 4.8). The bodies are moving relative to the O-XY system. At the point of contact, point P_m is fixed in the male surface and P_f is fixed in the female surface. The velocities of points P_m and P_f along the common normal to the

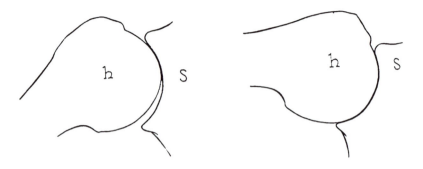

Figure 4.7   Loose-packed (left panel) and close-packed positions of the glenohumeral joint. h, humerus; s, scapula.
From MacConaill, M.A. & Basmajian, J.V. (1969). *Muscles and movements. A basis for human kinesiology.* Baltimore: Williams & Wilkins.

Table 4.3   Close-Packed and Least-Packed Positions of Some Joints

Joint	Close-packed	Least-packed
Shoulder	Abduction + external rotation	Semi-abduction
Ulnohumeral	Extension	Semiflexion
Radiohumeral	Semiflexion+semipronation	Extension+supination
Wrist	Dorsiflexion	Semiflexion
First carpometacarpal	Full opposition	Neutral position of thumb
Metacarpophalangeal (2-5)	Full flexion	Semiflexion+ulnar deviation
Interphalangeal	Extension	Semiflexion
Hip	Extension+medial rotation	Semiflexion
Knee	Full extension	Semiflexion
Ankle	Dorsiflexion	Neutral position
Tarsal joints	Supination	Semipronation
All toe joints	Dorsiflexion	Not given
Vertebral joints	Dorsiflexion	Not given

From Adams, L.M. (1993). The anatomy of joints related to function. In V. Wright & E.R. Ladin (Eds.), *Mechanics of Human Joints* (pp. 27-82) New York: Marcel Dekker, Inc.

surfaces at the point of contact, $\mathbf{V}_m^n$ and $\mathbf{V}_f^n$, are equal; otherwise the surfaces move out of contact or dig into each other. Because $\mathbf{V}_m^n = \mathbf{V}_f^n$, the relative velocity vector of P_m and P_f, $\mathbf{V}_{mf}$, lies in the tangent plane to the surfaces at the point of contact. The difference between the tangential components is the *skidding velocity*.

$$\mathbf{V}_{mf} = \mathbf{V}_m^t - \mathbf{V}_f^t = \mathbf{V}_m - \mathbf{V}_f \qquad (4.3)$$

where $\mathbf{V}_m^t$ and $\mathbf{V}_f^t$ are tangential components of the velocity of points P_m and P_f, respectively, and $\mathbf{V}_m$ and $\mathbf{V}_f$ are the resultant velocities of these points. For *pure rolling*, $\mathbf{V}_{mf} = 0$ but $\mathbf{V}_m \neq 0$ and $\mathbf{V}_f \neq 0$; for *pure skidding*, either $\mathbf{V}_m$ or $\mathbf{V}_f$ is equal to 0.

Rolling is the rotation of a bone around an axis located on the articular surface. During rolling, the instant center of rotation between the two bodies lies on their point of contact and the skidding velocity equals zero. The movement is similar to that seen when a wheel rolls over a road surface. If the conarticular surfaces were nondeformable, the center of rotation would be a shared point of the two surfaces. The rolling is a combined motion, both translation and rotation (Figure 4.9).

For rolling to occur the conarticular surfaces must be noncongruent. During rolling, the point of the common contact between articulating surfaces is not constant. When the joint angle changes, different points of both the male and female surface come near each other. In a joint, this prevents continuous stress on the same area of the articular cartilage. During rolling there are always

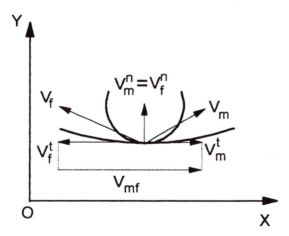

Figure 4.8 Surface-on-surface motion. Points P_m and P_f—not shown in the figure—are in contact with each other. $\mathbf{V}_m$ and $\mathbf{V}_f$ are the instant velocities of the points and $\mathbf{V}_m^n$ and $\mathbf{V}_f^n$ are their normal components, $\mathbf{V}_m^n = \mathbf{V}_f^n$. Tangential velocities of the points are $\mathbf{V}_m^t$ and $\mathbf{V}_f^t$. All the velocities are in the absolute reference frame. The relative velocity $\mathbf{V}_{mf} = \mathbf{V}_m^t - \mathbf{V}_f^t$ is skidding velocity.

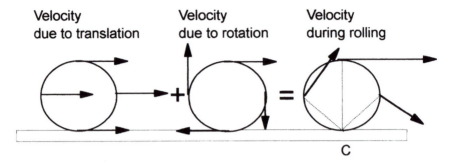

Figure 4.9 Rolling as a sum of the translation and rotation. The translational and rotational velocities of each point on the rim are summed. As a result, (1) the point of the common contact, C, has zero velocity (because the linear velocity of the point due to translation is equal in magnitude and opposite in direction to the velocity of this point due to translation); (2) the velocity of any point on the rim is perpendicular to a line joining that point with point C (dotted lines); and (3) the magnitude of the velocity of the point is proportional to its distance from point C. Hence, point C is the instantaneous center of rotation during rolling.

regions on both articular cartilages that are relieved of stress. Attrition of articular cartilages is often observed in the regions that are in close contact at the close-packed position. At this joint position, congruency of the conarticular surfaces is high and rolling does not occur. Rolling, as a mechanism of surface-on-surface movement, is also deemed to be important for joint lubrication. Rolling surfaces continually approach each other at the head end and move away from each other at the rear end. This motion helps squeeze the synovial fluid from the porous articular cartilage and create a lubricant film between the conarticular surfaces.

Skidding is a surface-on-surface movement in which the same region of an articular surface is in continuous contact with different regions of another surface. During skidding, the instant center of rotation does not lie on the contact point and the skidding velocity is not zero, $V_{mf} \neq 0$. A relative velocity exists between the two contact surfaces. Because during skidding one sector of an articular surface is in consistent contact with another surface, it is expected that the erosion of this area is greater than the attrition of those areas of the articular surface that do not slide. To visualize skidding, imagine two situations:

A. A car with locked brakes is gliding on ice. The tire does not rotate. The same point of the tire (male surface) is in constant contact with various areas of the road. This type of motion is called sliding. If the surface of the road is flat, the wheel and tire movements are rectilinear translation.

B. The tire of a stuck car is rotating without progression on an icy surface. The same point of the road (female surface) is in permanent contact with various areas of the tire. The surface-on-surface movement is called spinning. The wheel and tire movements are pure rotation around the wheel axle.

In the first condition, the maximal wearing down is expected on the male surface, whereas in the second wear is expected on the female surface. When skidding occurs, the rolling and sliding surface is called a *skidder* and the second mating surface is called the *skidding bed*. In human joints, the skidder is a male surface, and the skidding bed is a female surface.

In diarthroidal joints, female surfaces are usually curved, not flat. Hence, the skidding, which is a surface-on-surface motion, occurs because of translation or rotation of one bone with regard to another. The contribution of the translation and rotation to the skidding depends on the shape of the female surface and its congruency with the male surface. In the ensuing analysis, we assume that both surfaces are circular cylinders. If the mating surfaces are ideally congruent, the only possible bone movement is pure rotation and the surface-on-surface movement is pure gliding.

When the surfaces are incongruent, different cases are possible. Let us consider three cases shown in Figure 4.10. In case A, the skidder moves without rotation along the circular path prescribed by the skidding bed. Even though the skidder does not rotate, its orientation with regard to the juxtaposed region of the female surface changes. During the motion, the sundry regions of the skidder meet the regions of the skidding bed in sequence. For an observer positioned at the skidding bed, who does not know that the surface is curved, the skidder rolls and slides on the surface. The proportion of the rolling and sliding is determined by the ratio of the radii of curvature of the skidding bed, R, and the skidder, r (not shown in the figure). Hence, nonlinear translation of a skidder along a curved skidding bed results in rolling and gliding.

In case B, the skidder rotates counterclockwise around the geometric center, O, of the female surface. The surface-on-surface movement is pure sliding. The same point of the male surface is in constant contact with the female surface.

In case C, the skidder rotates 90° counterclockwise around the geometric center of the female surface and, in addition, turns 90° counterclockwise around its own center. The surface-on-surface movement is again a combination of rolling and sliding. The proportion of rolling and sliding evidently depends on the magnitude of both the rotations and on the r:R ratio. The relationship between the surface-on-surface motion and bone-on-bone motion is presented in Figure 4.11.

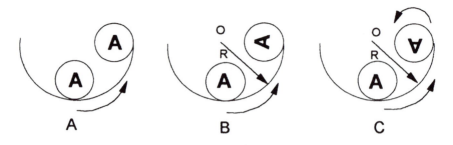

Figure 4.10 The motion of a circular skidder along a circular skidding bed. To make the kinematics visible, male and female surfaces of sharply contrasted size and curvature are drawn. The skidder starts at the bottom of the female surface and moves to the right along the surface.

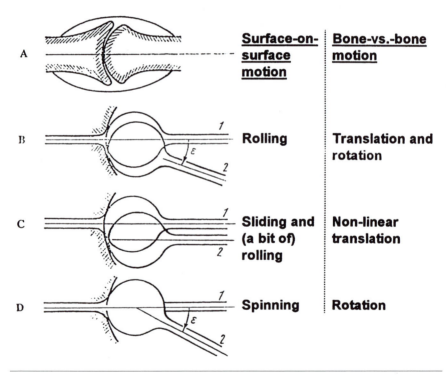

Figure 4.11 Relative motion of articular surfaces and adjoining bones. A. Joint geometry. B. Pure rolling. During articular motion, an instantaneous axis of rotation is at the point of surface-to-surface contact. The movement can be viewed as a combination of rotation and translation (see Figure 4.9). C. Gliding (mainly). The articular surfaces translate with respect to each other. Because of the curvature of the female surface this is not a pure gliding, however (see text for explanation). D. Spinning. The male bone rotates; the male surface slides in the opposite direction. 1 and 2 the are initial and final joint positions.

4.1.2.3 Slip Ratio

A joint surface-on-surface movement is, as a rule, a roll-and-slide movement. When a male surface rolls, it simultaneously slides in the opposite direction. For example, during arm abduction, the humeral head rolls upward and glides downward. In various joints, these movements are found together in different proportions. To characterize the contribution of rolling and gliding to joint motion, the *slip ratio* is used. To measure the slip ratio, any two contact points on the conarticular surfaces of a joint in a reference position are selected. After joint rotation, a new pair of contact points is chosen. Then, the distances between the initial and final points d_m and d_f are measured along the articular surfaces. Their ratio, $d_m{:}d_f$, is the slip ratio (Figure 4.12, left panel).

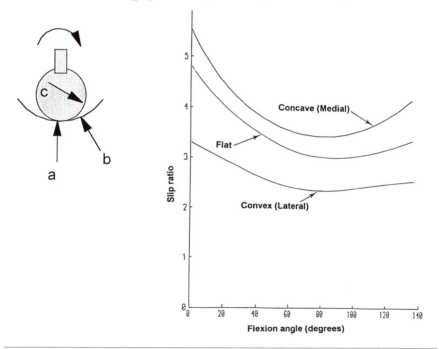

Figure 4.12 Calculating the slip ratio (left panel) and measured values of the slip ratio for the knee joint. On the left panel, point a is the point of initial contact for the male and female surface. Point b is the point on the female surface at which the contact occurs after the rolling and sliding, $b - a = d_f$. Point c is the point of the contact at the male surface after the joint motion, $c - a = d_m$. Slip ratio = $d_m{:}d_f$. On the right panel, the slip ratio is plotted versus anatomic knee angle for concave, flat, and convex tibial plateaus.

From O'Connor, J.J. & Zavatsky, A. (1990). Kinematics and mechanics of the cruciate ligaments of the knee. In V.C. Mow, A. Ratcliffe, & S. L.-Y. Woo (Eds.), *Biomechanics of diarthroidal joints* (Vol. 2, pp. 197-241). New York: Springer-Verlag. Reprinted by permission.

Three values of the slip ratio are of special interest. When a slip ratio equals 1, the joint motion is pure rolling. During pure rolling, the contact points on both the male and female surfaces shift by an equal amount. When the slip ratio is zero, the movement is pure sliding. During pure sliding, the contact point on the male surface remains constant, $d_m = 0$, and that on the female surface changes. During spinning rotation ("turning without progression"), the contact point of the female surface remains constant and the slip ratio is not defined because the denominator of the ratio equals zero, $d_f = 0$.

Slip ratios for the human knee joint are presented in Figure 4.12 (right panel). The slip ratios are different for various joint angular positions. The magnitude of the slip ratio depends on the shape (curvature) of the male and female joint surfaces and on constraints imposed on the joint motion by the ligaments.

■ ■ ■ *FROM THE SCIENTIFIC LITERATURE* ■ ■ ■

Use of Slip Ratio in Reconstruction of the Anterior Cruciate Ligament

Source: Thoma, W., Jäger, A., & Schreiber, S. (1994). Kinematic analysis of the knee joint with regard to the load transfer on the cartilage. In Y. Hirasawa, C.B. Sledge, & S.L.-Y. Woo (Eds.) *Clinical Biomechanics and Related Research* (pp. 96-102). Tokyo: Springer-Verlag.

During reconstruction of the anterior cruciate ligament (ACL), the ligament attachment location is important. The purpose of this study was to analyze the effect of small changes in the proximal insertion point of the ACL on knee joint kinematics and kinetics. In particular, the slip ratio was calculated for eight different proximal attachment points of the ACL. Figure 4.13B illustrates the eight attachment sites, numbered from 0 to 7. The radius r was equal to one-tenth of the total length of the ACL, which was equivalent to 3.9 mm in the specimen presented in Figure 4.13. The results for two positions, 2 and 6, are presented in Figure 4.13, C and D.

The relocation of the ACL attachment to position 2 does not change the slip ratio. The ratio changes dramatically after relocating the attachment point to position 6. If such a surgery were performed, it would result in spinning of the male bone with respect to the female and a high risk of wearing and tearing of the tibial articular surface; different parts of the femoral surface will be in a permanent skidding contact with the same part of the tibial surface. Note also that both the reconstructive techniques cause joint instability at some joint angles. Hence the best solution is to fix the torn ligament to its original place, i.e., exact isometric replacement.

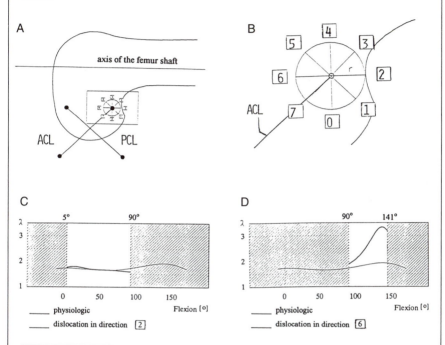

Figure 4.13   Geometry of ACL reconstruction (upper panels) and the slip ratio at various knee joint angles (bottom panels). Panel A shows the lateral femur condyle. ACL, anterior cruciate ligament; PCL, posterior cruciate ligament. On panel B, the proximal attachment point of the ACL and eight directions of dislocation (0-7) are shown. r, radius. The bottom panels show dislocation for directions 2 (panel C) and 6 (panel D). The original (physiologic) slip ratio compared with the slip ratio after a surgery.
From Thoma, W., Jäger, A., & Schrieber, S. (1994). Kinematic analysis of the knee joint with regard to the load transfer on the cartilage. In Y. Hirasawa, C.B. Sledge, & S.L.-Y. Woo (Eds.), *Clinical Biomechanics and Related Research* (pp. 96-102). Tokyo: Springer-Verlag. Adapted by permission.

4.1.2.4 Spin

Rolling and skidding are patterns of planar surface-on-surface motion. In a three-dimensional case, the third type of joint motion, *spin*, can also take place. The spin is the rotation of joint surfaces around an axis perpendicular to the surfaces at the point of contact. During spin, one point at the joint surface does not move and all of the others slide along circular paths. The examples of spin are external and internal rotation. Flexion and extension in the hip and shoulder (glenohumeral) joints are also spins (Figure 4.14). Any movement other than spin is called a *swing*. A swing without an accompanying spin is a *pure swing*.

4.1.3 Geometry and Algebra of Intra-articular Motion

Representative paper: MacConaill, 1966.

Geometric figures, such as triangles, rectangles, and circles, can be drawn either on a flat surface or on a curved one, for instance, on the earth surface. The geometry of curved surfaces differs from planar geometry studied in high school

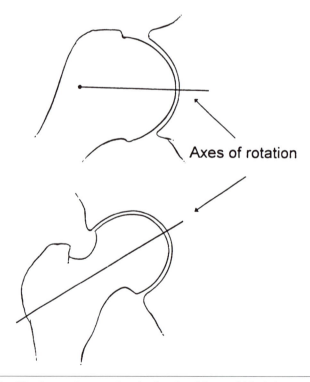

Axes of rotation

Figure 4.14   Flexion and extension in the shoulder and hip joints as spins. Joint rotation is performed around the axes that are perpendicular to the surface of contact.
Adapted from MacConaill, M.A. & Basmajian, J.V. (1969). *Muscles and movements. A basis for human kinesiology*. Baltimore: Williams & Wilkins.

classes. For example, one of the main concepts of the planar geometry is the notion of a straight line, which is drawn along the shortest distance between two points. On a curved surface, a straight line does not exist, any line is curved. Such curved surfaces are studied in *spherical geometry*. In this area of geometry, the shortest path between two points, for example, between New York and Tokyo, is not a straight line but a *geodesic* one. Likewise, the sum of three interior angles of a triangle is not equal to 180°. The properties of geometric figures depend not only on the figures themselves but also on the curvature of the surface. For instance, the difference between the sum of the internal angles of a triangle (S) and π, $r = S - \pi$, is positive for any ovoid surface, negative for a saddle surface, and zero for a flat surface. This difference is called a *residual*. The value of the residual on an ovoid or saddle surface depends on the area of the triangle: the greater the area the greater the residual.

Joint surfaces are generally not flat. Therefore, they can be studied invoking the ideas developed in nonlinear (i.e., spherical) geometry. This direction of research is called the geometry and algebra of articular kinematics. Because this is a special area of joint biomechanics, it is not the goal of this chapter to cover this field entirely. Only a brief account of the main ideas of the geometry and algebra of articular kinematics is given in the ensuing paragraphs. The propositions are given without rigorous proofs.

4.1.3.1 Main Notions of the Geometry and Algebra of Articular Surfaces

A line between any two points on a surface is called an *arc*. The shortest arc is a *chord* (a more general term *geodesic* is also acceptable). There is only one chord that connects the two points. When three points that are not on the same chord are connected by lines, the three-sided figure is called a *triangle* if all the sides are chords, and it is called a *trigone* if at least one side is not a chord. Similarly, a *multiangle* figure is a closed n-sided figure of which all the sides are chords; a *polygone* is a closed figure of which at least one side is not a chord.

Relative movement of articular surfaces is described by the motion of an imaginary infinitesimal rod perpendicular to the surface. The rod can either spin, or slide, or both. A spin of angle α is written as E^α. The expression $mE^\alpha m$ means a spin E^α at a point m. Using the equal sign to mean "is equivalent to" and the letter I to mean "no motion," we can write the following expressions (n is an integer):

$$E^\alpha + E^\beta = E^{\alpha+\beta}$$

$$E^0 = I \qquad (4.4)$$

$$E^{2\pi n} = E^0 = I$$

A slide along a path A from point m to point n is written as mAn. The reverse slide along the same path is $nA^{-1}m$. Two consecutive slides from m to n and then from n to p are represented as a product.

$$mAn \cdot nBp = mABp \qquad (4.5)$$

The symbol AB means "A followed by B." A round motion from m to n and then back to m is $mAn \cdot nA^{-1}m = mIm$.

A spin and a slide are commutative:

$$mAn \cdot nE^{\alpha}n = mE^{\alpha}m \cdot mAn = mE^{\alpha}An \qquad (4.6)$$

The final result of the spin and the slide would be the same whether these two motions occur simultaneously or in any sequence. In general, spin vectors commute with both spin and slide vectors. Slide vectors do not commute with each other.

4.1.3.2 Basic Theorem of Kinematics of Articular Surfaces

Any two consecutive slides—first from point m to point n and then from n to p—are equivalent to a single slide from m to p together with the spin. The magnitude and the sense of the spin are that of the residual of the trigone formed by the two consecutive slides, m→n and n→p, and the single slide, m→p.

To prove the theorem, consider a closed trigone of sliding motion (Figure 4.15). The theorem can be proved by inspection. Imagine an arrow beginning at point m and directed along (tangent to) A. Move the arrow to n while keeping it tangent to A. At n, spin it to B over the angle $(\pi - \gamma)$. Slide the arrow along B in such a way that it is tangent to B all the time. At p, rotate the arrow through the angle $(\pi - \alpha)$, making the arrow along C^{-1}. Ultimately, move the arrow along C^{-1} to the starting point m and rotate it over the angle $(\pi - \beta)$. The arrow is now in its starting position. The described three-stage motion is specified by the following equation:

$$mAE^{\pi-\gamma}n \cdot nBE^{\pi-\alpha}p \cdot pC^{-1}E^{\pi-\beta}m = mIm \qquad (4.7)$$

Replace $(\alpha + \beta + \gamma)$ by S and recall that $E^{3\pi} = E^{\pi}$ and we have

$$ABC^{-1}E^{3\pi-S} = ABC^{-1}E^{\pi-S} = ABC^{-1}E^{-r} = I$$
$$AB = CE^{r} \qquad (4.8)$$

The theorem is proved. The basic theorem of articular kinematics explains such phenomena of joint angular motion as induced twist (see Section **2.1.4**) and Codman's paradox (see Section **2.1.2** and **3.1.2.1.4**) in terms of intrajoint surface-on-surface movement.

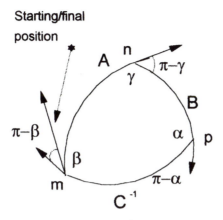

Figure 4.15 A trigone formed by two consecutive slides, A and B, and one single slide, C^{-1}. Slide A is from point m to point n, and slide B is from n to p. Slide C is directly from m to p (slide C^{-1} is from p to m). Internal angles opposite A, B, and C are α, β, and γ. The basic theorem states that $AB = CE^r$, where r is the residual of the trigone, $r = (\alpha + \beta + \gamma) - \pi$.

■ ■ ■ *FROM THE SCIENTIFIC LITERATURE* ■ ■ ■

Explaining Codman's Paradox in Terms of Intra-Articular Motion. Ergonomic Cycles

Source: MacConaill, M.A., & Basmajian, J.V. (1969). *Muscles and movements. A basis for human kinesiology* (pp. 36-51). Baltimore: Williams & Wilkins Company, Chapter 4.

A sequence of shoulder joint motions *forward flexion→horizontal extension→adduction* brings the arm to a laterally rotated position. The humerus has turned around its long axis, even though none of the consecutive movements by themselves involve axial rotation (Codman's paradox). This phenomenon is easy to explain using the basic theorem of joint kinematics. All of the three involved movements are at right angles to each other. Hence, the sum of the internal angles is S = 270° and that of the residual is r = 90°. Therefore, according to the basic theorem, the spin is also 90°. The sense of the spin, clockwise or counterclockwise, is the same as the sense of the movement sequence. As viewed by the subject, it is clockwise for the right limb and counterclockwise for the left limb.

See **CODMAN'S PARADOX,** *p. 250*

■ ■ ■ **CODMAN'S PARADOX,** *continued from p. 249*

Because the arm does not return to its original position after the described shoulder swing cycle, the cycle cannot be performed more than twice. Cycles of joint movement that bring the limb back to its starting position are called *ergonomic cycles*. The ergonomic cycles allow repetitive series of the same joint movements. For example, the sequence of the thumb movements in the first carpometacarpal joint is (a) *extension*, deviation in the plane of the palm; (b) *abduction,* and (c) *reposition*. This series constitutes the ergonomic cycle; it brings the thumb into its starting position.

4.1.4 Ligaments and Joint Motion: A Joint as a Mechanical Linkage

Representative paper: O'Connor & Zavatsky, 1990.

Joint motion is restricted not only by the geometry and conformity of the articular bones, but also by the soft tissues, especially the ligaments. If the ligaments were inelastic, they would impose one-sided constraints on joint motion; i.e., during motion the length of a ligament could not exceed its resting length. Fortunately, real ligaments elongate under the acting force. However, this elongation is disregarded in what follows.

With inelastic ligaments and incompressible articular surfaces that are not completely congruent, the joint surfaces touch each other only at a point. They are not separate nor do they interpenetrate during the joint motion. The joint can be considered a mechanical linkage with the bones and the ligaments as the links. In this spirit, the knee joint can be regarded a four-bar linkage (Figure 4.16). The links are the femoral and tibial articular surfaces and the anterior and posterior cruciate ligaments. This approach was described by Züppinger in 1904.

The four-link planar linkage ABCD allows knee joint flexion and extension. During joint angular motion, the geometry of the linkage changes. The ligaments rotate and change their orientation. Their length, however, is kept constant. The movement, constrained by both the conarticular surfaces and the cruciate ligaments, possesses several specific features. In particular, the instantaneous center of rotation is at the intersection of the neutral fibers of the cruciate ligaments, at the point I. During flexion and extension, the center moves with regard to each bone as well as with regard to the ligaments (Figure 4.17).

Such a complex movement results in concurrent rolling and sliding of the conarticular surfaces relative to each other with various slip ratios. The slip

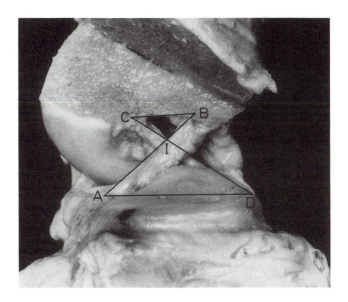

Figure 4.16   A sagittal section through a knee joint with the medial femoral condyle removed. The cruciate ligaments, together with the articular surfaces, make a four-bar linkage ABCD. The link AB represents the anterior portion of the anterior cruciate ligament and the link CD depicts the posterior portion of the posterior cruciate ligament. These portions of the cruciate ligaments represent the *neutral fibers* of the ligaments, those that do not change their length during flexion-extension. The tibial link, AD, of the linkage is between the attachments of the cruciate ligaments on the tibia; the femoral link, CB, connects their attachments to the femur.
From O'Connor, J.J. & Zavatsky, A. (1990). Kinematics and mechanics of the cruciate ligaments of the knee. In V.C. Mow, A. Ratcliffe, & S. L.-Y. Woo (Eds.), *Biomechanics of diarthroidal joints* (Vol. 2, pp. 197-241). New York: Springer-Verlag. Reprinted by permission.

ratio depends on the joint angular position (see Figure 4.12). In summary, kinematics of some joints, e.g., the knee joint, is determined by the ligaments to the same extent as it is prescribed by the conarticular surfaces. Although the assumption about inelasticity of the ligaments is not absolutely valid, it permits one to arrive at useful conclusions about joint kinematics.

4.2 CENTERS AND AXES OF ROTATION

Knowledge of the positions of the rotation axes is necessary for determining the action of muscles and tendons (muscle moment arms) as well as for other

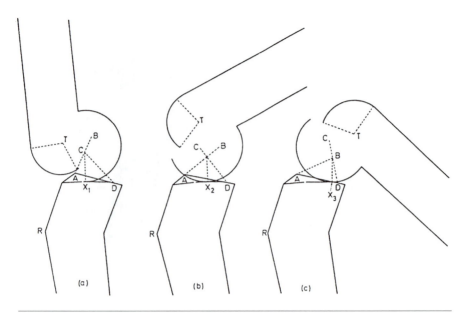

Figure 4.17 A planar model of the knee joint. At the various joint angles the four-link linkage ABCD changes its configuration, but the perpendicular to the surfaces at the contact points x_1, x_2, x_3 passes through the instant center of the cruciate linkage, which coincides with the center of joint rotation.

From O'Connor, J.J. & Zavatsky, A. (1990). Kinematics and mechanics of the cruciate ligaments of the knee. In V.C. Mow, A. Ratcliffe, & S. L.-Y. Woo (Eds.), *Biomechanics of diarthroidal joints* (Vol. 2, pp. 197-241). New York: Springer-Verlag. Reprinted by permission.

applications, such as joint replacement. Even relatively small errors in locating a joint center or axis may result in large inaccuracies when muscle moment arms and muscle moments need to be determined. For instance, a 2 cm superior displacement of the hip joint center decreases the force generated by the hip abductors by 44% and the muscle moment by 49% (Delp, Maloney, 1993).

Historically, research in joint kinematics has been progressing from simple models to complex. Anatomists and biomechanists of the 19th century postulated that the human joints have fixed axes of rotation that are perpendicular to each other. Biomechanical analysis of joint motion was limited to the planar case. In experiments, however, the researchers were not able to document the existence of the fixed axes. Even for a single human joint wide variations were reported. At the beginning of the 20th century, more complex approaches have been developed. The concept of instantaneous centers of rotation (ICR) has been taken up. The instantaneous centers are not fixed, rather they are specific to a given instant of motion. Still, research has been limited to planar movement.

Although in many cases the assumption of planar motion is a reasonable simplification, the use of planar models is bound up with a number of uncertainties. For joints with one DOF, this approach requires that the plane of motion be known in advance, before the experiment. To find the plane of motion, a three-dimensional analysis should be performed (see ensuing sections for details). In joints with two and three DOF, a body limb can move along an infinite number of paths. A "main" path, representing, for example, pure flexion without any abduction-adduction, can only be determined with three-dimensional analysis. Hence, planar analysis of joint kinematics is insufficient. Three-dimensional methods of recording and analysis are needed.

To describe a three-dimensional joint motion in detail, the accuracy of recording is of primary importance. Specifically, estimation of the axes of rotation in joints is highly susceptible to measurement errors. Special methods have been developed to extract reliable information from noisy data. It has been shown that, unlike technical devices, the position of the joint rotation axes is not firmly fixed. It depends on the forces acting on the limb and, during voluntary movement, on the subject's ability to reproduce the joint motion pattern. To examine that statement, the reader is invited to perform the following simple experiment. Stand before a mirror with your palm facing anteriorly and flex your elbow completely. Note the final position of your hand with regard to your shoulder. If you perform the flexion several times trying to vary the hand position at the destination along the mediolateral axis (do not move your upper arm), you will definitely find the hand at different places. Which of these movements is, then, a pure flexion? It can be a nontrivial question if the axes of rotation need to be determined with a high accuracy.

In summary, joint kinematics has grown up from an elementary description of planar joint gyration into a rather sophisticated and developed area of human biomechanics. It is not the goal of this chapter, however, to cover systematically the entire area. I will focus only on the main ideas and methods. For the reader interested in detailed description, references are provided.

4.2.1 Planar Joint Movement

Although joint motion generally occurs in three dimensions, some joint movements may be considered planar. Kinematic analysis of a planar motion is relatively simple. All points of a body move in parallel planes. In a plane parallel to the plane of motion, three-dimensional articular surfaces can be replaced by two curves, the contours of the sections. To describe a planar motion it is sufficient to know the movement of only two points on the body. Until recently, planar models of joint motion have been assumed in most analyses. Either fixed or instant centers of rotation have been presumed. The center of rotation is related to the axis of rotation. It is the point of the axis in the plane of motion (a piercing point). In the planar case, the axis of rotation always remains perpendicular to the plane of motion. Planes of many synovial joints do not coincide exactly with the reference planes, i.e., the anatomic planes. The observed motion is then seen as three-dimensional and can be resolved into its components in the three reference planes. When joint motion is assumed to be planar without validation and, in reality, the orientation of the helical axis changes, the joint motion deviates from a plane. Using planar models can result in large inaccuracies in this case.

Because any planar motion at each instant of time can be viewed as rotation around an instantaneous center (see Section **1.2.5.3**), an entire rolling-and-sliding motion in a joint can be viewed as a rotation around a center that changes its position. The joint movement can then be likened to a sequence of rotations around an instantaneous center that is itself moving along a certain pathway (centrode). Hence, the joint motion is completely determined by the ICR trajectory (centrode) and angles of rotation throughout the range of motion. We begin, however, with a more simple case, assuming that a joint center can be regarded as fixed.

4.2.1.1 Locating a Fixed Center of Rotation

Representative paper: Lewis & Lew, 1978.

When minute details of joint motion are not important, articular motion is usually considered rotation about a fixed center as if the motion were occurring about a single axis. A similar approach is used in clinical settings, where hinged joint prostheses and orthoses are often employed. They replace complex articular motion by simple rotation. To adjust a prosthesis or orthosis to a patient, the axis of rotation should be located. This task is not simple, even if the given joint motion is in fact pure rotation. It is analogous to finding the center of a circle of which only a part is known.

Roughly, the center of a joint can be located by visually inspecting the joint and following recommendations from the literature (suggestions regarding individual joints will be given in Section **5.1**). A more sophisticated technique is to locate a fixed axis using an optimization procedure. Recall (see Chapter **1**) that any real joint motion can be parameterized by either a 3×3 rotation matrix [R] and a displacement vector **P**, or by a 4×4 homogeneous transformation matrix [T]. If joint motion is considered pure rotation, angular joint motion is described by a 3×3 rotation matrix $[R_f]$, where subscript f refers to a given fixed axis of rotation. The idea of this method is to find an axis f to replicate the anatomic motion prescribed by [T] as exactly as possible. To do that, several markers on a body segment are selected and, after applying transformations [T] and $[R_f]$, the positions of the markers are compared. Axis f is

■ ■ ■ *From the Scientific Literature* ■ ■ ■

Optimal Hip Center Positioning During Arthroplasty

Sources: Delp, S.P. & Malloney, W. (1993). Effects of hip center location on the moment-generating capacity of the muscles. *Journal of Biomechanics, 26,* 485-499.

Lebar, A.M., Iglic, A., Antolic, V., Damjanic, F.B., Herman, S., Srakar, F., & Brajnik, D. (1995). Dependence of the hip forces and stem/cement interface stress distribution on the position of the hip joint rotation centre. *XVth Congress of the International Society of Biomechanics, Book of Abstracts*, July 2-6, 1995. Juväskylä, Finland, pp. 534-535.

Total hip replacement is performed to reduce pain and restore normal activity, mainly in aged patients. The surgery may alter the location of the hip rotation center and cause changes in the length and moment arms of the involved muscles. As a result, the capacity of the muscles to generate moments may decrease and the hip joint contact force may increase. Neither of these effects is desirable. These issues were addressed in the cited studies.

In the first study, the hip rotation center was displaced up to 2 cm from its anatomic position in three directions (Figure 4.18, drawing). The shift in the position changed the moment of force generated by the muscles (the four graphs). The largest changes were seen with the abduction moment (inferior-superior displacement), the adduction moment (medial-lateral displacement), the extension moment (anterior-posterior displacement), and the flexion moment (inferior-superior and anterior-posterior displacements). In particular, a 2-cm medial displacement decreased the adduction moment arm by 20%, force by 26%, and moment by 40%.

In the second study, the influence of the different positioning of the femoral head on the hip joint contact force was addressed (Figure 4.19). Shifting the joint rotation center 2 cm in the medial direction decreased the contact force from 3.0 to 2.0 body weights. Because large contact forces may prompt loosening of an implant, shifting the joint hip center medially was recommended. This recommendation should, however, be accepted with a restriction; too large a displacement may decrease the adduction moment in the joint to an unacceptable level.

See **ARTHROPLASTY,** *p. 256*

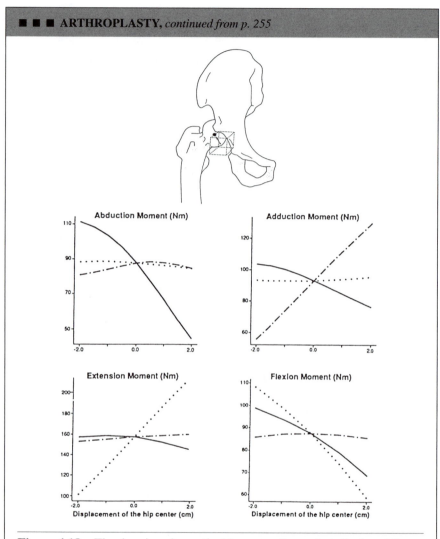

■ ■ ■ **ARTHROPLASTY,** *continued from p. 255*

Figure 4.18 The drawing shows the hip center (large dot) displaced 2 cm superiorly and 2 cm laterally from the anatomic hip center (small dot). The dashed cube, 4 cm on each side, shows the range of hip center location that was investigated. On the four graphs, moment-generated capacity versus displacement in each direction is shown. Positive displacements indicate anterior (dotted curve), superior (solid curve), and lateral (dot-dash curve). Negative displacements indicate posterior (dotted curve), inferior (solid curve), and medial (dot-dash curve).

Adapted from *Journal of Biomechanics, 26,* Delp, S.L. & Maloney, W., Effects of hip center location on the moment-generating capacity of the muscles, 495-499, 1993, with kind permission from Elsevier Science Ltd, The Boulevard, Langford Lane, Kidlington 0X5 1GB, UK.

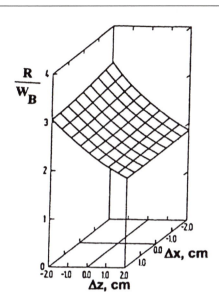

Figure 4.19 The magnitude of the hip contact force as a function of the femoral head shifts in the mediolateral (Δz) and superioanterior directions (Δx). The force is in body weights. Positive direction for Δz is medial and positive direction for Δx is inferior.

From Lebar, A.M., Iglic, A., Antolic, V., Damjanic, F.B., Herman, S., Srakar, F., & Brajnik, D. Dependence of the hip forces and stem/cement interface stress distribution on the position of the hip joint rotation centre. In *XVth Congress of the International Society of Biomechanics, Book of Abstracts* (pp. 534-535). Juväskylä, Finland, July 2-6, 1995. Reprinted by permission.

selected in such a way that the sum of the squared distances between the real positions of the markers (given by the transformation [T]) and the positions estimated from [R_f] is minimized.

For joints with a complex movement pattern, the fixed axis approach is useful for describing displacement between discrete angular positions. It does not, however, convey any detail of the movement during a motion.

4.2.1.2 Instant Rotation in Joints

The complex pattern of the joint movement is mostly the consequence of the complexity of the joint articular surfaces. Two factors contribute to this complexity: the shape of the joint surfaces and their congruity. If the mating joint surfaces were congruous cylinders, their geometric centers would coincide and the joint axis of rotation would coincide with the geometric centers and would remain stationary throughout the motion. Generally, the closer an articular surface is to a simple geometric shape with a constant radius of curvature, cylinder or sphere, the more stationary the axis. In the ankle joint, the radii of curvature are relatively uniform and the axis moves little during flexion and extension. In contrast, the radii of curvature in the femoral condyles vary substantially, and during knee motion the joint axis displaces considerably. In addition to the constant radii of curvature of both surfaces, male and female, the congruency of the surfaces is an important factor for determining the magnitude of the rotation center migration.

It is important to understand that neither a fixed nor a moving centrode (see Section **3.1.1.1.9** for explanation of the terms) represents a path of a real material point fixed to body segments. Unfortunately, this misinterpretation is very common in the anatomic and clinical literature. For instance, during mouth opening the mandible rotates and moves forward (and somewhat downward) (Figure 4.20). In some textbooks of anatomy the movement is described as shifting the momentary axis of rotation, located at the head of the mandible, plus rotation around this axis. This explanation is wrong. The ICR is not fixed at either the head of the mandible or at another point on the mandible or on the maxilla. Every point on the mandible—not only the head of the mandible—changes its position. Also, the mandible revolves around any point—not just the head of the mandible—at the same angle. The difference is that with regard

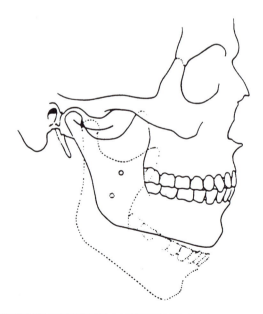

Figure 4.20 Movement of the mandible during mouth opening. The mandible slides forward and rotates.

to ICR the mandible only rotates, or changes its orientation, and with regard to all other points it changes both orientation and location, or the distance to the point. Hence, none of the anatomic points can be picked as the ICR.

Joint movement at any instant can be thought of as the male bone rotation about the center of curvature of the female surface plus the bone rotation about its own center of curvature (see Figure 4.10). When the geometric centers coincide, the center of rotation is fixed, the joint motion is pure rotation, and the surface-on-surface motion is pure gliding. Consider what happens when the geometric centers of the male and female surfaces, both circular, are different (Figure 4.21). In such a joint, the rotation is performed with regard to an ICR. To find the location of the ICR, think about the joint motion as a combined rotation around the female surface center plus rotation around the male surface center. This motion is equivalent to the motion of a two-link chain with the first joint in the geometric center of the female surface. If angular velocity around the female surface center is $\dot{\theta}_f$ and velocity with regard to the male surface center is $\dot{\theta}_m$, then rotational velocity of the male bone with regard to the female bone is $\dot{\theta}_c = \dot{\theta}_f + \dot{\theta}_m$, where $\dot{\theta}_c$ is the angular velocity of the male bone in its rotation around the ICR. According to Kennedy's theorem, the ICR lies on the line of centers. Its location with regard to the female joint center is given by equation 3.27, where radius-vector **A** determines the location of the male surface center. The position of the ICR depends on the ratio of the two angular velocities. When the ratio $\dot{\theta}_f : \dot{\theta}_m$ is negative, the ICR is on the extension of the line connecting the geometric centers of the male and female surfaces. When

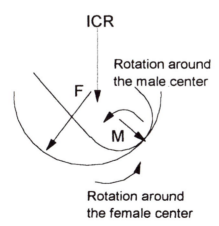

Figure 4.21 Joint rotation as a compound rotation around the male (M) and female (F) centers. The difference between the curvature of the surfaces is greatly exaggerated. In this example, the male and female rotation are of the same direction, counterclockwise. This motion is kinematically equivalent to the motion of a two-link chain. Hence, equation 3.27 can be applied.

$A = 0$, the ICR coincides with the geometric center of the mating surfaces. For pure rolling, the ICR is at the point of common contact.

For example, the articular surfaces of the glenohumeral joint are not absolutely congruent. The curvature of the glenoid cavity is smaller than that of the humerus. The humeral center is shifted from the glenoid center. The rotation in the joint is made up of two elementary rotations, one about the humeral joint center and one about the glenoid joint center. The humerus rotates around the two centers in opposite directions. As a result, during joint motion the ICR is not fixed; it migrates.

■ ■ ■ *FROM THE SCIENTIFIC LITERATURE* ■ ■ ■

Movement in the Glenohumeral Joint

Source: Kirsch, S., Nägerl, H. & Kubein-Meesenburg, D. (1993). Kinematics and statics of the human shoulder. In *XIVth Congress of the International Society of Biomechanics, Abstracts* (pp. 692-693). Paris, July 4-8, 1993.

The authors found that the glenoid cavity and the head of the humerus are not ideally congruent. Specifically, the glenoid radii are larger than humeral ones. The average difference is 2.6 ± 1.7 mm in the frontal plane and 2.8 ± 2.5 mm in the horizontal plane. Hence, the glenoid center and the humeral center do not coincide. The glenoid center, M_g, is shifted from the humeral center, M_h, toward the articular surface by $R = 2.7$ mm (Figure 4.22). As a result, the rotation in the glenohumeral joint is a combination of two elementary rotations, around M_g and around M_h. The position of the resultant center was found from the ratio of the two angular velocities, $\dot{\theta}_g / \dot{\theta}_h$, and the distance R between M_g and M_h (see Figure 4.22).

$$\ell_g = R\left(\ell + \frac{\dot{\theta}_g}{\dot{\theta}_h}\right) \quad \text{and} \quad \ell_h = -R\left(\ell + \frac{\dot{\theta}_h}{\dot{\theta}_g}\right)$$

The maximal upward dislocation of M_h was observed at the resultant rotation angle $\alpha = 60°$. To arrive at this position, the humerus rotated about M_g by $\alpha_g = -70°$ and around M_h by $\alpha_h = 130°$. In this study, the distance between the glenoid center and the ICR, ℓ_g, was estimated on average 5.7 mm, and the distance between the humeral center and the ICR, ℓ_h, was on average 3.1 mm. Poppen and Walker (1976) found instant center values $\ell_h = 6 \pm 2$ mm in 12 uninjured subjects and greater values in some patients with shoulder lesions.

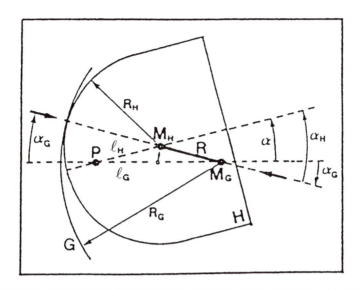

Figure 4.22 Structure of humeral movement. G, glenoid; H, humerus; M_G, glenoidal center; M_H, humeral center; R_H and R_G, radii of curvature; P, resultant center of rotation. Because in the figure a finite rather than instantaneous rotation is shown, P does not lie on the line of centers, $M_H M_G$, as it is prescribed by Kennedy's theorem. α, α_H, and α_G are the angles of rotation with respect to P, the humeral and glenoidal center. In abduction, the humerus rotates around M_G and M_H in opposite directions, M_H migrates upward, and P translates toward the scapula by 3.1 mm. From Kirsch, S., Nägerl, H., & Kubein-Meesenburg, D. (1993). Kinematics and statics of the human shoulder. *XIVth Congress of the International Society of Biomechanics, Abstracts* (pp. 692-693). Paris, July 4-8, 1993. Adapted by permission.

4.2.1.3 Locating an Instant Center of Rotation (The Reuleaux Method)

Representative papers: Panjabi et al., 1982, a, b; Soudan et al., 1979; Zatsiorsky et al., 1984.

When a body rotates, the velocity vector of any point fixed with the body is perpendicular (tangent) to the radius drawn from the center of rotation to the point. Therefore, if the velocity of any two points is known, it is possible to draw the perpendicular lines to them (radii). The point of the intersection of the two perpendiculars is, evidently, the ICR (Figure 4.23). This method, called

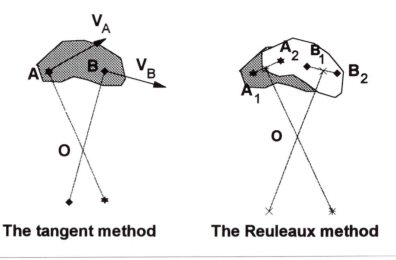

The tangent method   **The Reuleaux method**

Figure 4.23   The tangent method and the Reuleaux method of locating an instantaneous center of rotation.

the *tangent method*, is by itself absolutely correct, but it can almost never be applied in practice because the instantaneous velocity is usually not known.

In the tangent method, an instantaneous position and the velocity of two points, A and B, must be known. In contrast, in the Reuleaux method, two consecutive positions of two points, $A_1 \to A_2$ and $B_1 \to B_2$, must be known. In both methods, the point of intersection of two perpendiculars represents a center of rotation.

The Reuleaux method is a classic method of planar kinematics developed in the 19th century. The method is based on a simple idea: a line connecting two consecutive positions of a point, P_1 and P_2, approximates the direction of velocity between sequential positions P_1 and P_2. Hence, the perpendicular erected from the midpoint of the line $P_1 \to P_2$ approximates the radius. In the Reuleaux method, two points on the moving segment are tracked and two perpendiculars from the midpoints are erected. The point of intersection of the perpendiculars represents the location of the mean center of rotation during this time. As a result, the continuous motion is replaced by the discrete coordinates of a point(s). Unfortunately, the Reuleaux method, as well as other methods of planar joint kinematics, is very susceptible to measurement errors (Figure 4.24).

One of these errors is due to a possible misalignment of the plane of movement and the plane of a registering device (e.g., the TV camera, X-ray plate). When joint motion is a pure rotation, trajectories of individual markers fixed to the rotating body segment are concentric circles. The projections of the circles on a parallel plane are also circular. Normals of these circles intersect at one point, the center of rotation. If the image plane is not perfectly parallel to the

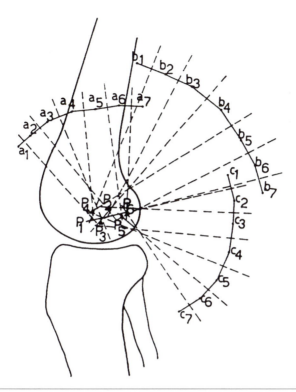

Figure 4.24 Instant center pathway as found by the Reuleaux method. Instead of an expected continuous centrode, a collection of dispersed points was detected. Similar erratic shifts in axial position have been reported by many other researchers.

Reprinted from *Journal of Biomechanics, 12,* Soudan, K., Auderkercke, R.V., & Martens, M., Methods, difficulties and inaccuracies in the study of human joint mechanics and pathomechanics by the instant axis concept, example: The knee joint, 27-33, 1979, with kind permission from Elsevier Science Ltd, The Boulevard, Langford Lane, Kidlington 0X5 1GB, UK.

plane of motion, a deformed view of the motion is registered on the film plane. Instead of circles, ellipses are seen. Normals of the ellipses do not intersect at one center. The researcher, not being aware of the planar misalignment, is prone to explain the movement along elliptical paths as a combination of rotation (which is correct) and translation of the centrode (which is not correct in this case). To avoid this mistake, the image plane must be put perfectly parallel to the motion plane. This is not simple, especially when only small parts of the circles or ellipses are registered. Tiny deviations are hard to visualize and can be recorded and corrected exclusively with three-dimensional measurement techniques.

4.2.1.4 Finite and Instantaneous Centrodes

Representative papers: Woltring et al., 1985; Woltring, 1990.

The Reuleaux method is based on a discrete, stepwise estimation of the center of rotation. Although historically the method was designed for locating an instantaneous center of rotation, it in fact provides an estimation of a finite center position rather than an instantaneous one.

Consider a finite displacement of two points anchored to a body, A and B (Figure 4.25). The initial and final coordinates of point A are (X_1, Y_1) and (X_2, Y_2), and the initial and final coordinates of point B are (X_3, Y_3) and (X_4, Y_4). Vectors **A** and **B** are given as $\mathbf{A} = (X_2 - X_1, Y_2 - Y_1)$ and $\mathbf{B} = (X_4 - X_3, Y_4 - Y_3)$. Coordinates of their median points are

$$\overline{X'} = \frac{X_1 + X_2}{2}; \ \overline{Y'} = \frac{Y_1 + Y_2}{2}$$

$$\overline{X''} = \frac{X_3 + X_4}{2}; \ \overline{Y''} = \frac{Y_3 + Y_4}{2} \tag{4.9}$$

Equations of the lines that are normal to the vectors and pass through the median points are

$$(X - \overline{X'})(X_2 - X_1) + (Y - \overline{Y'})(Y_2 - Y_1) = 0$$

$$(X - \overline{X''})(X_4 - X_3) + (Y - \overline{Y''})(Y_4 - Y_3) = 0 \tag{4.10}$$

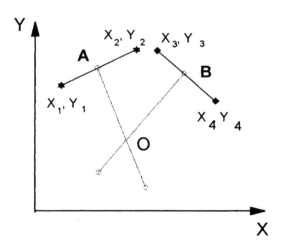

Figure 4.25 Finding a finite center of rotation with the Reuleaux method.

The point at which the lines intercept can be found as a solution to the linear equation 4.7. For convenience, rewrite equation 4.10 as

$$X(X_2 - X_1) + Y(Y_2 - Y_1) = \overline{X'}(X_2 - X_1) + \overline{Y'}(Y_2 - Y_1)$$
$$X(X_4 - X_3) + Y(Y_4 - Y_3) = \overline{X''}(X_4 - X_3) + \overline{Y''}(Y_4 - Y_3) \tag{4.11}$$

Represent the right part of equation 4.11 as vector **V**:

$$\mathbf{V} = \begin{bmatrix} \overline{X'}\,(X_2 - X_1) + \overline{Y'}(Y_2 - Y_1) \\ \overline{X''}\,(X_4 - X_3) + \overline{Y''}(Y_4 - Y_3) \end{bmatrix} = \frac{1}{2}\begin{bmatrix} X_2^2 - X_1^2 + Y_2^2 - Y_1^2 \\ X_4^2 - X_3^2 + Y_4^2 - Y_3^2 \end{bmatrix} \tag{4.12}$$

and introduce a matrix [M] with a determinant det [M] = Δ:

$$[\mathbf{M}] = \begin{bmatrix} X_2 - X_1 & Y_2 - Y_1 \\ X_4 - X_3 & Y_4 - Y_3 \end{bmatrix} \tag{4.13}$$

Then, the coordinates of the point of interception of the two normals can be found from

$$\begin{bmatrix} X \\ Y \end{bmatrix} = \frac{1}{\Delta}\begin{bmatrix} Y_4 - Y_3 & Y_1 - Y_2 \\ X_3 - X_4 & X_2 - X_1 \end{bmatrix} \cdot \mathbf{V} \tag{4.14}$$

as

$$X = \frac{1}{2}\Delta\left[(Y_4 - Y_3)(X_2^2 - X_1^2 + Y_2^2 - Y_1^2) + (Y_1 - Y_2)(X_4^2 - X_3^2 + Y_4^2 - Y_3^2)\right]$$
$$Y = \frac{1}{2}\Delta\left[(X_3 - X_4)(X_2^2 - X_1^2 + Y_2^2 - Y_1^2) + (X_2 - X_1)(X_4^2 - X_3^2 + Y_4^2 - Y_3^2)\right]$$

$$(4.15)$$

The concept of a finite center presumes that between two consecutive data observation points the body rotates from position 1 to position 2 about a short-lived fixed center, the finite center. In reality, during this period the ICR is moving. The difference between the real instant center pathway and its estimation deduced from the finite axis location can be too large to accept. Evidently the similarity is better the nearer the successive positions are to each other. Hence, the estimation improves with more data collection points and it seems natural to recommend increasing their number. Nothing is perfect, however. If the rotation is small, the estimation of the center of rotation position is highly susceptible to measurement errors. Recall (see Section **1.2.5.3**) that a center of rotation is not defined for pure translation. When rotation is small, the center of rotation approaches infinity and even the smallest measurement errors can generate a large inaccuracy in the estimation of the center coordinates. Therefore a trade-off solution should be found. The motion steps should be small enough to get a good similarity between the finite and instantaneous center and they should be large enough to estimate location of the finite center with acceptable accuracy. The accuracy also depends on the pattern of the motion, i.e., its affinity with pure rotation. In short, in many joints, because of the complex pattern of articular motion and inherent inaccuracies in the measurement, the finite center of rotation is an unreliable estimator of the true instantaneous center. One of the suggested solutions to this problem is approximating successive positions of the markers by continuous curves and then computing a centrode. If the fitting curve is by itself accurate, the direction of the normals on the fitting curve can be computed with good precision. In studies on articular kinematics, several procedures for smoothing the sampled position data have been suggested (e.g., Holzreiter, 1991; Panjabi, 1979; Panjabi et al., 1982a, b; Soudan et al., 1979; Woltring et al., 1985; Woltring, 1990) and the methods for optimal marker placement have been suggested (Crisco et al., 1994).

In spite of all the technical difficulties associated with determining the joint centrode, the procedure, if accurate, is still useful for diagnostic purposes. All the points of the rotating bone are instantly moving perpendicular to the radius drawn from the ICR to the point—this is the idea on which the Reuleaux method is based. If the ICR is located such that the direction of displacement of the contact point is along the articular surface, the point is gliding on the surface (Figure 4.26). When the ICR location is not perfectly matched with the contour of the joint surfaces, the articulating surfaces can either distract or impinge each other. Location of the joint centrode is also important for determining the muscle moment arm, i.e., the shortest distance from the line of muscle action to the axis of rotation. Even a small, say 5 mm, displacement of the instant center can change the estimated magnitude of the moment arm dramatically.

4.2.2 Three-Dimensional Joint Movement

In accordance with the Chasles' theorem (see "Euler's and Chasles' Theorems" on page 52), any three-dimensional, rigid-body motion can be viewed as a rotation around and translation along an axis, called a screw or helical axis. The values of the helical rotation and translation are invariant to coordinate transformation, that is, they are independent of the chosen reference frame. During a joint motion the helical axis, which is the instantaneous axis about and along which a human body segment is thought to be moving, can itself move (it can translate [change place] and rotate [change orientation]) (Figure

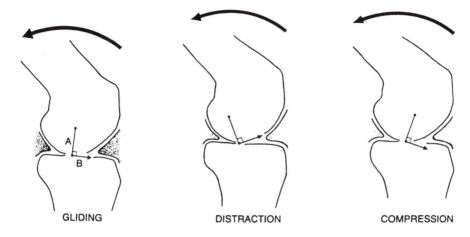

GLIDING   **DISTRACTION**   **COMPRESSION**

Figure 4.26   Location of the instant joint center in the normal knee (left figure) and deranged knees. In a normal knee a radius from the ICR to the tibiofemoral contact point is perpendicular to a line tangential to the tibial surface (line B). As a result, the contact point on the femur glides on the tibial condyle. In the knees with displaced ICR, the direction of the instant rotation is not along the joint surface—the femoral contact point will be either distracted (central figure) or compressed (right figure) during increased flexion.

From Frankel, V.H., Burstein, A.H., & Brooks, D.B. (1971). Biomechanics of internal derangement of the knee. Pathomechanics as determined by analysis of the instant centers of motion. *Journal of Bone and Joint Surgery, 53A,* 945-962. Adapted by permission.

■ ■ ■ *From the Scientific Literature* ■ ■ ■

Joint Centrode in Pathology

(A) Source: Soudan, K., Auderkercke, R.V., & Martens, M. (1979). Methods, difficulties and inaccuracies in the study of human joint mechanics and pathomechanics by the instant axis concept, example: The knee joint. *Journal of Biomechanics, 12,* 27-33.

Earlier, these authors determined the ICR location in the knee joint during joint flexion using the methods described in the literature and arrived at rather bizarre results (see Figure 4.24). Later, the accuracy was improved. Specifically, great attention was given to the proper alignment of the plane of measurement with the plane of joint motion. Then when the ICR was determined with all the precautions taken, the centrodes for the healthy and impaired joints showed obviously different patterns (Figure 4.27).

See **PATHOLOGY,** *p. 268*

■ ■ ■ **PATHOLOGY,** *continued from p. 267*

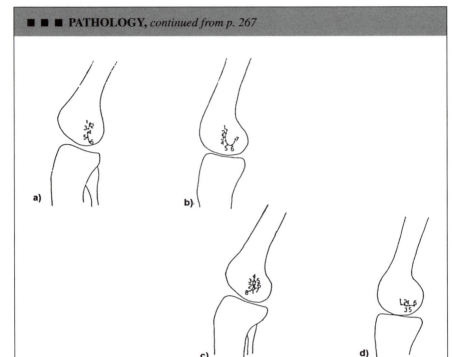

Figure 4.27   Centrodes of the knee joints: (a) normal knee, (b) deranged knee with a torn meniscus, (c) deranged knee with an anterior cruciate ligament tear, (d) deranged knee with postoperative control after meniscectomy. The numbers are for the sequential positions of the centrode.

Adapted from *Journal of Biomechanics, 12,* Soudan, K., Auderkercke, R.V., & Martens, M., Methods, difficulties and inaccuracies in the study of human joint mechanics and pathomechanics by the instant axis concept, example: The knee joint, 27-33, 1979, with kind permission from Elsevier Science Ltd, The Boulevard, Langford Lane, Kidlington 0X5 1GB, UK.

(B) Source: Amevo, B., Aprill, C., & Bogduk, N. (1992). Abnormal instantaneous axes of rotation in patients with neck pain. *Spine, 17,* 748-756.

Instantaneous axes of rotation for cervical joint motion segments C2-3 to C6-7 were determined in 109 patients with uncomplicated neck pain. Unequivocally abnormal ICRs were found in 46% of patients, and marginally abnormal ICRs were found in a further 26% of patients. The authors concluded that an abnormal ICR significantly correlates with the presence of neck pain.

4.28). The path of the instantaneous screw axis is called the *axode* and a surface produced by the migration of the instantaneous axis during motion is called the *helical axis surface*. The axodes are generalization for spatial motion of the centrodes associated with the planar motion.

Similar to the planar case and depending on the method of measurement, the helical axes can be either finite or instantaneous. A finite axis describes a motion step; an instantaneous axis describes a motion at infinitesimal time. The finite axis only approximates the position of the instantaneous helical axis.

4.2.2.1 Joints with One Degree of Freedom

Joints with one DOF that are represented as hinge or pivot joints are by definition planar. These joints allow motion in only one plane, about a single axis of rotation. However, when the axis of rotation of such a joint, such as the tibiotalar joint, is not perpendicular to the cardinal anatomic planes, articular motion occurs in all three planes. To employ the methods of planar kinematics, the motion plane should be known and the image plane should be positioned strictly parallel to the plane of the articular motion. After that, the methods described in Section **4.2.1** can be applied. For the planar joints with instant axes of rotation, the axodes are two rolling cylinders. The curves formed by the intersection of these cylinders with the motion planes are the centrodes of the planar motion. For hinge joints, the axodes degenerate to a single axis.

Some one-DOF joints are not planar. In these joints, the axis of rotation changes its orientation during articular motion. Such joints should be analyzed by the methods of three-dimensional kinematics.

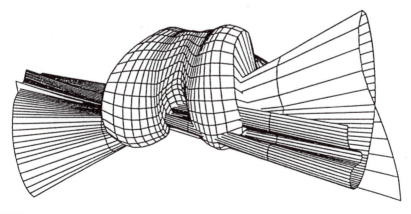

Figure 4.28   Helical axis surface of the knee joint with respect to the femur (distal view).
From Fuss, F.K. (1993). Helical axis surface of the knee joint. In *XIVth Congress of the International Society of Biomechanics, Abstracts* (pp. 438-439). Paris, July 4-8, 1993. Reprinted by permission.

4.2.2.2 Joints with Two and Three Degrees of Freedom

In joints with two and three DOF, articular motion can be performed around an infinite number of axes. The rotation axes are not unique. They depend on the particular motion pathway. For instance, in a hip motion, flexion-extension and abduction-adduction can be combined in many different ways. Because of the unlimited variety of motion, kinematic studies of joints with more than one DOF have addressed either a given articular motion or all of the expanse of possible joint motion.

To describe a three-dimensional articular motion, the methods presented in Chapters **1** and **2** (specifically direction cosines, Euler's rotations, and their variant [joint configuration convention]), as well as the helical axes method, are commonly used. Each of the techniques has its own advantages and disadvantages. The values of the Euler's angles depend on the specified rotation sequence. Position of one of the axes influences predicting the other axes. Therefore, all of the axes of rotation for a joint must be solved at the same time. The method is very sensitive to a precise orientation of the coordinate frame with regard to the anatomic landmarks and the rotation plane. If the rotation plane and the coordinate plane are offset, the registered joint rotation will include all three movements: flexion-extension, abduction-adduction, and internal-external rotation. In contrast to that, the values of rotation about and translation along the helical axes are invariant to coordinate transformations. However, if the rotation is small, measurement errors become very large; also, for pure translation the helical axis is not defined at all. Nevertheless, in the current literature, the screw method is becoming more and more popular. For spheroid joints, the axodes are two rolling cones with shared apexes at the center of the sphere.

Although all of the mentioned methods of spatial kinematics are valid and well established, the reported results, especially from different laboratories, are not always consistent. The poor reproducibility of the results cannot be attributed entirely to the inaccuracy of measurements. The main reason is a high sensitivity of joint motion to the small changes in applied forces (Figure 4.29) and the difficulties in standardizing a given motion.

From the example presented in Figure 4.29 it follows that an exact prescription of a "given motion" is the crucial issue in determining the helical axes. If a joint motion is prescribed verbally, for example, as "flexion," different subjects and experimenters will follow the instruction in a slightly different way. The location of the helical axes will also be different. This is a "Catch-22" situation: to find the helical axis a movement should be prescribed unambiguously; to assign the articular motion the axis of rotation should be specified. The solution depends on the required accuracy. If the requirement is not very strong, the axes for the main joint movements—flexion-extension, abduction-

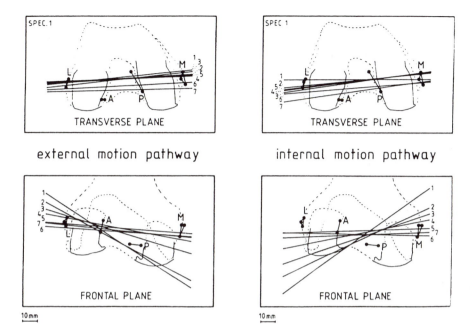

Figure 4.29 Helical axes of an external (left panel) and internal (right panel) flexion pathway of one knee specimen. In the experiment, the knee specimens were tested in a number of flexion positions from extension to approximately 95°. Either external or internal torque of 3 Nm about the long axis of the tibia was applied. Correspondingly, two flexion pathways, external and internal, were recorded. A, anterior cruciate ligament; P, posterior cruciate ligament; L, lateral collateral ligament; M, medial collateral ligament. The helical axes are not parallel to the main anatomic axes and, also, they change their orientation during the motion. If flexion is defined as a rotation around a horizontal lateromedial axis, the registered knee motion (intended to represent the knee flexion) cannot be regarded as a (pure) flexion because the movement is performed around the oblique helical axes. Rather, the motion is a combination of the pure flexion and pure axial rotation. Even small changes in applied load induce large changes in the articular motion pattern.

Reprinted from *Journal of Biomechanics, 23,* Blankevoort, L., Huiskes, R., & de Lange, A., Helical axes of passive joint motions, 1219-1229, 1990, with kind permission from Elsevier Science Ltd, The Boulevard, Langford Lane, Kidlington OX5 1GB, UK.

adduction, and internal-external rotation—can be found under the assumption that various subjects can execute these motions in the same way. If this assumption does not stand, precise determination of the helical axes for the main joint motions does not make much sense. The same is true for joint kinematics during specific activities, such as walking and running. People change their movement patterns from attempt to attempt and they move differently by habit; high variability is inherent to the system. As a result, various gross motor patterns produce different articular motions.

4.2.2.3 Envelopes of Passive Joint Motion

Representative paper: Blankevoort, Huiskes, & de Lange, 1988.

Joint motion is restricted by elastic structures and bone-on-bone contacts. However, within a certain region, the joint moves without perceptible resistance from the surrounding joint structures. When the range of motion increases, so does the resistance. Therefore, two parts of the articular motion can be discerned: the part with low resistance from the joint supporting structures and the part with high resistance from these structures. Within the first region, the motion pathway is very sensitive to small loads. The border between the first and the second regions defines the *envelope of passive joint motion*. Within the envelope, small changes in the magnitude and direction of the applied forces have large effects on the location of rotational axes. The envelope itself depends on the magnitude of what is defined as a "small resistance." The magnitude is selected arbitrarily, but it should be kept constant across the different studies. To decide on the envelope of passive motion, first a "torque-angular displacement" curve is determined and then a threshold value of the "low resistance" is selected (Figure 4.30).

When the resistance limits are resolved, the envelope can be found registering a range of motion in various directions (Figure 4.31). Natural joint motion typically occurs within the limits prescribed by the envelope of passive joint motion. Articular motion beyond these limits imposes a high internal load and may damage the joint structures.

In some joints, the limits of angular motion are imposed by the bone structures rather than by the soft tissues. In these joints, the range of motion can be calculated as a difference between the arcs of the male and female surfaces (Figure 4.32).

4.3 SUMMARY

Although joint motion is often considered a simple rotation in an anatomic plane, e.g., joint flexion in the sagittal plane, real joint movements are usually much more complex. Joint axes are usually not fixed, oblique to the anatomic planes and to each other, and may not intersect. This chapter contains a thorough theory developed for studying joint motion. Restrictions on the shape of joint surfaces are not imposed, but the surfaces and ligaments are presumed to be nondeformable.

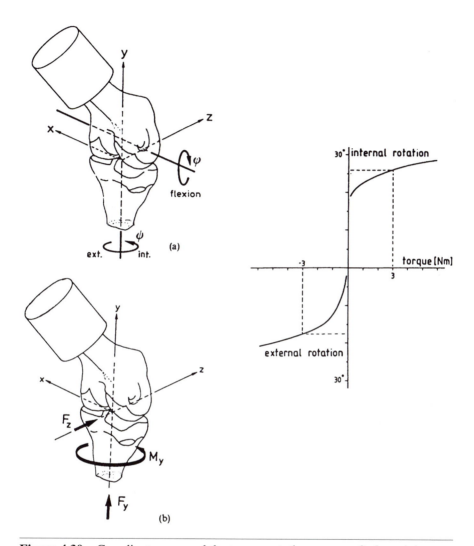

Figure 4.30 Coordinate axes and the torque-rotation curves of a knee joint specimen at 25° of flexion. The threshold value for the envelope of passive motion was chosen at the tibial rotation torques $M_y = \pm 3$ Nm.
Reprinted from *Journal of Biomechanics, 21,* Blankevoort, L., Huiskes, R., & de Lange, A., 1988, The envelope of passive knee joint motion, 705-720, 1988, with kind permission from Elsevier Science Ltd, The Boulevard, Langford Lane, Kidlington OX5 1GB, UK.

The chapter starts from elementary examination of joint surfaces and corresponding joint motions. The surfaces are classified as flat, synclastic (wholly concave or wholly convex), anticlastic, or sellars (convex in one direction and concave in the perpendicular direction). The joints with synclastic surfaces are

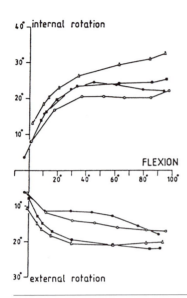
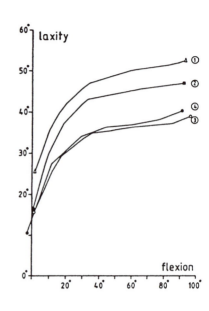

Figure 4.31 The envelope of passive knee motion between extension and 100° of flexion with external and internal tibial torques of ±3 Nm (left panel). The right panel shows the total rotatory laxity, i.e., the total amount of tibial rotation between the limits. The data are for four specimens.

Reprinted from *Journal of Biomechanics, 21,* Blankevoort, L., Huiskes, R., & de Lange, A., 1988, The envelope of passive knee joint motion, 705-720, 1988, with kind permission from Elsevier Science Ltd, The Boulevard, Langford Lane, Kidlington 0X5 1GB, UK.

spheroidal, ellipsoidal, hinge, and pivot joints. Convex surfaces are called male, and concave surfaces female. Joints with spheroidal (approximately spherical) articulating surfaces are ball-and-socket joints with three DOF. Joints with ellipsoidal and sellar mating surfaces have two DOF. Hinge and pivot joints have one DOF.

To characterize the shape of joint surfaces in detail several parameters of differential geometry, in particular joint congruence, are used. The joint congruence affects a pattern of surface-on-surface motion as well as the contact areas in the joints. The close-packed position is the position of complete congruency in a joint. Other positions are the loose-packed positions.

Two basic patterns of planar surface-on-surface motion are rolling and skidding. During rolling, the skidding velocity is zero; the male bone rotates about an instant axis located on the articular surface and the point of the common contact is not constant. In contrast, during skidding one of the points of contact is constant: the same region of an articular surface is in continuous contact with different regions of another surface. The slip ratio characterizes the rela-

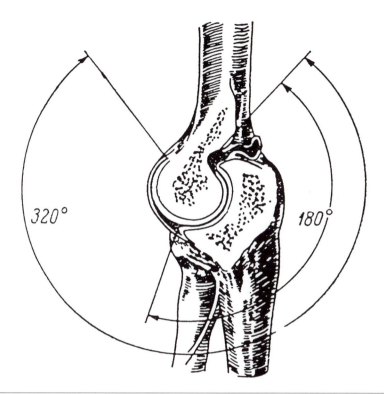

Figure 4.32 Calculation of the range of motion from the bony geometry, the portion of the arc subtended by the articular surface. In the elbow joint, the arc of the male surface, the trochlea of the humerus, is 320°; the arc of the female surface, the trochlear fossa of the ulna, is 180°. The range of motion limited by the bony surfaces is 320° − 180° = 140°.

tive contribution of rolling and skidding to joint motion. Spin is rotation of joint surfaces around an axis normal to the surfaces at the point of contact.

In the algebra of articular kinematics, the algebra based on the ideas of spherical geometry, relative motion of joint surfaces is represented as motion of an imaginary rod normal to the surface. The rod can slide, or spin, or both. According to the main theorem of joint algebra any two consecutive joint motions are equivalent to a single slide combined with a single spin. Available methods allow a compact description of several joint motions performed in order.

When ligaments are included in the problem, and they are assumed to be inelastic, the joints are analogous to mechanical linkages. In this spirit, the knee joint is a four-bar mechanism.

Kinematic models of various complexity have been used to analyze joint motion. The models range from the simplest, i.e., planar rotation around a fixed axis parallel to an anatomic axis, to rather complex, such as three-dimensional rotation about and translation along arbitrarily oriented changeable axes. To apply a planar model, the plane of rotation should be either known in advance or established through three-dimensional analysis. Inaccuracies in determining the plane of rotation lead to errors in describing even the most simple joint movements, such as planar rotation about a fixed axis. If an optical method is used, an elliptical rather than circular trajectory of a marker is recorded.

Location of an instant center of rotation in a joint can be found using Kennedy's theorem. Joint motion can be represented as a combined rotation around the female joint center plus rotation around the male joint center. According to the theorem, the ICR lies on the line of centers and its position is defined by the ratio of the two angular velocities. Other methods include the tangent method and the Reuleaux method. In the tangent method, the ICR location is found at the intersection of the two normals to the instant velocity vectors of any two points on the body of interest. In the Reuleaux method, the normals are erected from the midpoints of the lines connecting two consecutive positions of the points. The mathematical foundation of the Reuleaux method is described (see equation 4.15).

In joints with two and three DOF, articular motion can be performed around an infinite number of axes. Kinematic studies of these joints have addressed either a given articular motion or all possible joint motions. To study a given joint motion, e.g., flexion, it should be standardized beforehand. After that, to find the ICR and the joint centrode the helical axis method is recommended. The envelope of joint motion is defined as a border between two regions of joint motion—with a small and a large passive resistance to an applied force. This is a useful method for describing joint kinematics.

4.4 QUESTIONS FOR REVIEW

1. Define geometrically ideal joints.
2. What is the difference between synclastic and anticlastic joint surfaces?
3. Name the articular surfaces of the joints with one, two, and three DOF. What are these joints called?
4. Name the male and female surfaces in the glenohumeral joint.
5. Define joint congruence and calculate the overall joint congruence at a point with the principal curvatures $k^e_{min} = 3$ cm^{-1} and $k^e_{max} = 7$ cm^{-1}.
6. Define the skidding velocity. What is the skidding velocity for pure rolling? For pure skidding?
7. Define the slip ratio. What is the slip ratio for pure rolling?

8. Give an example of spin in a human joint.
9. What is a geodesic?
10. Discuss the basic theorem of kinematics of articular surfaces.
11. Discuss the four-bar model of the knee joint.
12. Discuss the concept of fixed centers of rotation and the method used to locate them.
13. Using Kennedy's theorem, find the ICR location in a joint.
14. Describe the Reuleaux method of locating the ICR in a joint.
15. What is the helical axis surface?
16. Discuss envelopes of passive joint motion.

BIBLIOGRAPHY

Adams, L.M. (1993). The anatomy of joints related to function. In V. Wright & E.R. Ladin (Eds.), *Mechanics of human joints* (pp. 27-82). New York: Marcel Dekker, Inc.

Amevo, B. (1991). Instantaneous axes of the typical cervical motion segments. 1. An empirical study of technical errors. *Clinical Biomechanics, 6,* 31-37.

Amevo, B., Aprill, C., & Bogduk, N. (1992). Abnormal instantaneous axes of rotation in patients with neck pain. *Spine, 17,* 748-756.

Ateshian, G.A., Rosenwasser, M.P., & Mow, V.C. (1992). Curvature characteristics and congruence of the thumb carpometacarpal joint: Differences between female and male joints. *Journal of Biomechanics, 25,* 591-607.

Ateshian, G.A., Soslowsky, L.J., & Mow, V.C. (1991). Quantitation of articular surface topography and cartilage thickness in knee joints using stereophotogrammetry. *Journal of Biomechanics, 24,* 761-776.

Barnett, C.H., Davies, D.V., & MacConaill, M.A. (1961). *Synovial joints, their structure and mechanics.* London: Longman.

Bell, A.L., Pedersen, D.R., & Brand, R.A. (1990). A comparison of the accuracy of several hip center location prediction models. *Journal of Biomechanics, 23,* 617-621.

Blankevoort, L., Huiskes, R., & de Lange, A. (1988). The envelope of passive knee joint motion. *Journal of Biomechanics, 21,* 705-720.

Blankevoort, L., Huiskes, R., & de Lange, A. (1990). Helical axes of passive joint motions. *Journal of Biomechanics, 23,* 1219-1229.

Bogduk, N., Amevo, B., & Pearcy, M. (1995). A biological basis for instantaneous centres of rotation of the vertebral column. *Proceedings of the Institution of Mechanical Engineers. Part H. Journal of Engineering in Medicine, 209*(3), 177-183.

Bryant, J.T., Wevers, H.W., & Lowe, P.J. (1984). One parameter model for error in instantaneous centre of rotation measurements. *Journal of Biomechanics, 17,* 317-323.

Chao, E.Y.S. (1980). Justification of triaxial goniometer for the measurement of joint rotation. *Journal of Biomechanics, 13,* 989-1006.

Crisco, J.J. III, Chen, X., Panjabi, M.M., & Wolfe, S.W. (1994). Optimal marker placement for calculating the instantaneous center of rotation. *Journal of Biomechanics, 27,* 1183-1187.

Delp, S.L. & Maloney, W. (1993). Effects of hip center location on the moment-generating capacity of the muscles. *Journal of Biomechanics, 26,* 495-499.

Dempster, W.T. (1955). *Space requirements of the seated operator.* (Technical Report WADC-TR-55-159). Wright-Patterson Air Force Base, OH: Aerospace Medical Research Laboratory.

Dimnet, J. (1980). The improvement in the results of kinematics of in vivo joints. *Journal of Biomechanics, 13,* 653-661.

Dimnet, J., Carret, J.P., Gonon, G., & Fischer, L.P. (1976). A technique for joint center analysis using a stored program calculator. *Journal of Biomechanics, 9,* 771-778.

Dimnet, J. & Guingard, M. (1984). The finite displacement vector's method. *Journal of Biomechanics, 17,* 387-394.

Engin, A.E. (1979). On the biomechanics of major articulating human joints. In N. Akkas (Ed.), *Progress in biomechanics* (pp. 157-188). Alphen aan den Rijn, The Netherlands: Sijthoff & Noordhoff.

Fick, R. (1908). *Handbuch der Anatomie und Mechanik der Gelenke unter Berücksichtigung der bewegenden Muskeln. Spezielle Gelenk und Muskelmechanik.* Jena, Germany: Verlag von Gustav Fisher.

Fioretti, S., Jetto, L., & Leo, T. (1990). Reliable in vivo estimation of the instantaneous helical axis in human segmental movements. *IEEE Transactions on Biomedical Engineering, 37,* 398-409.

Frankel, V.H., Burstein, A.H., & Brooks, D.B. (1971). Biomechanics of internal derangement of the knee. Pathomechanics as determined by analysis of the instant centers of motion. *Journal of Bone and Joint Surgery, 53A,* 945-962.

Fuss, F.K. (1993). Helical axis surface of the knee joint. In *XIVth Congress of the International Society of Biomechanics, Abstracts* (vol. 1, pp. 438-439). Paris.

Gerhardt, J.J. & Russe, O.A. (1975). *International SFTR Method of Measuring and Recording Joint Motion.* Chicago: Year Book Medical.

Gertzbein, S.D., Chan, K.H., Tile, M., Seligman, J., & Kapasouri, A. (1985). Moiré patterns: An accurate technique for determination of the locus of the centres of rotation. [Technical note] *Journal of Biomechanics, 18,* 501-509.

Grant, P. (1973). Biomechanical significance of the instantaneous center of rotation: The human temporomandibular joint. *Journal of Biomechanics, 6,* 109-113.

Gysi, A. (1910). The problem of articulation. *Dental Cosmos, 52,* 1-19.

Holzreiter, S. (1991). Calculation of the centre of rotation for a rigid body. *Journal of Biomechanics, 24,* 643-647.

Huiskes, R. & Blankevoort, L. (1990). The relationship between knee motion and articular surface geometry. In V.C. Mow, V.C. Ratcliffe, & S.L.-Y. Woo (Eds.), *Biomechanics of diarthroidal joints* (pp. 269-286). New York: Springer-Verlag.

Huiskes, R., Kremers, J., de Lange, A., Woltring, H.J., Selvik, G., & van Rens, T.H.J.G. (1985). An analytical stereo-photogrammetric method to determine the three-dimensional geometry of articular surfaces. *Journal of Biomechanics, 18,* 559-570.

Kapandji, I.A. (1974). *The physiology of the joints* (2nd ed.). Edinburgh: Churchill-Livingstone.

Keats, T.E., Teeslink, R., Diamond, A.E., & Williams, J.H. (1966). Normal axial relationships of the major joints. *Radiology, 87,* 904-907.

Kirsch, S., Nägerl, H., & Kubein-Meesenburg, D. (1993). Kinematics and statics of the human shoulder. *XIVth International Congress of Biomechanics* (pp. 692-693). Abstracts, Paris.

Kirstukas, S.J., Lewis, J.L., & Erdman, A.G. (1992). 6R instrumental spatial linkages for anatomical joint motion measurement. Part 1: Design. *Journal of Biomechanical Engineering, 114,* 92-100.

Koff, D.G. (1995). Joint kinematics: Camera-based approach. In R.L. Craik & C.A. Oatis (Eds.), *Gait analysis* (pp. 183-204). St. Louis: Mosby.

Kubein-Meesenburg, D., Nägerl, H., & Fanghänel, J. (1990). Elements of a general theory of joints. 1. Basic kinematics and static functions of diarthroses. *Anatomische Anzeiger, 170,* 301-308.

Kubein-Meesenburg, D., Nägerl, H., & Fanghänel, J. (1991a). Elements of a general theory of joints. 4. Coupled joints as simple gear systems. *Anatomische Anzeiger, 172,* 309-321.

Kubein-Meesenburg, D., Nägerl, H., & Fanghänel, J. (1991b). Elements of a general theory of joints. 5. Basic mechanics of the knee. *Anatomische Anzeiger, 173,* 131-142.

Lange, A. de, Huiskes, R., & Kauer, J.M.G. (1990). Effects of data smoothing on the reconstruction of helical axis parameters in human joint kinematics. *Journal of Biomechanical Engineering, 112,* 107-113.

Lebar, A.M., Iglic, A., Antolic, V., Damjanic, F.B., Herman, S., Srakar, F., & Brajnik, D. (1995). Dependence of the hip forces and stem/cement interface stress distribution on the position of the hip joint rotation centre. In *XVth Congress of the International Society of Biomechanics, Abstracts* (pp. 534-535). Juväskylä, Finland.

Lewis, J.L. & Lew, W.D. (1977). A note on the description of articulating joint motion. *Journal of Biomechanics, 10,* 675-678.

Lewis, J.L. & Lew, W.D. (1978). A method for locating optimal "fixed" axis of rotation for the human knee joint. *Journal of Biomechanical Engineering, 100,* 187-193.

MacConaill, M.A. (1946). Studies in the mechanics of synovial joints. *Irish Journal of Medical Science, 49,* 223-235.

MacConaill, M.A. (1953). The movements of bones and joints. *Journal of Bone and Joint Surgery, 35B,* 290-297.

MacConaill, M.A. (1966). The geometry and algebra of articular kinematics. *Bio-Medical Engineering, 1,* 205-212.

MacConaill, M.A. & Basmajian, J.V. (1969). *Muscles and movements. A basis for human kinesiology.* Baltimore: Williams & Wilkins.

Mow, V.C, Ratcliffe, A., & Woo S.L.-Y. (Eds.). (1990). *Biomechanics of diarthroidal joints* (Vols. I and II). New York: Springer-Verlag.

Nägerl, H., Kubein-Meesenburg, D., Becker, B., & Fanghänel, J. (1991). Elements of a general theory of joints. 3. Elements of theory of synarthrosis: Experimental examinations. *Anatomische Anzeiger, 172,* 187-195.

Nägerl, H., Kubein-Meesenburg, D., & Fanghänel, J. (1990). Elements of a general theory of joints. 2. Introduction into a theory of synarthrosis. *Anatomische Anzeiger, 171,* 323-333.

Nägerl, H., Kubein-Meesenburg, D., Fanghänel, J., Thieme, K.M., Klamt, B., & Schwestka-Polly, R. (1991). Elements of a general theory of joints. 6. General kinematical structure of mandibular movements. *Anatomische Anzeiger, 173,* 249-264.

Norkin, C.C. & Levangie, P.K. (1992). *Joint structure and function. A comprehensive analysis* (2nd ed.). Philadelphia: F.A. Davis Company.

O'Connor, J.J. & Zavatsky, A. (1990). Kinematics and mechanics of the cruciate ligaments of the knee. In V.C. Mow, A. Ratcliffe, & S.L.-Y. Woo (Eds.), *Biomechanics of diarthroidal joints* (Vol. 2, pp. 197-241). New York: Springer-Verlag.

Panjabi, M.M. (1979). Centers and angles of rotation of body joints: A study of errors and optimization. *Journal of Biomechanics, 12,* 911-920.

Panjabi, M.M., Goel, V.K., & Walter, S.D. (1982a). Errors in kinematic parameters of a planar joint: Guidelines for optimal experimental design. *Journal of Biomechanics, 15,* 537-544.

Panjabi, M.M., Goel , V.K., Walter, S.D., & Schick, S. (1982b). Errors in the center and angle of rotation of a joint: An experimental study. *Journal of Biomechanical Engineering, 104,* 232-237.

Persson, T., Lanshammar, H., & Medved, V. (1995). A marker-free method to estimate joint centre of rotation by video image processing. *Computer Methods and Programs in Biomedicine, 46,* 217-224.

Poppen, N.K. & Walker, P.S. (1976). Normal and abnormal motion of the shoulder. *Journal of Bone and Joint Surgery, 58A,* 195-201.

Ramakrishnan, H.K. & Kadaba, M.P. (1991). On the estimation of joint kinematics during gait. *Journal of Biomechanics, 24,* 969-977.

Reuleaux, F. (1866). *The kinematics of machines: Outlines of a theory of machines* (A.B.W. Kennedy, Trans.). New York: Dover.

Scherrer, P.K. & Hillberry, B.M. (1979). Piecewise mathematical representation of articular surfaces. *Journal of Biomechanics, 12,* 301-311.

Shiavi, R., Limbird, T., Frazer, M., Stivers, K., Strauss, A., & Abramovitz, J. (1987). Helical motion analysis of the knee. 1. Methodology for studying kinematics during locomotion. *Journal of Biomechanics, 20,* 459-469.

Shoup, T.E. (1976). Optical measurement of the center of rotation for human joints. *Journal of Biomechanics, 9,* 241-242.

Sommer, H.J. & Miller, N.R. (1980). A technique for kinematic modeling of anatomical joints. *Journal of Biomechanical Engineering, 102,* 311-317.

Soudan, K., Auderkercke, R.V., & Martens, M. (1979). Methods, difficulties and inaccuracies in the study of human joint mechanics and pathomechanics by the instant axis concept, example: The knee joint. *Journal of Biomechanics, 12,* 27-33.

Spiegelman, J. & Woo, S. (1987). A rigid-body method for finding centers of rotation and angular displacements of planar joint motion. *Journal of Biomechanics, 20,* 715-721.

Spoor, C.W. & Veldpaus, F.E. (1980). Rigid body motion calculated from spatial coordinates of markers. *Journal of Biomechanics, 13,* 391-393.

Suntay, W.J., Grood, E.S., Hefzy, M.S., Butler, D.L., & Noyes, F.R. (1983). Error analysis of a system for measuring three-dimensional joint motion. *Journal of Biomechanical Engineering, 105,* 127-135.

Thoma, W., Jäger, A., & Schrieber, S. (1994). Kinematic analysis of the knee joint with regard to the load transfer on the cartilage. In Y. Hirasawa, C.B. Sledge, & S.L.-Y. Woo (Eds.). *Clinical Biomechanics and Related Research* (pp. 96-102). Tokyo: Springer-Verlag.

Veldpaus, F.E., Woltring, H.J., & Dortmans, L.J.M.G. (1988). A least squares algorithm for the equiform transformation from spatial marker co-ordinates. *Journal of Biomechanics, 21,* 45-54.

Walter, S.D. & Panjabi, M.M. (1988). Experimental errors in the observation of body joint kinematics. *Technometrics, 30,* 71-78.

Woltring, H.J. (1990). Estimation of the trajectory of the instantaneous centre of rotation in planar biokinematics. *Journal of Biomechanics, 23,* 1273-1274.

Woltring, H.J. (1991). Representation and calculation of 3-D joint movement. *Human Movement Science, 10,* 603-616.

Woltring, H.J., Huiskes, R., De Lange, A.L., & Veldpaus, F.E. (1985). Finite centroid and helical axis estimation from noisy measurements in the study of human joint kinematics. *Journal of Biomechanics, 18,* 379-389.

Wongchaisuwat, C., Hemami, H., & Hines, M.J. (1984). Control exerted by ligaments. *Journal of Biomechanics, 17,* 525-532.

Youm, Y. (1978). Kinematics theory and evaluation of articular joints. *Perspectives in Biomechanics, 1B,* 831-868.

Zatsiorsky, V.M. (Saziorski, W.M.), Aruin, A.S., & Selujanow, W.N. (1984). *Biomechanik des menschlichen Bewegungsapparates.* Berlin: Sportverlag.

Züppinger, H. (1904). *Die aktive Flexion in unbelasten Kniegelenk.* Zurich: Habilitationschrift.

5

CHAPTER

KINEMATICS OF INDIVIDUAL JOINTS

In Chapter **5** I will present a general introduction to the kinematics of particular joints. The discussion is limited to the major joints. Section **5.1** is intended to familiarize the students with the concept of nominal (fixed) joint axes. For the joint axes to fall on a point, the articular movement should be perfectly circular. Although this requirement is usually too restrictive, the abstraction of nominal joint axes appears to be the only way to carry out kinematic analysis of complex human movements. In this section, nominal axes in the main joints are specified. The specification is based on suggestions taken from the literature. Unfortunately, for some joints the recommendations do not agree with each other. Because there is no reason to prefer one author or technique over another, several available recommendations are presented in the text (properly marked in case of evident controversy). It is up to a researcher or student to choose a technique that he or she prefers. I apologize for this inconvenience. The biomechanics of human motion is still a young science with many unresolved issues.

The ensuing sections address kinematics of particular joints in greater detail. The joints are described in a bottom-to-top sequence: from toes of the foot to the fingers of the hand and then to mandibular motion. Individual joints have been investigated to different extents. Some joints have attracted much attention from researchers, and others have not been studied thoroughly. This disproportion is reflected in the level of details in describing individual joints.

Sections **5.2** through **5.5** concentrate on the joints of the leg. Section **5.2** describes the joints of the foot. In natural movements, the foot is considered a two-speed construction with coupled dorsiflexion in the metatarsophalangeal joints and ankle joint complex performed with reference to the parallel axis, either transverse or oblique. Section **5.3** is devoted to the ankle joint complex with its two joints, talocrural and subtalar. The kinematics of the knee joint—one of the most complex joints in the human body—is analyzed in Section **5.4**. Despite all the efforts of numerous researchers, many aspects of knee joint kinematics are still a matter of contention. Although the exact point in question may vary, the real issue of the controversy is the constraints acting on the joint, whether they are fixed or variable. If knee joint motion is guided completely by the articular surfaces and the inelastic ligaments the pattern of the knee motion—disregarding its complexity—is fixed and can be revealed. For example, a unique axis of joint flexion-extension exists and can be documented. However, if the muscles surrounding the joint are acting not only as the movers but also as stabilizers, the pattern of the knee motion depends on a specific combination of muscle activity and is changeable. Many axes of rotation will be discovered. Section **5.5** discusses the hip joint. The necessity to locate the joint center precisely during total hip replacement has stimulated a series of research on the kinematics of this joint.

Section **5.6** discusses kinematics of the spine, which is an extremely complex structure containing more than 100 articulations. The section begins with an explanation of movement in synarthroses (the preceding text addressed only diarthroidal joints). Vertebral movement depends on the deformability of the intervertebral disks and cannot be comprehended without using the concepts of strength of materials, a field of mechanics that studies the statics and dynamics of deformable elastic bodies. Therefore, some notions of strength of materials are used in this subsection to explain the motion of vertebrae. Next, the kinematics of two regions of the spine, the thoracolumbar and cervical is discussed. Each subsection discusses movement of individual motion segments and then the regional movement as an entirety. The section ends with a short subsection devoted to the rib cage movement.

Sections **5.7** to **5.10** are concerned with the kinematics of arm movement. The kinematics of the shoulder complex—the mobility of this complex exceeds the mobility of any other joint in the body—is addressed in Section **5.7**. Individual joints composing the shoulder complex are considered and then movement of the shoulder complex as a whole. Section **5.8** contains an examination of the elbow joint complex. The kinematics of the wrist is examined in Section **5.9** and kinematics of the joints of the hand in Section **5.10**. Finally, Section **5.11** concentrates on the temporomandibular joint.

This chapter is different from others in style. Because the chapter is based on experimental and anatomical data, there are no illustrations from the literature. Compared with the preceding chapters, there are many more references in Chapter **5**. In particular, the original quantitative data are referenced. Nevertheless, it was impossible to mention the contribution of all the authors without being involved in lengthy discussions not appropriate in a basic textbook. Instead, a list of sources is presented at the end of the chapter.

5.1 NOMINAL JOINT AXES

When studying gross human motion, researchers habitually ignore details of the joint kinematics and assume the existence of geometrically ideal joints. In ideal joints, the axes are (a) fixed, (b) orthogonal, (c) parallel to the main anatomic axes, and they (d) intersect at one fixed center. Those axes are called the *nominal* axes. The nominal axes are defined by the location of the joint centers with regard to skeletal landmarks. Typically, the landmarks are bony prominences that need to be accurately identified. Subjective error in determining the joint centers is almost unavoidable. It depends on the tracer's training and knowledge of anatomy. Simple accurate methods for locating the joint centers and nominal axes are not available at this time (see Section **4.2.1.1**). Because real (instantaneous) axes usually differ from the nominal, the selection of the nominal axes presumes a certain latitude and is based on a convention. One such convention is described in Table 5.1. The second convention is presented in Figure 5.1. Note that for some joints, for instance the hip joint, the recommendations are not in good agreement.

To locate the nominal joint centers, it is best to first draw the planes of the joint centers on the skin. After that, with the segment out of the vertical, the

Table 5.1   Nominal Joint Centers Relative to Bony Landmarks

Joint	Bony landmarks	Comments
Ankle (the talocrual joint)	Level of a line between the tip of a lateral malleolus of the fibula and a point 5 mm distal to the tibial malleolus.	(1) The talocrual joint axis is often determined by using the distal tip of the medial malleolus, rather than the lateral malleolus, as a reference bony landmark. (2) The axis can be estimated by palpating the malleoli.
Knee	Midpoint of a line between the centers of the posterior convexities of the femoral condyles.	(1) When this line is used for dissecting human body segments, a chunk of a thigh is added to the shank mass. An alternative way is to dissect along the knee joint slit. (2) According to Plagenhoef (1971), the knee center should be located by determining the center of the flat portion of the femoral condyles (Figure 5.1B).
Hip	(a) A point at the tip of the femoral trochanter 12 mm anterior to the most lateral projecting part of the trochanter.	(1) Dempster gives two slightly different landmarks. The first recommendation is for people with a normal angle of inclination of the femoral neck to the femoral shaft, which is on average about 125°.

(continued)

Table 5.1   *(continued)*

Joint	Bony landmarks	Comments
	(b) The lateral projection of the hip joint center is (i) within a 1.2 × 1.5 mm ellipse around the greater trochanter, or (ii) on a midhalf of the vertical line dropped from the anterior superior spine to the seat in a seated person (a standard pelvic tilt of 40° is assumed), the accuracy is ±3°.	Deviation of the neck-to-shaft angle in either direction shifts the greater trochanter with respect to the center of the hip joint along the vertical line. (2) According to Plagenhoef (1971) the hip joint center is located 3 cm above the most lateral bony prominence of the greater trochanter (Figure 5.1C).
Gleno-humeral	Midregion of the palpable bony mass of the head and tuberosities of the humerus; with the arm abducted about 45° relative to the vertebral margin of the scapula. A line dropped perpendicular to the long axis of the arm from the outer-most margin of the acromion will approximately bisect the joint.	Accuracy improves if the location of the joint center is determined in three joint positions rather than in one (Figure 5.1D).
Clavi-scapular	Midpoint of a line between the coracoid tuberosity of the clavicle (at the posterior border of the bone) and the acromioclavicular articulation (or the tubercle) at the lateral end of the clavicle. The point, however, should be visualized as on the underside of the clavicle.	
Sterno-clavicular	Midpoint position of the palpable junction between the proximal end of the clavicle and the sternum at the upper border (jugular notch) of the sternum.	
Elbow	Midpoint of a line between (1) the lowest palpable point of the medial epicondyle of the humerus, and (2) a point 8 mm above the radiale (radiohumeral junction).	Plagenhoef (1971) recommended another procedure (Figure 5.1E).

Joint	Bony landmarks	Comments
Wrist	On the palmar side of the hand, the distal wrist crease at the palmaris longus tendon, or the midpoint of a line between the radial styloid and the center of the pisiform bone.	
	On the dorsal side of the hand, the palpable groove between the lunate and capitate bones, on a line with the metacarpal bone III.	

According to W.T. Dempster (1955).

planes, bony landmarks, and skin creases are used together to locate the joint centers (Figure 5.2 and Table 5.2).

The planes used to locate the joint centers are also employed to divide the human body into separate links. The links span from one joint center to another.

Another approach to determine the nominal axes and joint centers—instead of marking the joint center planes—is to estimate their location from the measurements of skeletal bones. In particular, the joint center longitudinal positions can be estimated from the segment lengths (Figure 5.3). In the study by Leva (1996), the joint centers were assumed to lie on the longitudinal axes of the respective segments. The following nomenclature was used in that study:

- Acromion (ACR) was defined as the most lateral point on the lateral margin of the acromial process of the scapula.
- Ankle joint center (AJC) was the center of a transverse section of the talus, approximately at the level of the distal tip of the fibula (fibular sphyrion).
- The ball of humerus (BHUM) was the midpoint of the most proximal portion of the intertubercular groove of the humerus.
- The elbow joint center (EJC) was the center of the transverse section of the humerus at the level of the greatest projection of the medial humeral epicondyle (near the level of the skin crease at the anterior surface of the elbow).
- The hip joint center (HJC) was the center of the femoral head.
- The knee joint center (KJC) was the midpoint between the maximal protrusions of the femoral epicondyles (approximately at the level of the lower third of the patella when the knee is extended).

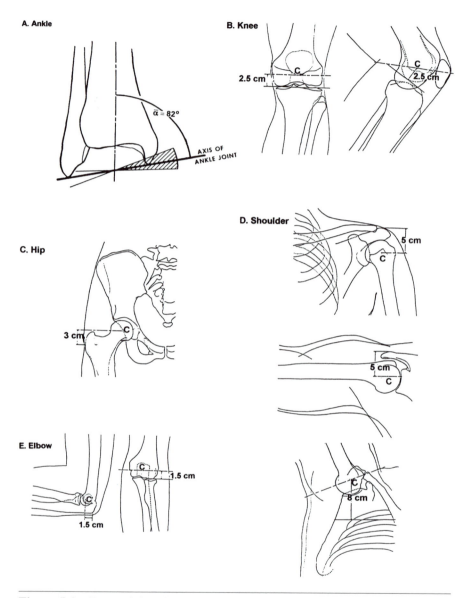

Figure 5.1 Determining the nominal joint centers. A. The position of the ankle joint axis is estimated by palpating the malleoli. The joint axis is under the tip of each malleolus. B. The nominal axis of the knee joint passes through the center of the flat portion of the condyles of the femur. C. The center of the hip joint is located approximately 3 cm above the most lateral bony prominence of the greater trochanter. D. The position of the center of the glenohumeral joint when the arm is down, sideways, and raised overhead. E. The position of the axis of the elbow joint is estimated by palpating the lateral and medial epicondyle of the humerus.

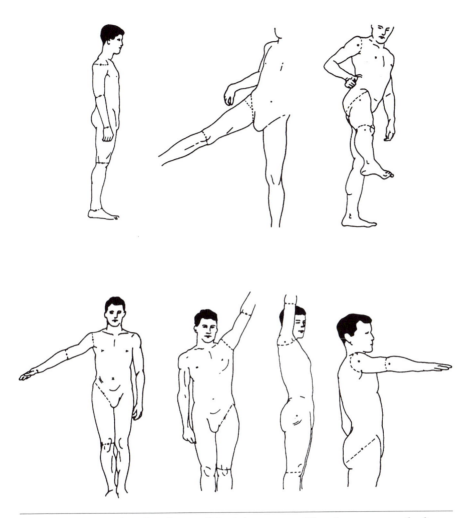

Figure 5.2 Locating the planes of the joint centers. The planes are marked with skin pencils using bony prominences and skin creases. The center is then located when the body segment is out of the vertical plane.

From Plagenhoef, S, *Patterns of human motion.* Copyright © 1971. All rights reserved. Adapted by permission of Allyn & Bacon.

- The lateral malleolus (LMA1) was the most lateral point on the lateral malleolus.
- The third metacarpal (MET3) was a point on the dorsal sulcus between the tip of the third metacarpal (knuckle) and the base of the third finger.
- The radiale (RAD) was the most proximal point on the lateral edge of the radius.
- The shoulder joint center (SJC) was the center of the humeral head.
- The sphyrion (SPHY), or tibial sphyrion, was the distal tip of the tibia.

Table 5.2   Location of the Planes of the Joint Centers Relative to Bony Landmarks

Dissected segments	Bony landmarks
Foot/shank	(1) The superior border of the calcaneum anterior to the Achilles tendon as palpated medially and laterally.
	(2) The upper border of the head of the talus.
	(3) The lower tip of the fibula.
Shank/thigh	(1) The midpoint of the posterior curvature of the medial condyle of the femur as palpated.
	(2) The same for the lateral condyle.
Thigh/trunk	(1) The anterior superior spine of the ilium.
	(2) The ischiopubic sulcus between the thigh and the scrotum.
	(3) The line from the ischium to above the femoral trochanter.
Trunk/neck	(1) The top of the sternoclavicular joint.
	(2) Between the C-7 and T-1.
Neck/head	(1) Decapitate the skull from the atlas.
Trunk/arm	(1) The palpable sulcus above the acromioclavicular joint.
	(2) The anterior axillary fold at the projection of the thoracic contour.
	(3) The posterior axillary fold at the projection of the thoracic contour.
Upper arm/forearm	(1) The lower border of the medial epicondyle of the humerus.
	(2) Eight millimeters above the radiale.
Forearm/hand	(1) The midpoint of the pisiform bone.
	(2) The distal wrist crease at the palmaris longus tendon of the midpoint on the volar surface of the navicular bone.
	(3) The palpable sulcus dorsally between the lunate and capitate bones.

According to W.T. Dempster (1955) and adapted from S. Plagenhoef et al. (1983).

- The stylion (STYL) was the distal tip of the styloid process of the radius.
- The tibiale (TIB) was the most proximal point on the medial margin of the head of the tibia.
- The wrist joint center (WJC) was the center of the transverse section of the capitate bone, at the level of the palpable groove between the lunate and capitate bones.

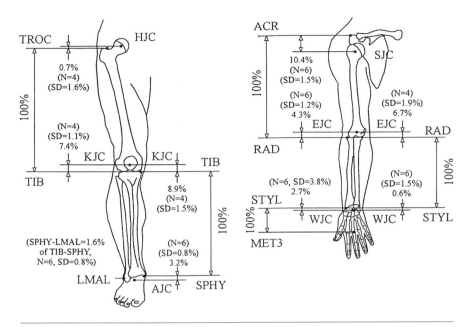

Figure 5.3   The percentage of longitudinal distances of joint centers from neighboring bony landmarks. Each percentage is relative to the closest 100% distance. N, number of subjects; SD, standard deviation.

Reprinted from *Journal of Biomechanics, 29,* Leva, P. de, Joint center longitudinal positions computed from a selected subset of Chandler's data, 1231-1233, 1996, with kind permission from Elsevier Science Ltd, The Boulevard, Langford Lane, Kidlington 0X5 1GB, UK.

The original data are from the studies of:

Chandler, R.F., Clauser, C.E., McConville, J.T., Reynolds, H.M., & Young, J.W. (1975). Investigation of inertial properties of the human body (AMRL TR 74-137). Wright-Patterson Air Force Base, Ohio (NTIS No. AD-A016485).

Clauser, C.E., McConville, J.T., & Young, J.W. (1969). Weight, volume and center of mass of segments of the human body (AMRL TR 69-70). Wright-Patterson Air Force Base, Ohio (NTIS No. AD-710 622).

Although the statistics presented in Figure 5.3 have not been validated by independent researchers at this time, the data seem to be more accurate than figures from other studies. The use of this method is recommended.

5.2 THE JOINTS OF THE FOOT

The two main functions of the foot are support and propulsion. Poor foot mechanics is one of the most common ailments in modern society, probably due to insufficient and inadequate physical activity and wearing improper shoes.

••• *ANATOMY REFRESHER* •••

The Foot

The foot includes 26 bones plus two sesamoids that form an arched construction (Figure 5.4). The bones are held together by ligaments. The bones are classified as the seven tarsal bones (the talus, calcaneus, navicular, cuboid, and three cuneiform bones), five metatarsal bones, and 14 phalanges (three for each toe except the first, which has two). Depending on the way of counting, there are 33 to 57 joints in the foot.

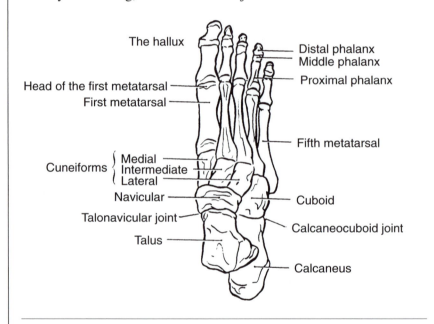

Figure 5.4   Foot skeleton, top view.

5.2.1 Metatarsophalangeal Joints: The Foot as a Two-Speed Construction

Representative papers: Bojsen-Møller, 1978; Sammarco, 1989.

Natural movement in the metatarsophalangeal (MTP) joints, specifically pushoff, is performed about either a transverse axis through the MTP joints of the first and second toe or about an oblique axis through the MTP joints of the second to the fifth toe (Figure 5.5). Toe dorsiflexion usually occurs with an accompanying plantar flexion about a secondary axis in the ankle joint complex. Hence, these two articular motions are coupled. The secondary axis is parallel to the primary axis and can also be either transverse or oblique.

The resistance arm for the transverse MTP joint axis is longer than the arm for the oblique axis (the ratio averages 6:5). Hence, if a certain moment of force is developed in the ankle joint, it will result in a larger displacement of the metatarsals. In other words, rotation about the transverse axis is similar to selecting a high-speed gear. Joint motion about the oblique axis matches a low-speed gear. When high velocity is desired, for example, in sprinting, people perform the pushoff about the transverse axis. When a large force is required, for example, in lifting heavy loads or in the first steps of a sprint, the pushoff is performed about the oblique axis.

The oblique axis through the five MTP joints is called the *metatarsal break,* and it is a generalization of the rotation axes of all five joints (Figure 5.6). The orientation of the metatarsal break with regard to the long axis of the foot—the

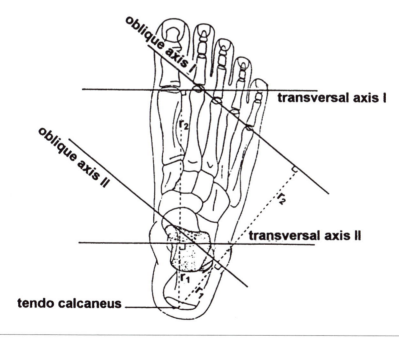

Figure 5.5 Coupled axes of foot rotation in the MTP joints and ankle joint complex. r_1 is the moment arm of the triceps surae relative to the ankle joint. The moment arms for the transverse and oblique axes are equal. r_2 is the resistance arm of the foot, i.e., the distance between the primary and the secondary axis. The distance is larger for the transverse axis.

Adapted from Bojsen-Møller, F. (1978). The human foot—a two speed construction. In E. Asmussen & K. Jørgensen (Eds.). *Biomechanics VI-A* (pp. 261-266). Baltimore: University Park Press.

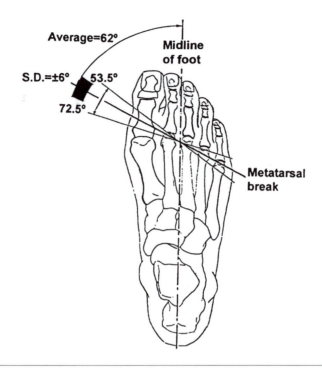

Figure 5.6 Metatarsal break (the line connecting the heads of the second and the fifth metatarsals).
Data from Isman, R.I. & Inman, V.T. (1968). Anthropometric studies of the human foot and ankle. Technical Report (May 1968). Biomechanics Laboratory, University of California, Berkeley.

line connecting the midpoint of the posterior aspect of the calcaneus with the gap between the second and the third toe—varies among individual persons from 50° to 70°.

Large MTP joint mobility is required not only for some specific skills, such as the crouched position of a baseball catcher, but also for ordinary walking. Normal range of motion in extension for all the five MTP joints is 90°. The joints, with the exception of those in the great toe, can flex up to 50°. Normal range of flexion for the great toe is 30°. Hyperextension of the toes raises the longitudinal arch of the foot (Figure 5.7). Wearing tight-fitting shoes over a period of years may provoke a deformity of the great toe, *hallux valgus,* which is a lateral angulation of the first toe more than 15°. In patients with hallux valgus the bony protuberances just proximal to the toes, the bunions, emerge. Bunions change the joint kinematics. In normal MTP joints, the articular surfaces glide with regard to each other throughout most of the joint motion—except at the limit of the extension, such as during squatting, where the joint is being compressed. In patients with hallux valgus and bunions, range of motion

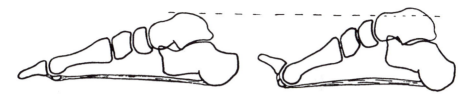

Figure 5.7 Raising the longitudinal arch of the foot due to hyperextension of the toes.

is limited and joint compression and distraction may occur even in the median range of the joint excursion.

5.2.2 Joints of the Midfoot

Representative papers: Elftman, 1960; Gershman, 1988.

The intertarsal joints have flat articulating surfaces. The joint mobility is limited because of the tight ligamentous structures surrounding the joints. Limited motion occurs also in the tarsometatarsal joints, between the distal tarsal and the metatarsal bones. The tarsometatarsal joints when considered together are called *Lisfranc's joint* (Figure 5.8). This joint is important for the shape of the arch of the foot. Total range of motion in the midfoot in flexion-extension is about 15°, mainly from plantar flexion.

The calcaneocuboid and talonavicular joints have two axes of rotation, one for flexion-extension and one for mediolateral rotation. These are two-degrees of freedom (DOF) joints. The joints share the same axes. Because of this relationship, the calcaneocuboid and talonavicular joints are often considered as

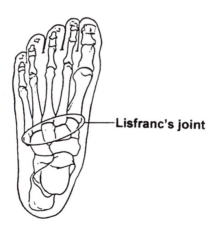

Figure 5.8 Lisfranc's joint.

one midtarsal joint, *Chopart's joint*. The joint includes the saddle-shaped articulation between the calcaneus and the cuboid and the articulation of the navicular with the head of the talus. The axis of mediolateral rotation in the joint is inclined about 15° and directed away from the midline of the foot approximately 9°. Flexion and extension in the joint occurs around two parallel axes—one through the talus and the second through the calcaneus (Figure 5.9). Direction of the axes varies substantially among subjects. On average, these two axes point anteromedially at an angle of 57° to the midline of the foot and are inclined from the floor 52°.

5.3 THE ANKLE JOINT COMPLEX

Representative papers: Inman, 1976; van den Bogert et al., 1994.

The ankle joint complex is a combination of two joints: the ankle, or talocrural (TC), joint between the tibia-fibula and the talus; and the subtalar, or talocalcaneal, joint between the talus and the calcaneus. The complex permits three rotations that are usually labeled as plantar flexion and dorsiflexion, inversion-eversion, and external-internal rotation, or abduction-adduction. Plantar flexion is the movement of the sole downward; inversion is the turning of the sole inward (weight on the outer edge of the foot), and abduction is the movement of the toes sideways in a medial direction.

5.3.1 The Talocrural Joint

The talocrural joint is a hinge, mortise joint with the tibia as the female surface and the talus as the male surface. The joint has one DOF permitting rotation

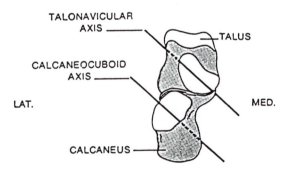

Figure 5.9   The flexion-extension axes of the calcaneocuboid and talonavicular joints, projection on the frontal plane. The foot is in the neutral position.
From Mann, R.A. & Inman, V.T. (1964). Phasic activity of intrinsic muscles of the foot. *Journal of Bone and Joint Surgery, 46A*, 469-476. Adapted by permission.

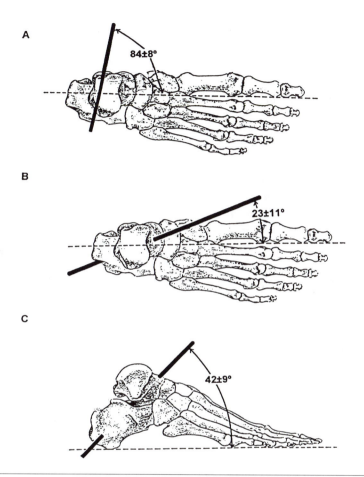

Figure 5.10   Rotation axes in the ankle joint complex. A. The angle between the talocrural joint axis and the midline of tibia in the frontal plane. B. The angle between the subtalar joint axis and the midline of the foot. C. The angle between the subtalar joint axis and the horizontal axis. The talocrural joint axis runs just distal to the tips of the malleoli (on average, 5 ± 3 mm on the medial side and 3 ± 2 mm on the lateral side [Inman, 1976]). The axis is obliquely oriented to the long axis of the leg as well as to the cardinal anatomic planes. Viewing from medial to lateral, the axis is directed posteriorly (on average 6° as reported by Inman, 1976; 6.8° ± 8.1° according to Bogert et al., 1994) and inclined downward (8° according to Inman, 1976; 7.0° ± 5.4° according to Bogert et al., 1994). (See also Figure 5.1A.) The subtalar joint axis intersects the joint and runs into the calcaneus posteriorly and the talus anteriorly. It deviates toward the medial side (16° according to Manter, 1941; 23° ± 11° with range 4°-47° according to Inman, 1976; 18.0° ± 1.6° according to Bogert et al., 1994) and inclines from the horizontal plane on average 42° according to Manter (1941), 42° ± 9° with range 20.5°-68.5° according to Inman (1976), or 37.4° ± 2.7° according to Bogert et al. (1994). The data are for an unloaded foot.

Data from Inman, V.T. (1976). *The joints of the ankle.* Baltimore: Williams & Wilkins.

along a predetermined path. The axis of rotation in this joint lies between the tips of the malleoli. The axis is offset in two anatomic planes, frontal and transverse (Figure 5.10). Because the axis is not perpendicular to the limb segments, the joint produces motion in all three anatomic planes. Hence, the movement in the talocrural joint can be described either as plantar flexion-dorsiflexion around an axis of rotation that is neither in the frontal plane nor the horizontal, or as a conjunct rotation involving adduction (valgus) and pronation associated with the plantar flexion and abduction (varus) and supination coupled with the dorsiflexion (Figure 5.11). In the latter case, plantar flexion and dorsiflexion are defined as rotations around a lateromedial horizontal axis.

For studying gross motor movement, a fixed axis of plantar flexion-dorsiflexion is usually estimated by palpation of the malleoli (see Figure 5.1A). When details of joint kinematics are not very important, the nominal ankle joint axis is estimated by using the tip of the lateral malleolus as a reference and directing the joint axis horizontally in the frontal plane (see Table 5.1). In this case, the offset of the axis from the frontal plane of about 6° (see Figure 5.10) is neglected.

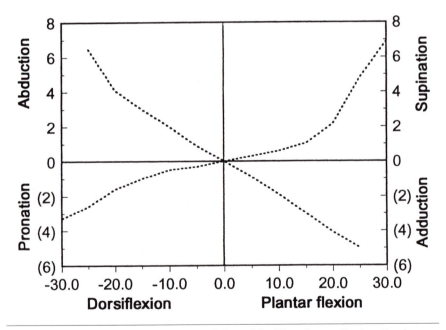

Figure 5.11 Contingent movement of the ankle. The plantar flexion (horizontal axis) correlates with both the pronation and adduction (vertical axis).
Data from Dempster, W.T. (1955). *Space requirements of the seated operator.* (Technical Report WADC-TR-55-159). Courtesy of the Wright-Patterson Air Force Base, OH: Aerospace Medical Research Laboratory.

5.3.2 The Subtalar Joint

The subtalar joint has a single oblique axis. Because the axis is slanting, internal and external leg rotation within a limited range results in inversion and eversion of the foot. Some authors model the subtalar joint as a miter hinge (Figure 5.12). The subtalar joint also behaves as a screw: during pronation the talus advances along the joint axis with regard to the (fixed) calcaneus. Thus, the joint was modeled as a spiral of Archimedes, a screw that is right-handed in the right foot and left-handed in the left foot (Figure 5.13). Forward advancement of the talus is, however, small (approximately 1.5 mm per 10° of rotation).

Because movement of the talus cannot be registered in living people using external markers, the kinematics of the subtalar joint is hard to study. Furthermore, movements in the talocrural and the subtalar joints are usually coupled.

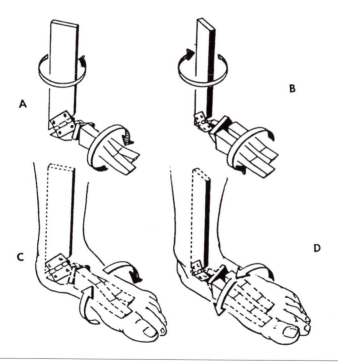

Figure 5.12   Modeling of subtalar joint as a miter (oblique) hinge. Rotating of one member causes rotation of the other member. (A and C) Outward rotation of the tibia causes elevation of the medial border of the foot. (B and C) Inward rotation of the tibia results in elevation of the lateral border of the foot.
From Mann, R.A. (1985). Biomechanics of the foot. In: *Atlas of Orthotics: Biomechanical Principles and Applications*. 2d ed. American Academy of Orthopaedic Surgeons. St. Louis: Mosby. Reprinted by permission.

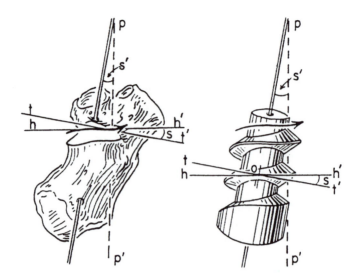

Figure 5.13   Modeling of the subtalar joint as a spiral of Archimedes.
Movements of the subtalar and transverse tarsal joints, Manter, J.T., *Anatomical Record,*
80, 397-410, Copyright © 1941. Adapted by permission of Wiley-Liss, Inc., a subsidiary
of John Wiley & Sons, Inc.

Separating the contributions of the individual joints is not easy. Accordingly,
total mobility of the foot with regard to the shank is studied—especially in
applied research—instead of studying motion in the single joints. Mobility in
the mediolateral direction is estimated through the Achilles tendon angle (Figure 5.14). The angle, if measured between the heel and the leg, can be 12° to
13° during sideward, cutting movements. The angle can reach 30° if measured
between the shoe and the leg.

 Because the axes in the talocrural and the subtalar joints are not perpendicular to each other, rotation of the foot with regard to the shank, as observed from
a fixed frame of reference, requires three Euler's angles for description. In this
sense, the ankle joint complex is analogous to joints with three DOF. The angular coordinates are not independent, however, and if the axes of coordinates
are chosen along the joint rotation axes only two angular coordinates are sufficient. Because of good agreement with the clinical terminology, the three angles
are more convenient for a pure kinematic description. However, when dynamic
analysis is desired and force moments are to be determined, the use of a ball-
and-socket model may lead to large inaccuracies. The three-DOF joint model
assumes one fixed center of rotation in the joint. When the moments are calculated with regard to this center, the force moments transmitted by the bone
surfaces and the ligaments are neglected. As a result, the three-DOF joint model
overestimates the muscle forces and underestimates the loading of the bones

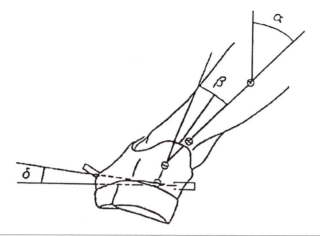

Figure 5.14   The angles used to characterize lateral mobility of the foot: the leg angle, α; the Achilles tendon angle, ψ; and the forefoot angle, δ.
From Stacoff, A., Steger, J., & Stussi, E. (1993). Lateral stability in sidewalk cutting movements. In *XIVth Congress of the International Society of Biomechanics, Abstracts* (pp. 1278-1279). Paris, July 4-8, 1993. Adapted by permission.

and ligaments. This is especially seen in the torsional moments about the longitudinal axis of the shank and bending moments in the mediolateral direction.

5.4 THE KNEE

The knee joint consists of the tibiofemoral and the patellofemoral joints.

5.4.1 The Tibiofemoral Joint

Representative paper: Hollister et al., 1993.

In the tibiofemoral joint, articular motion occurs in all three planes, with the greatest range of motion in flexion-extension (about 145°). Internal and external rotation, as well as abduction-adduction, depend on the level of knee flexion. Full extension of the knee precludes almost completely both internal-external rotation and abduction-adduction. In this position, the femoral condyles and tibial plateaus are interlocked. When the knee is flexed, the range of the other two joint motions increases. It reaches its maximum at 30° of flexion for abduction-adduction (a few degrees only) and at 90° for internal-external rotation (up to 30° in internal rotation and up to 45° in external rotation).

Knee extension is conventionally combined with external rotation of the tibia. This combination is called the *screw-home mechanism* (Figure 5.15). The conjunct axial rotation occurs mainly during the last 30° of extension. When the knee is flexed 30° to 150°, the axial rotation can be performed independently of the knee flexion-extension.

The direction of the coupled rotatory movement is not firmly fixed, however. It depends on the position of the tibia before knee extension. If the tibia is intentionally rotated externally in flexion, the tibia rotates internally rather than externally during the knee extension. This is opposite from what would be expected from the screw-home mechanism.

Despite the long history of research, the location of the knee joint axes is still a matter of discussion. At least four points of view exist in the literature.

1. The instantaneous axis of flexion-extension lies in the intersection of the frontal and the transverse planes and displaces during the motion. As a result, the centrode of the joint is located in the sagittal plane and can be found if all the precautions necessary for accurate measurements are taken. An example of this approach was given in Figure 4.27.

2. A fixed, rather than movable, axis exists (Figure 5.16). However, the axis is offset in both the frontal and transverse planes by 3.3° on average. From medial to lateral, the axis is directed posteriorly and distally. Slices from the

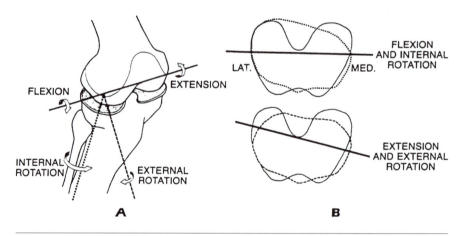

Figure 5.15 Screw-home mechanism of the knee motion. The displacement of the tibial plateau is shown. A. The tibial plateau (shaded area) and the screw-home mechanism. B. The position of the tibial plateau (broken outlines) with regard to the femoral condyles (solid outlines) in knee flexion (top) and extension (bottom). The tibia rotates internally during knee flexion and it rotates externally during knee extension.

From Helfet, A.J. (1974). Anatomy and mechanics of movement in the knee joint. In A.J. Helfet (Ed.), *Disorders of the knee* (pp. 1-17). Philadelphia: J.B. Lippincott Co. Adapted by permission.

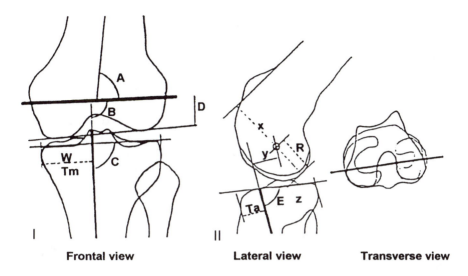

| Frontal view | Lateral view | Transverse view |

Figure 5.16   Axes of rotation of the knee joint. The axes are fixed but do not coincide with the cardinal anatomic axes. I. Representation in the plane parallel to the flexion-extension axis (pseudofrontal plane). The following angles are shown: A, the angle between the flexion-extension axis and the long axis of the femur (mean ± SD, 84° ± 2.4°); B, the angle between the flexion-extension axis and the axis of internal-external rotation (88° ± 1.2°); C, the angle between the axis of internal-external rotation and the tibial plateau (89° ± 2.1°). Linear dimensions are W, width of the tibia in the pseudofrontal plane; Tm, distance between the medial tibia and the axis of internal-external rotation (the ratio Tm/W = 47.5% ± 4.1%). II. Representation of the plane perpendicular to the flexion-extension axis (pseudosagittal plane). E is the angle between the axis of internal-external rotation and the tibial plateau (85° ± 3.5°). Also, the following linear dimensions are shown: z, width of the tibia in the pseudosagittal plane; Ta, distance of the axis of internal-external rotation to the anterior tibia (Ta/z = 31.8% ± 10.6%); y, perpendicular distance between the two axes of rotation (y/W = 31.6% ± 12.3%); x, distance between the anterior femoral shaft and the posteromedial femoral condyle; R, distance between the flexion-extension axis and the posteromedial femoral condyle. The percentage ratio locating the femoral axis in the pseudosagittal plane is R/x = 35.3% ± 5.1%.

From Hollister, A.M., Jatana, S., Singh, A.K., Sullivan, W.S., & Lupichuk, A. (1993). The axes of rotation of the knee. *Clinical Orthopaedics and Related Research, 290,* 259-268. Adapted by permission.

posterior femoral condyles made perpendicular to the flexion-extension axis have circular contours; hence, the anatomic data confirm the kinematic findings. In a sagittal plane projection, the path of the joint axis is an ellipse. For a perpendicular projection of a single axis, it should be a point. The ellipse, which is due to the projection error, is mistakenly interpreted as a displacement of the instantaneous axis. The (single) axis is in the femur and goes through the origins of the medial and lateral collateral ligaments. The axis proceeds through the notch, slightly superior to the crossing of the cruciate ligaments. If the knee joint were an ideal planar four-bar mechanism with inelastic ligaments and noncompressible surfaces, the axis should pass exactly through the intersection of the cruciate ligaments (see Figures 5.16 and 4.17). Because the joint axis does not coincide with the lateromedial cardinal axis, knee flexion and extension also produce angular motion with regard to all three cardinal planes. The screw-home mechanism is an example of such coupled movement. There also exists an independent rotation about a longitudinal axis.

The longitudinal rotation axis does not intersect the flexion-extension axis. The longitudinal axis is anchored with the tibia, anterior to the flexion-extension axis, not perpendicular to it, and moves about it. The axis passes near the anterior cruciate ligament insertion on the tibia and is directed posteromedially near the posterior cruciate ligament insertion on the femur. Knowledge of the positions of these axes and the amounts of rotation allows for a full description of the relative position of the tibia on the femur.

3. Knee flexion-extension takes place about an axis that changes its orientation during the motion. This is a helical motion rather than simple rotation as in a hinge joint (see Figure 4.28). The screw-home mechanism is evidence of that. The mechanism helps to stabilize the joint.

4. Mechanically, knee motion may take place about an infinite number of axes. The motion depends on acting forces and the subject's ability to reproduce a requested movement. For example, when asked to flex the knee, subjects may combine or not combine this movement with internal-external rotation at their will. This approach results in determining the knee motion envelopes (see Figure 5.17 compared with Figure 4.31) and multiple axes of joint rotation (see Figure 4.29).

The selection of the most suitable approach depends, at least in part, on the required accuracy. If very high accuracy is needed, the fourth method seems to be the best one. When a total joint prosthesis is installed in the human knee, the complex anatomic movement is frequently replaced by a pure rotation about a fixed axis. To decrease the negative consequences of replacing the complex natural movement by a simple rotation, the geometry of the prosthesis should be carefully selected. If requirements for high accuracy are not very strict, easy tests of the knee derangements that are based on simplifying assumptions are useful. The first approach is accepted: the knee joint axis is assumed to lie (approximately) in the frontal and transverse planes and displaces during knee motion. In people with normal knees, a radius from the instant center of rotation to the contact point is perpendicular to the tibial surface. It indicates that the surface motion is tangential gliding. In deranged knees, the instant center is displaced during some portion of the angular motion. The joint surfaces do

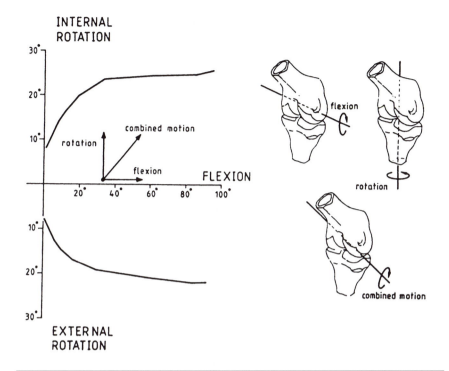

Figure 5.17 The envelope of passive knee joint motion. The external and internal joint torques of 3 Nm were applied. A combination of flexion (occurring about a horizontal axis) and axial rotation (occurring about a tibial longitudinal axis) results in an oblique helical axis.

Reprinted from *Journal of Biomechanics, 21,* Blankevoort, L., et al., The helical axes of passive knee joint motion, 1219-1226, 1990, with kind permission from Elsevier Science Ltd, The Boulevard, Langford Lane, Kidlington OX5 1GB, UK.

not glide smoothly throughout the range of motion but are either distracted or compressed. Such a joint is like a bent hinge that does not fit into the door jamb any longer. The articular movement results in either forcible ligament stretching or in abnormally high pressure imposed on the joint cartilage (see Figure 4.26).

5.4.2 The Patellofemoral Joint

Representative papers: Goodfellow et al., 1976; van Eijden et al., 1986.

The patellofemoral joint does not fit into any class of joints described in the preceding chapter (see Table 4.1). In the joint, both the patellar and femoral articular profiles are convex. Hence, the patellofemoral joint is formed by two male surfaces. The surface movement in the joint is gliding and rolling. For the femur position shown in Figure 5.18A, the gliding motion is clockwise. The mean amount of patellar gliding on the femoral condyles is approximately 6.5 mm per 10° of flexion between 0° (full knee extension) and 80° and 4.5 mm per 10° of flexion between 80° and 120°. In total, up to 7 cm of patellar gliding is possible. The direction of rolling is not constant. It is counterclockwise between 0° and 90° and clockwise between 90° and 120°. Because of the complex pattern of patellar movement, the slip ratio changes from negative to positive during knee flexion (Figure 5.18B). At the time of knee flexion, the patella also translates slightly mediolaterally and rotates externally.

During knee flexion, the contact point between the patella and femur moves in a clockwise direction along the femoral profile. Its position on the patellar surface changes in a more complex way. Between 0° and 90° of flexion, the contact point moves upward, and between 90° and 120° it moves in the opposite direction (see Figure 5.18 for details). Because of the shift of the contact point on the patella and also because of its rocking, the lever arms of the quadriceps tendon and the patellar ligament change during knee motion. Thus, the patella cannot be treated as a simple pulley that serves for redirecting the force exerted by the quadriceps muscle around the distal femur. Contrary to early conjecture, the tension in the patellar ligament is not equal to the force exerted by the quadriceps muscle. Hence, the patella acts not only as a spacer to increase the moment arm of quadriceps tendon but also as a variable gear.

5.5 THE HIP JOINT AND THE PELVIC GIRDLE

Representative papers: Neptune & Hull, 1995.

The male surface of the hip joint is represented by the femoral head and the female surface by the acetabulum, a concave surface at the lower aspect of the pelvis. The femoral head is almost spheroidal, with the greater deviation from circularity in the coronal plane. Because of its unique sphericity, the joint axes do not displace during joint motion. The joint is usually modeled as a ball-and-socket joint with a fixed center and with restraints only at the terminal ranges. The hip joint center (HJC) coincides with the geometric centers of the femoral head and acetabulum. With the pelvis fixed, the HJC is the center of the sphere described by the three-dimensional rotation of a point anchored to the femur. This assumption is approximately valid in the normal case. It is not valid for people with various hip diseases, such as developmental dysplasia. Surface movement in a normal hip joint is gliding. Flexion in the hip joint is spin (see Figure 4.14).

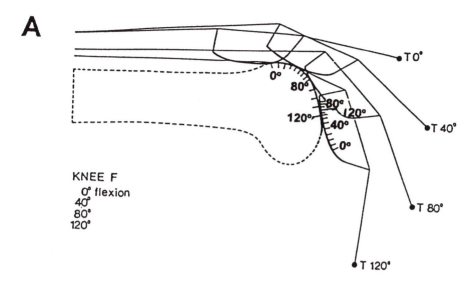

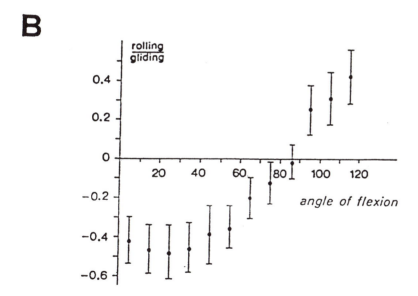

Figure 5.18 A. The position of the patellar ligament, patella, and quadriceps tendon and the location of the contact points as a function of the flexion-extension angle. Point T is the point of attachment of the patellar ligament to the tibial tuberosity. B. Slip ratio (rolling:gliding ratio) as a function of the flexion-extension angle.

From Eijeden, T.M.G.J. van, Kouwenhoven, E., Verburg, J., & Weijes, W.A. (1986). A mathematical model of the patello-femoral joint. *Journal of Biomechanics, 19,* 219-229. Reprinted by permission.

The orientation of the hip joint movement with regard to the cardinal anatomic planes is influenced by the geometry of the femur. Two angles are important: the neck-to-shaft angle (Figure 5.19) and the anteversion angle (Figure 5.20).

The ranges of motion in the hip joint in different directions are interrelated. The amplitude of abduction-adduction and axial rotation depends on the hip position in the sagittal plane. It is minimal when the hip is in full extension. In this position, the fibers of the iliofemoral, pubofemoral, and ischiofemoral ligaments are fully tensed and prohibit movement in the frontal and transverse planes.

The hip joint is not accessible; therefore, locating the HJC is not an easy task. Finding the precise location is especially important when the goal is to calculate forces acting on the femoral head because even a small inaccuracy in locating the HJC gives rise to large errors in estimating the muscle moment arms (see Figure 4.18). In addition to the stereoroentgenography method, which is the most accurate but cannot be recommended for everyday use because of the radiation hazard, several methods have been suggested to locate the HJC. The methods are divided into two groups: functional and morphological. The functional methods are based on kinematic analysis of hip joint movement, i.e., the HJC is assumed to be at the center of the sphere described by the markers located on the thigh. While a subject with the pelvis firmly fixed performs several circumductory movements of the leg, the trajectory of the markers is registered. Then, the pivot point of the joint rotation is determined (Cappozzo, 1984). The method cannot be used in patients with limited hip mobility, however.

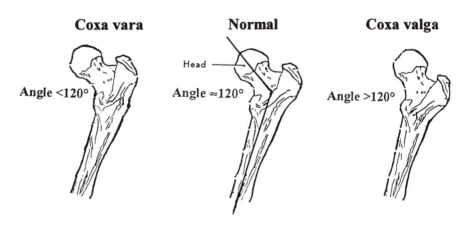

Coxa vara **Normal** **Coxa valga**

Head

Angle <120°   Angle ≈120°   Angle >120°

Figure 5.19 The neck-to-shaft angle. Normally, the angle is about 125°. An angle less than 125° brings about a condition known as *coxa vara*; an angle greater than 125° results in *coxa valga*.

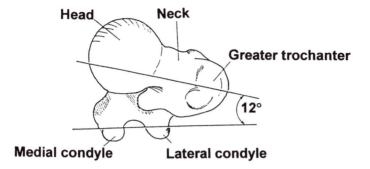

Figure 5.20 The anteversion angle (top view of the right femur). The angle is formed by the long axis of the femoral head and the transverse axis of the femoral condyle. In adults, the angle is about 12°. An angle of more than 12° results in the internal rotation of the leg during gait. When the angle is less than 12° (this condition is called retroversion), the leg usually rotates externally during walking.

Morphological methods rely on external landmarks. For the flexion-extension axis, the superior aspect of the greater trochanter is often used as a landmark. The HJC during a natural movement is located by placing a skin marker over the superior aspect of the greater trochanter. This is not, however, a generally accepted practice (see Table 5.1 and Figures 5.1C and 5.3 for comparison). Placing an external marker over the greater trochanter is not a reliable procedure because the soft tissues displace during the movement. A higher accuracy is achieved when the position of the greater trochanter during motion is calculated using palpable pelvis landmarks: the anterior superior iliac spine (ASIS); posterior superior iliac spine (PSIS); and pubic symphysis (PS), the midpoint between the pubic tubercles (Seidel et al., 1995). To find the HJC, the following distances are first determined: the pelvic width (ASIS to ASIS distance), pelvic height (perpendicular from PS to the inter-ASIS line), and pelvic depth (ASIS to PSIS distance). With regard to the relevant ASIS, the HJC lies 14% of pelvic width medial (mean error, 0.58 cm), 79% of pelvic height inferior (mean error, 0.35 cm), and 34% of pelvic depth posterior (mean error, 0.30 cm). There is no difference in the HJC estimation between men and women. Other methods of computing the HJC location from palpable bony landmarks have also been suggested (Bell et al., 1989, 1990; Davis et al., 1991).

The pelvis can move with regard to the thigh and the spine. The movement is called the *pelvic tilt*. Position of the pelvis is defined by using the anterior and posterior iliac spines as reference points. In a neutral position the angle between the line connecting the ASIS and PSIS and the horizontal line is approximately 10°—ASIS lower than PSIS (Day et al., 1984). Forward pelvic tilt is a combination of hip flexion and lumbosacral hyperextension and backward pelvic tilt is the opposite. A pelvic inclination in the frontal plane is called lateral pelvic tilt.

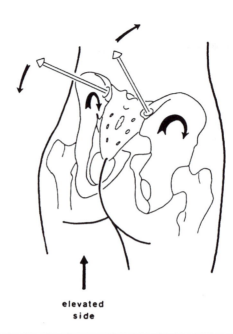

**elevated
side**

Figure 5.21   Hip movement during walking. The outline arrows are normals drawn from the dimples. Heavy arrows are movement directions. During unilateral elevation of the hip, the ipsilateral pelvic bone rotates downward and the contralateral bone rotates upward. The whole movement is referred to as *pelvic torsion*. The torsion axis runs parallel to the line connecting the dimples. The torsion angle is the angle between the dimple normals.
Reprinted from *Journal of Biomechanics, 20,* Drerup, B. & Hierholzer, E., Movement of the human pelvis and displacement of related anatomical landmarks on the body surface, 971-977, 1987, with kind permission from Elsevier Science Ltd, The Boulevard, Langford Lane, Kidlington 0X5 1GB, UK.

During natural movements, small motion at the sacroiliac joints is observed (Figure 5.21). This motion, which is difficult to register, is described by the torsion angle. To measure the angle, the so-called dimples of Venus, i.e., the tips of the scapulae and the lumbar fossulae, are located and the normals to them are restored (Figure 5.21, light arrows). Assuming that the torsion axis is parallel to the line connecting the dimples, the torsion angle is defined as the angle between the dimple normals about this axis. The angle amounts to 6.4°.

5.6 THE SPINE

Any change in spine posture involves the joined movement of several motion segments. People cannot move the individual motion segments independently.

••• ANATOMY REFRESHER •••

Spine

The skeletal system of the trunk comprises the pelvis, spine, and rib cage. The spine, or vertebral column, is divided into four regions: cervical; thoracic, which serves as an essential part of the rib cage; lumbar; and sacrum (Figure 5.22). The spine is a flexible rod with 7 mobile segments in the cervical region, 12 segments in the thoracic region, and 5 segments in the

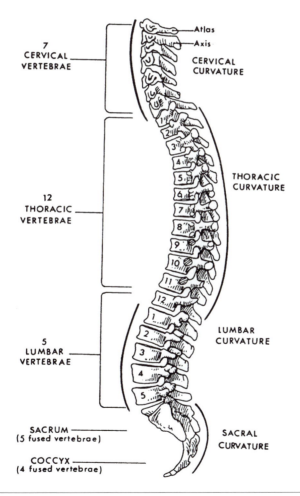

Figure 5.22 The regions of the spine.

See **ANATOMY,** *p. 312*

●●● **ANATOMY,** *continued from p. 311*

lumbar region. The fourth region of the spine, the sacral-coccyx region, includes nine fused vertebrae that together with the right and left ilia form the pelvis.

Two adjacent vertebrae and their interposed intervertebral disk form a *motion segment* of the spine (Figure 5.23). Each segment has six DOF. In the vertebral column, there are two types of joints: the *intervertebral joints* between the vertebrae and adjacent disks and the *facet joints* between the facets (articular processes) of the neighboring vertebrae. The intervertebral joints are synarthroses and the facet joints are flat synovial joints. Movement at the intervertebral and facet joints of the same motion segments is coupled.

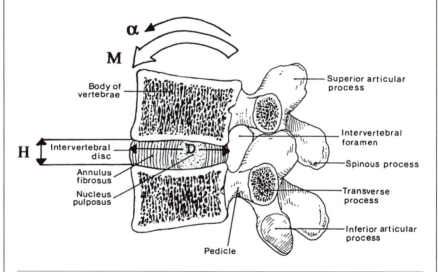

Figure 5.23   Lumbar spine motion segment sectioned in the midsagittal plane.

The intervertebral disks are flexible spacers between adjacent vertebrae. The disk consists of a central *nucleus pulposus*, which is a ball of hydrophilic jelly, and the outer *annulus fibrosus*, a series of laminae formed by collagen fibers. The disk height (thickness) increases from the cervical to the lumbar region from about 3 mm to 9 mm.

The neck joints are classified as the joints of the upper cervical spine, or occipital-atlanto-axial complex, and the joints of lower cervical spine.

The upper cervical vertebrae, C-1 and C-2, also known as the *atlas* and *axis*, differ markedly from the other vertebrae. The centrum of C-2, called the *dens*, forms a longitudinal axis about which the atlas and the head (occiput) rotate. The occipital-atlanto-axial complex is a unique structure formed by two synovial joints (they are the only synovial joints of the spine). Because the joints are synovial, muscle forces are always needed to position and steady the head. As a whole, the complex provides large skull mobility, especially in axial rotation, and at the same time protects the spinal cord and blood transport through the vessels to the brain. The lower cervical spine motion segments, C-3 through C-7, posses intervertebral disks.

The spinal column contains more than 100 articulations, both synarthroses and diarthroses.

Hence, movement of the motion segments is coupled. Kinematics of the spine deals either with the specific motion segments or with the entire region of the spine or both.

Each of the motion segments has six DOF: because the intervertebral disks can deform, the vertebrae, in addition to being able to rotate, can translate. The spine as a whole can, however, produce only three movements: flexion-extension, lateral flexion, and axial rotation. The spine movement results from concurrent rotation and translation of the vertebrae. The amount of motion available at various motion segments depends mainly on the size of the disks, while the orientation of the facet joint surfaces, which changes from region to region, defines the direction of the allowable movement.

5.6.1 Movement in Synarthroses

All the intervertebral joints, except the two joints of the occipital-atlanto-axial complex, are synarthroses. In synovial joints, which were discussed in the preceding text, the relative motion of adjacent segments occurs by rolling and sliding on two cartilage surfaces. The relative movement of the motion segments of the spine is allowed by an elastic connection provided by the intervertebral disks. Vertebral flexion and extension causes compression in one part of the disk and traction in another part. This is similar to relative movement of two pieces of solid material glued to a piece of rubber between them. Such a motion is constrained by the rubber layer. When a relative motion between two consecutive vertebrae is analyzed, the lower vertebra is customarily considered a fixed body and the upper vertebra is treated as a moving body.

The coordinate axes are taken along the inferior and posterior margins of the stationary, lower vertebra.

Because the disks can deform, the same resultant force and torque being applied to various vertebrae produces different movements depending on the disk stiffness and its dimensions, i.e., height and diameter. Hence, vertebral motion cannot be understood apart from the disk geometry, stiffness, and the acting forces.

••• STRENGTH OF MATERIALS REFRESHER •••

Torsion and Bending of Circular Shafts (Intervertebral Disks)

Stress is the force per unit cross-sectional area (in newtons per square meter). *Shear stress* is the stress in the plane of the cross section (also in newtons per square meter). The *shear modulus*, G, is the proportionality constant between the shear stress and shear angle (in newtons per square meter). The shear modulus represents the property of the material. In what follows, the disk is modeled as a straight cylinder made of an elastic material.

Torsion (axial rotation). Consider a straight shaft (disk) having a circular cross section. The disk is subjected to equal and opposite twisting couples at the two adjacent vertebrae. The *angle of twist,* or *axial rotation,* α_a, is the angle by which the top vertebra turns with respect to the bottom vertebra (in radians). For a disk of a height H, radius r, and cross-sectional area A, on which a torque M_a is acting, the angle of twist is:

$$\alpha_a = \frac{M_a H}{G I_p} \tag{5.1}$$

where I_p is the *polar moment of inertia,* and

$$I_p = \int r^2 dA = \frac{\pi r^4}{2} = \frac{\pi D^4}{32} \tag{5.2}$$

where the integration covers the entire area.

Combining equations 5.1 and 5.2, we get

$$\alpha_a = \frac{32 M_a H}{\pi G D^4} \tag{5.3}$$

Bending. When the line of force does not coincide with the symmetry axis of the column, bending stress occurs. The equation relating angular displacement of the disk, α_b, with the applied bending moment, M_b, is

$$\alpha_b = \frac{M_b H}{EI_d} = \frac{64 M_b H}{\pi E D^4} \tag{5.4}$$

where E is the *modulus of elasticity* or *Young's modulus* (in newtons per square meter) and I_d is a *diametral moment of inertia*, $I_d = 0.5 I_p$.

Consider the example in Figure 5.24. The column is loaded with two forces, F and –F, acting in opposite directions along the same line of action. The line of action does not pass through the center of gravity (G), however. In this context, the center of gravity is the location of the resultant force acting in compression and proportional to the elements of area. The col-

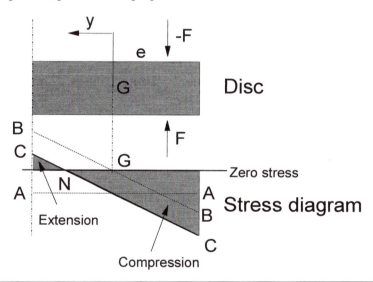

Figure 5.24 A model of intervertebral disk bending: the stress distribution is in a vertical cross section of a column loaded by the off-center forces, F and –F. Compressive stress in isolation is illustrated by the line AA. The bending stress alone has the center of gravity G for a neutral point and is shown as the line BB. The total stress is $S = \dfrac{F}{A} + \dfrac{Fey}{I}$ and is illustrated by the line CC passing through zero at a neutral point N. To the right of the neutral point the disk experiences compression stress, to the left tension stress.

See **STRENGTH**, *p. 316*

● ● ● **STRENGTH,** *continued from p. 315*

umn is under a joint action of compression and bending and is stressed correspondingly.

The compressive stress is F/A, where A is the area of the horizontal section. The bending stress is My/I, where M is the bending moment; y is the distance from the "neutral line" of the column, the line that experiences neither compression nor tension during bending; and I is the area moment of inertia, $I = \int y^2 dA$. The bending moment equals the product of force F times e, the eccentricity (i.e., the distance from the line of force to the center of the column). This distance is crucial in provoking bending stress.

From equations 5.1 and 5.4 it follows that rotatory movement of the vertebra varies inversely with the disk stiffness and its diameter in the fourth power and it varies directly with the disk height.

Because the disk deforms under load, the pattern of the vertebral movement depends not only on the torque applied to the vertebra but also on the resultant force acting on the disk. If a pure torque were applied (the resultant force is zero), the rotation would take place around an axis of symmetry. When compression and torsion are applied simultaneously, as is usually the case, the axis of rotation is not necessarily located on the axis of symmetry. To locate the axis of rotation analytically, i.e., solving equations, both the compression load and the torque should be known. The same spine movement, bending for example, is performed differently in loaded and unloaded conditions. Thus, the location of the instantaneous axis of rotation cannot be fixed precisely. It depends very much on the type of loading. However, if the loading is almost constant, for example, during slow neck bending, the location of the instantaneous axis of rotation can be estimated.

The bending axis of a motion segment is located in the lower vertebra. The vertebra moves along an arc during bending (Figure 5.25). The radius of curvature characterizes the steepness of the arc.

The relative movement of a vertebra with regard to the vertebra immediately below it can also be viewed as a combination of rotation and translation. As was mentioned in Section **3.1.1.1.9**, an infinite number of combinations of rotation and translation can describe the same movement. Hence, a pure kinematic approach is not productive. However, a useful insight can be gained if it is assumed that a vertebra rotates around a pole which (1) is located on the inferior end plate of the moving vertebra and (2) moves parallel to the inferior margin of the reference vertebra, axis X of the coordinate system (Figure 5.26).

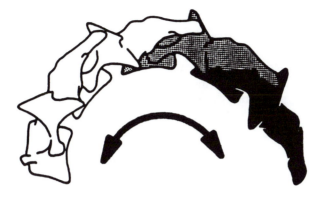

Figure 5.25 The arcs described by a C-7 vertebra during flexion-extension.

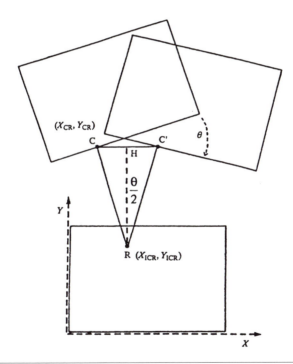

Figure 5.26 An illustration of a vertebra that has rotated and translated from an initial position to a final position about an ICR located at R. The center of rotation C translates to C'. H is the midpoint of interval CC'. θ is the angle of rotation.

This material has been reproduced from *Proceedings of the Institution of Mechanical Engineers, Part H, Vols. 209 (177-183),* Bogduk, N., Amevo, B., & Pearcy, M., A biological basis for instantaneous centres of rotation of the vertebral column, 1995, by permission of the Council of the Institution of Mechanical Engineers.

In the framework of this model, the vertebral motion is viewed as translation along the axis X and rotation around the pole, the center of rotation. Because a rotation about the ICR is equal to the rotation about any parallel axis, the angle subtended at the ICR by the arc of motion of C is equal to the angle of rotation (θ) undergone by the whole vertebra (see Figure 5.26). Because triangle CC'R is an isosceles triangle, HR = CC'/2tan (θ/2), where HR and CC' are the height and base of the triangle. If the location of the ICR is experimentally determined, location of C can be found as

$$X_{CR} = X_{ICR} - CC'/2$$
$$Y_{CR} = Y_{ICR} - \frac{CC'}{2\tan(\theta/2)} \tag{5.5}$$

The whole procedure is as follows:

1. the ICR is located using, for example, the Reuleaux method;
2. angle θ is measured as the angle formed by any of the vertebral sides in the two consecutive positions;
3. a line is constructed from the ICR parallel to axis Y;
4. two rays are constructed at an angle of θ/2 to this line;
5. the points of intersection of these rays with the inferior borders of the vertebra are selected as the centers of rotation.

From the preceding analysis it follows that the position of the ICR is determined by (a) location of the center of rotation C, (b) translation of the vertebra in parallel with the axis X, and (c) rotation of the vertebra. In some patients, as compared with healthy people, the ICR is displaced. The displacement can be explained in terms of the three mentioned factors. For instance, elevation of the ICR can occur only when the translation is decreased, the rotation is increased, or both. The three mentioned mechanisms, in turn, can be explained by a combination of biomechanical changes, such as increased muscle pull. Unfortunately, at this time such an examination can only be performed in an informal, qualitative way. The results are presented in Table 5.3. It is believed that, in spite of a somewhat superficial analysis, these results are still useful for clinicians.

5.6.2 Lumbar and Thoracic Spine

Single motion segments do not move separately. Any spine movement combines displacement of several motion segments.

Table 5.3 Biomechanical Factors Influencing Vertebral Kinematics[1]

Change	Change in biomechanical property							
	Disk stiffness		Muscle force		Shear stiffness	Ligament tension		Facets
	Anterior	Posterior	Anterior	Posterior	Annulus fibrosus	Anterior	Posterior	Facet impactio
CR↑	←	→	←	→				Impaired
CR↓	→	←	→	←				
θ↑	→	→	→			→	→	
θ↓	←	←	←			←	←	
CC'↑					→			Impaired
CC'↓					←			

[1]Symbol ↑ denotes increase in translation or rotation or a forward shift of the center of rotation. Symbol ↓ denotes a reduction in translation or rotation or a posterior shift of C.

Adapted by permission of the Council of the Institution of Mechanical Engineers from Bogduk, N., Amevo, B., & Pearcy, M. (1995). A biological basis for instantaneous centres of rotation of the vertebral column. *Proceedings of the Institution of Mechanical Engineers. Part H. Journal of Engineering in Medicine, v. 209 (3), 177-183.*

5.6.2.1 Individual Joints

The vertebrae, when flexing and extending, move with regard to each other along circular arcs. The curvature of the arcs, as compared with that in the cervical spine, is small, especially for flexion-extension. Movement of the motion segments takes place by compressing and stretching the intervertebral disks (Figure 5.27).

In principle, the kinematics of the thoracic vertebrae are analogous to the movement of the lower spine motion segments. However, movement of the thoracic motion segments is relatively limited. The thoracic cage, thin intervertebral disks, configuration of the articular facets, and apposition of the spinous processes (in extension) all contribute to the restricted mobility of the thoracic spine region in flexion-extension and lateral bending. The lumbar intervertebral disks are relatively thick and allow large flexibility in these movements. Axial rotation in the lumbar region is, however, restricted by the articular facets.

Moving in the caudal direction along the spine from T1-2 to L5-S1, the range of motion of the individual motion segments increases in flexion-extension; decreases, on average, in axial rotation; and is almost constant in lateral flexion (except for the L5-S1 joint where the flexibility is low) (Figure 5.28).

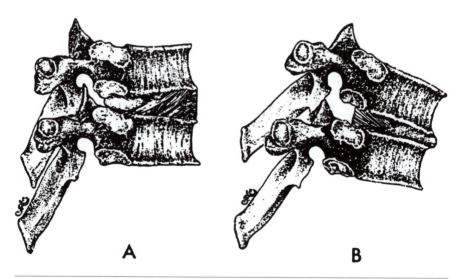

A   **B**

Figure 5.27 Spine extension (A) and flexion (B) cause compression and extension in different parts of the intervertebral disks. Hyperextension of the spinal column is limited by the shape of the articular facets.
From MacConaill, M.A. & Basmajian, J.V. (1969). *Muscles and movements. A basis for human kinesiology.* Baltimore: Williams & Wilkins.

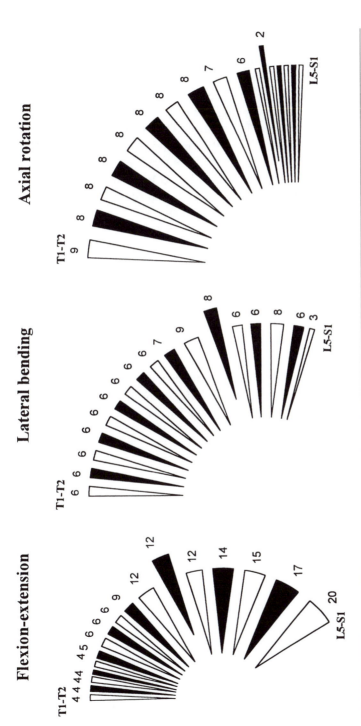

Figure 5.28 Range of motion of the spinal segments, angular degrees. The pulled out segments are for the T12-L1. In axial rotation, the range at motion at the lumbar spine otion units is 2 degrees.
Data from White, A.A. & Panjabi, M.M. (1975). Spinal kinematics. In M. Goldstein (Ed.), *The research status of spinal manipulative therapy* (pp. 93-102). Bethesda, Maryland: U.S. Department of Health, Education, and Welfare.

Lateral movement of the individual motion segments is conjunct with their axial rotation. Because of the spine curvature, the motion segments in a neutral position are not oriented perfectly vertically; they are inclined. When lateral flexion is performed, the movement is performed in the frontal plane of the spine, which does not coincide with the frontal plane of the individual vertebrae. As a result, during lateral flexion, the motion segment also rotates about its frontal and longitudinal axes. During lateral bending, the spinous processes of the thoracic region—especially of its upper portion—rotate to the convex side. In the middle and lower portions of the thoracic spine the coupling is rather weak. The spinous processes of the lumbar region turn toward the spine convexity. This pattern of movement, which is opposite of that in the cervical spine, is only partly explained by spine geometry, i.e., the oblique orientation of the longitudinal axes of the vertebrae with regard to the longitudinal axis of the trunk. Kinetic factors contribute to that. The ligaments at the rear of the spine (ligamenta flava and interspinous ligaments) provide resistance to the lateral bending. As a result, the spinous processes do not move as far away from the vertical as do the bodies of the vertebrae. Coupled translations and rotations of the lumbar spine segments resulting from imparted forces and moments are summarized in Figure 5.29. In healthy subjects, the coupled movements of the lumbar motion segments are inconsistent.

In healthy people, the instant axis rotation (IAR) lies within the adjacent disks but does not usually pass through the disk centers (Figure 5.30, upper panel). For instance, in the L3-4 disk the IAR is anterior to the facet joints and in the region of the posterior part of the nucleus. The IAR tends to move toward the side of rotation (Cossette et al., 1971). When measured from the posterior skin surface along the anteroposterior diameter, the L5-S1 joint center of rotation is estimated to be at 34% of the diameter (McNeill et al., 1980). The estimation is valid for young, nonobese subjects. In some patients with pathologic disorders, the IAR is located quite far from the disk (Figure 5.30, bottom panel, A and B).

5.6.2.2 The Spine as a Whole: The Lumbar-Pelvic Rhythm

Trunk flexion involves, as a rule, spine flexion and pelvic tilt. Commonly, the first 30° to 40° of flexion take place in the lower motion segments of the spine and then pelvic tilting occurs. In opposite movement, from full flexion to upright posture, the movement sequence is reversed: first the pelvis rotates backward and then the spine extends. This coupled movement of the spine and pelvis in the sagittal plane is called the *lumbar-pelvic rhythm* (Figure 5.31). The contribution of the thoracic spine to trunk bending is minimal.

The pelvis and spine exhibit different movement patterns. In bending, for instance, the pelvis rotates around an axis through the hip joints, while the

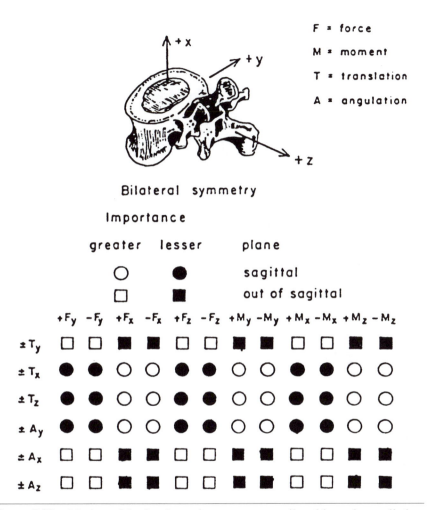

Figure 5.29 Motion of the lumbar spine segments attributable to the applied forces and force moments. Note the coupling movements and bilateral symmetry.

Reprinted from *Journal of Biomechanics, 12,* Frymoyer, J.W., et al., The mechanical and kinematic analysis of the lumbar spine in vivo, 165-172, 1979, with kind permission from Elsevier Science Ltd, The Boulevard, Langford Lane, Kidlington 0X5 1GB, UK.

spine is a flexible rod with many axes of rotation. Nevertheless, in modeling lifting activities, the spine is commonly considered one inflexible link connected with a rotating pelvis. (A three-link model of the spine, as compared with the one-link model, provides less information about lumbar motion. [Kippers, Parker, 1989]). The first 30° of trunk flexion is principally achieved by rotation in the lumbar spine. After the first 30°, pelvis tilt contributes substantially to trunk bending (Figure 5.32).

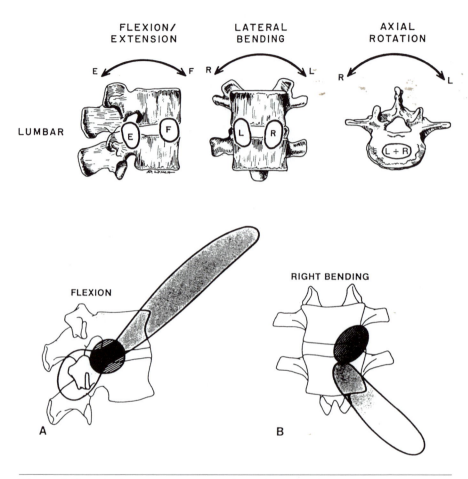

Figure 5.30 The instantaneous axes of rotation in the lumbar spine. Upper panel. In the normal spine, E, F, L, and R designate, correspondingly, extension, flexion, movement to the left and to the right from a neutral position. Lower panel. Location of the IAR in normal spine (dark areas) versus the spine with the degenerated disks, gray areas.

From White, A.A. & Panjabi, M.M. (1975). Spinal kinematics. In M. Goldstein (Ed.), *The research status of spinal manipulative therapy* (pp. 93-102). Bethesda, Maryland: U.S. Department of Health, Education, and Welfare.

Two types of kinematic models, linear and nonlinear, have been developed for predicting L-5 and S-1 orientation at a given body posture. The independent variables of the models are trunk inclination and knee flexion. According to the linear model, beyond approximately 30° the pelvis rotates forward at the rate of 2° per each 3° of trunk flexion; the spine contributes the other 1° (Figure 5.33). To the opposite effect, when the thighs are flexed more than 10° to 15°, the pelvis rotates backward at a rate of 1° per 3° of thigh flexion. In com-

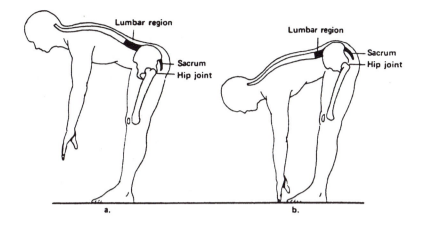

Figure 5.31 In the lumbar-pelvic rhythm (a) the lumbar spine flexes and (b) the pelvis tilts anteriorly.
From Norkin, C.C. & Levangie, P.K. (1992). *Joint structure and function. A comprehensive analysis* (2nd ed.) Philadelphia: F.A. Davis Company. Reprinted by permission.

plex movement involving both trunk bending and knee flexion, these two effects are subtracted from each other.

The nonlinear model gives more detailed estimations. In the model, sacral rotation is expressed in degrees and the lumbosacral angle as a percentage of the maximal rotation in this joint (to accommodate for individuals of different flexibility) (Figure 5.34).

Lateral trunk flexion involves the motion of the thoracic and lumbar segments to various degrees. Axial rotation is produced at the thoracic and lumbosacral levels. In stooped postures, the range of axial rotation decreases. This has been explained by the increased compression of the joint surfaces in this position.

The range of spinal motion in live subjects is smaller than the full range shown by cadaveric motion segments. For instance, at a fully flexed toe-touching posture, each lumbar motion segment reaches approximately only 75% of the full range allowed by the ligaments (Adams, Dolan, 1991). Evidently, mobility is limited by the back muscles. During fast lunging movements, the range of motion often exceeds the static limits of spine mobility. Because during spinal flexion the distance between skin markers located above the vertebrae prominences, for instance T-1 and L-5, increases, it has been suggested that tape measurements be used for estimating the amount of spine flexion ("skin-stretching techniques"). These methods have been shown to be unreliable, however. Spine mobility is often overestimated by the failure to account for hip movement and the movement in the atlanto-occipital joint. In particular, a popular reach test, during which a fingertip-to-ground distance is

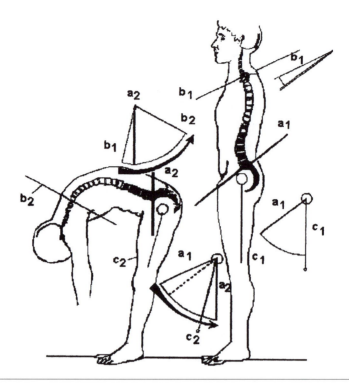

Figure 5.32   Contribution of the spinal column and hip joint flexion to the performance in the ground-reach test. The top sector on the left represents the contribution of the spine flexion into the performance. The bottom sector stands for the contribution of the hip joints. Symbols: a, intervertebral disk level between L-5 and S-1; b, intervertebral disk level between C-7 and T-1; c, femur support axis; subscript 1 refers to the initial upright, and subscript 2 to the final bent posture.

From Erdman, H. (1979) *Die körperlichen Untersuchung. Die Wirbelsäule in Forschung und Praxis* (vol. 83). Stuttgart: Hippokrates. Adapted by permission.

recorded, is not a very accurate measure of the spine's ability to bend; the test outcome depends on the hip movement and the arm length (see Figure 5.32).

5.6.3 The Cervical Region: Head and Neck Movement

Representative paper: Woltring, et al., 1994.

The head-neck motion system has been the focus of many studies. Under-standing head-neck movement is important for both motor control and injury prevention. A proper head position is paramount for gaze control and body orientation in space. The response of the head and neck structures to dynamic

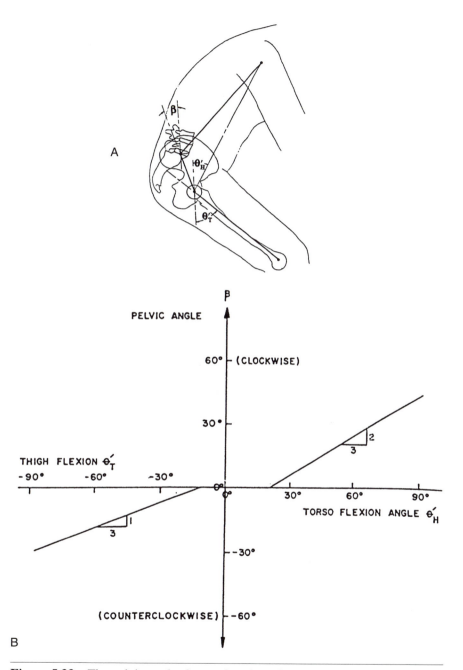

Figure 5.33 The pelvic angle, β, as a function of torso flexion, θ_H', and thigh flexion, θ_T'.

A is from *Occupational Biomechanics,* Chaffin, D.B. & Andersson, G.B.J., Copyright © 1984. Adapted by permission of John Wiley & Sons, Inc. B is reprinted courtesy of Wright-Patterson Air Force Base, OH.

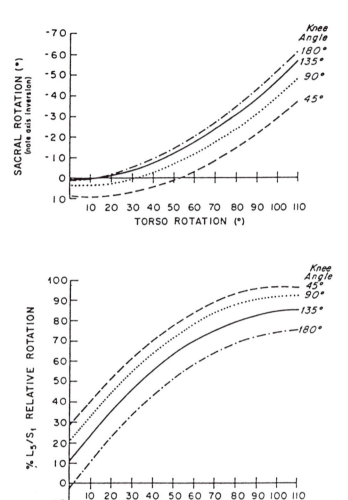

Figure 5.34 Sacral rotation (top panel) and percentage of maximal L5-S1 range of motion as functions of the trunk inclination and the included knee angle.

From Anderson, C.K., Chaffin, D.B., Herrin, G.D., & Matthews, L.S. (1985). A biomechanical model of the lumbosacral joint during lifting activities. *Journal of Biomechanics, 18,* 571-584. Reprinted by permission.

loading is important for understanding crash accidents, for instance, whiplash trauma.

The cervical spine allows the head a large range of motion. Kinematically, the head-neck system consists of eight joints of complex geometry. The head-neck motion has been analyzed as the movement of the individual joints and as an entity, i.e., the head movement with regard to the trunk. We begin from the analysis of the individual joints.

5.6.3.1 Individual Motion Segments

To analyze the individual motion segments, systems of coordinates should be defined. In a standing position, only some motion segments have their disks horizontal. Individual disks can be inclined up to 45°. Hence, the axial rotation of the head or the trunk does not correspond to the axial rotation of the motion segments. An uncomplicated axial rotation of a segment may induce rather complex head-trunk movement. Inversely, the simple head-trunk rotation may result from a more or less perplexing movement of the spinal motion segments. Hence, one should discern the movement described in the reference frames imbedded in an adjacent vertebra or anchored to the trunk as a whole. The following discussion of the kinematics of specific joints makes use of the local reference systems defined for each joint individually. In the neutral position, the disk is assumed to be horizontal.

5.6.3.1.1 The Upper Cervical Spine Joints

Representative paper: Werne, 1957.

The head can rotate in the occipital-atlanto-axial complex around three anatomic axes. In these movements, all three constituent parts of the complex, the occiput, C-1, and C-2, are involved. Flexion and extension take place both in the occipital-atlantal joint and in the atlanto-axial joint. On average, the range of motion is 13° in the occipital-atlantal joint and 10° in the atlanto-axial joint. Lateral bending takes place entirely at the occipital-atlantal joint (the range of motion is only about 8°) and axial rotation occurs exclusively at the atlanto-axial joint (the range is on average 47°). The absence of axial rotation in the occipital-atlantal joint has been used during radiological evaluation of the upper cervical spine joints. When a truly lateral view of C-1 is desired, a correct picture can be obtained by placing the film at the lateral side of the skull, disregarding the position of the neck and shoulders.

The axial rotation is a screw-like motion: the rotation of C-1 is coupled with vertical translation. This "telescoping" mechanism is called a *double-threaded screw*. The topmost vertebra of the neck, C-1, assumes its highest point in the neutral position and its lowest point at the extremes of axial rotation to the right or the left. The instantaneous axes of rotation for occipito-atlantal motion are offset with regard to each other; the sagittal axis is about 2 to 3 cm above the transverse axis. In the C1-2 joint, the axis of axial rotation goes through the central portion of C-2, the dens.

In summary, each of the joints forming the occipital-atlanto-axial complex has two rotational DOF, and the complex as a whole has three DOF. The atlas can move independently, but all the lower vertebrae, starting from C-2, are attached to each other and can only move together. Motion of the occipital-atlanto-axial complex can occur in isolation; however, this rarely takes place during natural movements. The cervical spine functions as a unit.

5.6.3.1.2 The Lower Cervical Spine Joints
Representative paper: Lysell, 1969.

The motion segments of the lower cervical spine allow for flexion-extension, lateral flexion, and axial rotation between all of the vertebrae. Representative values of rotation of the lower cervical spine are given in Figure 5.35. Despite the small height of the cervical intervertebral disks, the mobility, as compared with other spine regions, is relatively large.

In a normal cervical spine, the axes of rotation lie in the vertebral bodies below the moving vertebra (Figure 5.36). The location allows for a sliding movement of the mating surfaces of the facet joints. In cases of pathologic abnormalities, the center of rotation may be displaced, which, in turn, may lead to the compression or distraction of the facet joint surfaces.

Movement in individual lower cervical joints is coupled. Because the facets guide the motion, the flexion-extension is coupled with sagittal (horizontal and vertical) translation of about 2.5 to 4.0 mm, and lateral flexion with axial rotation (when flexing to the right the spinous processes turn to the left). The spinous processes rotate to the convexity of the arc formed by the neck. The movement in individual joints is also interdependent.

5.6.3.2 The Entire Head-and-Neck Movement

Kinematically, the head-neck system is highly redundant. For example, in a planar movement, the head has three DOF relative to the trunk. The head-neck system as a whole has eight planar DOF, according to the number of joints. The maneuverability of the cervical spine is then $8 - 3 = 5$. One can reach a conclusion that the end effector, or head, could assume the same position in an infinite number of ways. It does not happen, however. Only the occipital-atlanto-axial complex can move independently. Movement of the remaining cervical joints is closely coupled. The coupling takes place because the same muscles serve several joints, and the forces are transmitted through the compliant disks. Hence, the vertebrae C-3 through C-7 move as a unit. The movement, however, occurs around a distributed set of axes of rotation (Figure 5.37). Various approaches have been used to analyze the head-neck movement.

The simplest and still practical model of the planar neck-head movement assumes two fixed axes of rotation. The first axis is at the level of C7-T1 (some authors place it at the level C6-7, where range of motion is larger than

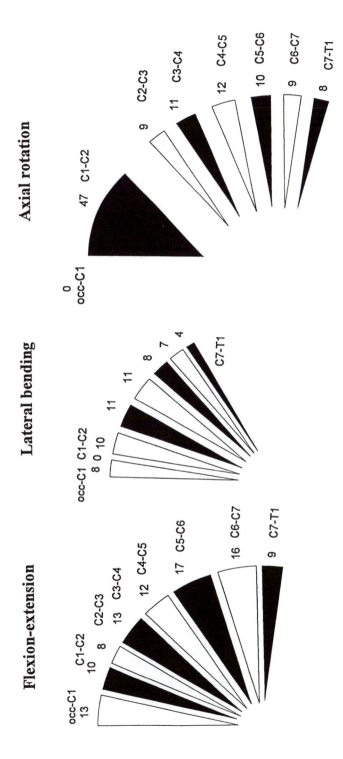

Figure 5.35 Range of motion of the cervical spine segments.
Data from White, A.A. & Panjabi, M.M. (1978). The basic kinematics of the human spine: A review of past and current knowledge. *Spine, 3,* 12-20.

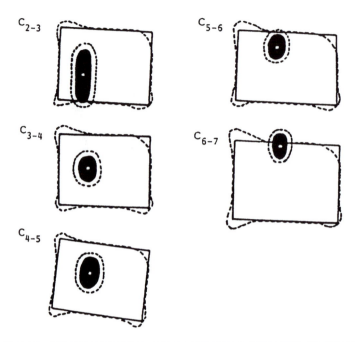

Figure 5.36 Instantaneous axes of rotation (flexion) of the cervical motion segments in normal subjects. The white dots are the mean locations of the IAR. The solid areas represent the intersubject variability and correspond to plus or minus one standard deviation from the mean. The dotted lines indicate the superimposed range of technical errors. The data are normalized with regard to the disk dimensions.

Reprinted from *Clinical Biomechanics, 6,* Amevo, B., et al., Instantaneous axes of rotation of the typical cervical motion segments. 1. An empirical study of technical errors, 31-37, 1991, with kind permission from Elsevier Science Ltd, The Boulevard, Langford Lane, Kidlington 0X5 1GB, UK.

that at C7-T1). The second axis is selected either at the occipital-atlantal joint (occ-C1) or, combining movement in the occ-C1 with the motion in the atlanto-axial joint, at the C-1 level. In the framework of this dual-joint model, the neck is considered a solid body and the whole trunk-neck-head system is regarded as a three-link planar kinematic chain. This chain was analyzed in detail in Chapter 3 (see Section **3.1.1.2** and Figure 3.13).

According to Kennedy's theorem, the ICR for the head in the reference frame anchored to the trunk is located on the line of centers of the two joints. Exact positioning of the ICR depends on the ratio of the two angular velocities (see equation 3.27). When the head and the neck are rotating in the same direction, the ICR is located between the two joints. However, when the ratio $\dot{\theta}_1/\dot{\theta}_2$ is negative (the neck and the head are moving in the opposite directions), the

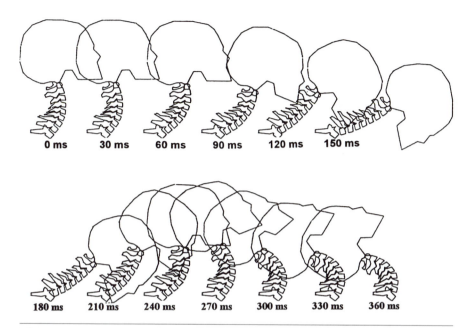

Figure 5.37 An example of a plane head-neck motion. Whiplash movement due to an acceleration pulse applied posteriorly at the base.
Reprinted from *Journal of Biomechanics, 17,* Merrill, T., Goldsmith, W., and Deng, Y.C., Three-dimensional response of a lumped-parameter head-neck model due to impact and impulsive loading, 2:81-95, 1984, with kind permission of Elsevier Science Ltd, The Boulevard, Langford Lane, Kidlington 0X5 1GB, UK.

ICR is located outside of the neck, but still on the line of centers. This situation occurs when people flex and extend the lower cervical spine and extend and flex the upper spine.

The dual-hinge model can be generalized to a multiple chain or even to a continuous model (Figure 5.38). This model includes either several vertebrae or infinitely many segments with vanishing length but totaling a fixed neck length. When the model assumes equal contribution of all the vertebrae to the total motion, the ICR is found to be very close to the midpoint of the bending arc. The model does not differentiate, however, between equal participation of all the joints and dominant involvement of some joints only. Also, the model cannot describe positions with opposite head and neck movement, such as neck extension with concurrent head flexion. To get a more detailed picture, *basic arcs* are used. In various neck movements, several adjacent vertebrae form an arc with a curvature shared by all the vertebrae in the group (basic arc). Any given position of the cervical spine can be described by two or, maximally, three basic arcs (Figure 5.39).

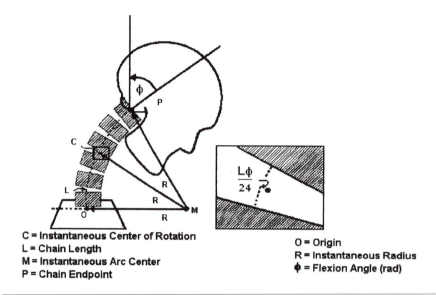

C = Instantaneous Center of Rotation
L = Chain Length
M = Instantaneous Arc Center
P = Chain Endpoint

O = Origin
R = Instantaneous Radius
φ = Flexion Angle (rad)

Figure 5.38 The neck modeled as an arch: the instantaneous arc center, M, and the instantaneous center of rotation of the head with regard to the trunk, C, are different points.

Adapted from *Journal of Biomechanics, 27,* Woltring, H.J., et al., Instantaneous helical axis estimation from 3-D video data in neck kinematics for whiplash diagnostics, 1415-1432, 1996, with kind permission from Elsevier Science Ltd, The Boulevard, Langford Lane, Kidlington 0X5 1GB, UK.

The cervical spine as a whole possesses a large range of motion. This is explained mainly by the relatively large size of the intervertebral disks as compared with the size of the vertebrae. The ratio between disk thickness and the cervical vertebrae height is large. Although spinal mobility is restrained not only by the intervertebral disks but also by articular facets, spinous processes, and ligaments, the rule still holds: the greater the ratio, the greater the range of motion. The movement of the cervical spine as a whole is also limited by outside obstacles. For instance, mobility in flexion is usually restricted not by the cervical spine itself but by the chin in contact with the breastbone. On average, the flexibility is estimated at about 145° of flexion and extension, 90° of lateral bending, and almost 180° of axial rotation, left and right motions combined. Wide variations exist, however, among individual people. During head-and-neck forward rotation (nodding), the first 8° are, as a rule, performed in the occipital-atlantal joint. About 50% of neck axial rotation takes place in the atlanto-axial joint, C1-2. Habitually, the first 45° of rotation occurs at C1-2 and then the lower cervical spine becomes involved. The motion in each plane is nearly equally distributed among the C-3 through C-7 motion segments.

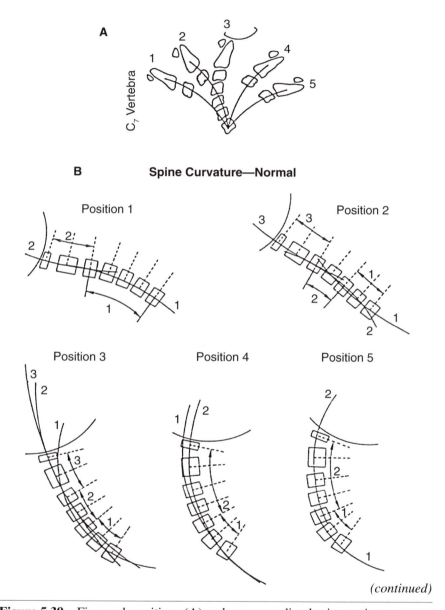

Figure 5.39 Five neck positions (A) and corresponding basic arcs in two subjects with normal (B) and abnormal (C) spine. 1, 2, 3 are basic arcs. For the second subject, in positions 3 and 4 the curvature of the upper cervical spine is close to a straight line (infinite radius). Therefore, no curve is indicated. (In this publication, the term "mean circles" instead of "basic arcs" is used.)

Adapted from *Journal of Biomechanics, 12,* Dimnet, J., et al., Cervical spine motion in sagittal plane: Kinematic and geometric parameters, 959-969, 1982, with kind permission from Elsevier Science Ltd, The Boulevard, Langford Lane, Kidlington 0X5 1GB, UK.

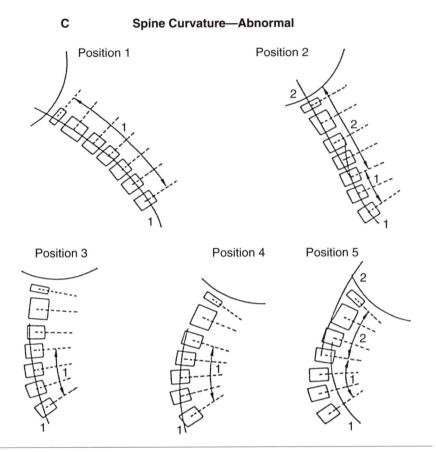

Figure 5.39 *(continued)*

Normal mobility of the whole spinal column in young people, including lumbar, thoracic, and cervical spine segments, is presented in Figures 5.40 and 5.41. Ideally, the norms should be classified according to age. However, norms for older patients are not available. Instead, the qualitative estimations are offered (Table 5.4).

5.6.4 The Rib Cage

Representative papers: Minotti & Lexcellent, 1991; Primiano, 1982.

The rib cage forms a closed kinematic chain. Some links of the chain, such as the shafts of the 11th and 12th ribs and the costal cartilages, are markedly deformable. The rib cage moves as a whole. The main movements are *elevation* and *depression*. The 1st ribs move upward and forward (a "pump-handle

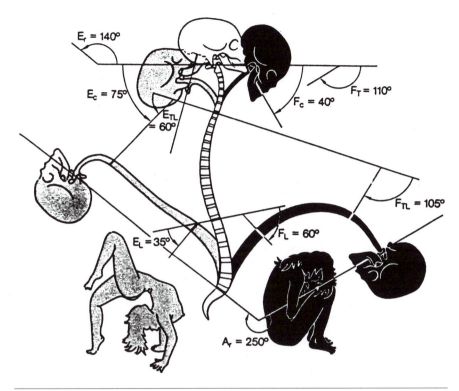

Figure 5.40 Normal flexion and extension potential of the spinal column. A, a total amplitude of the spine motion, the flexion and extension combined; F, flexion; E, extension; C, cervical; T, thoracic; L, lumbar; r, resultant.
From Kapandji, J.A. (1972). *Physiologie articulaire* (vol. 3). Paris: Maloine. Reprinted by permission.

motion"). The lower vertebrospinal ribs, from 2nd to 7th, move upward and laterally (a "bucket-handle motion"). During elevation of the 8th to 10th ribs their anterior ends open in a caliper-like fashion (Figure 5.42). The floating ribs, 11th and 12th, being connected with the skeleton only at their heads, move in any direction depending on the forces acting on them.

5.7 THE SHOULDER COMPLEX

The shoulder girdle acts as a movable but steady support for the motions of the humerus. The complex construction of the shoulder complex provides a unique mobility that surpasses the mobility of any other joint in the body.

Describing kinematics of the shoulder mechanism is not trivial. Usually, articular motion of a segment is defined with regard to the adjacent proximal segment, which is deemed fixed. For example, the foot movement in the ankle

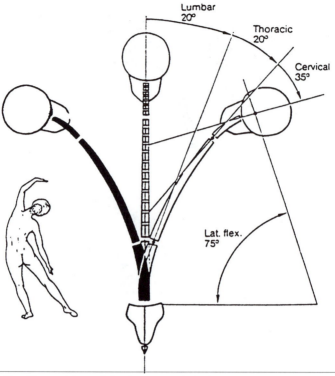

Figure 5.41 Normal lateral flexion potential of the spinal column.
From Kapandji, J.A. (1972). *Physiologie articulaire* (vol. 3). Paris: Maloine. Reprinted by permission.

Table 5.4 Differences in Mobility in the Three Spinal Column Sections*

	Cervical	Thoracic	Lumbar
Flexion-extension	+++	+	+++
Lateral bending	++	+	++
Axial rotation	++	+	++
Lateral bending with rotation	+++	+	+++

* +++ large, ++ intermediate, + slight extent of mobility

From Junghanns, H. (1990). *Clinical implications of normal biomechanical stresses on spinal function.* Rockville, MD: An Aspen Publication.

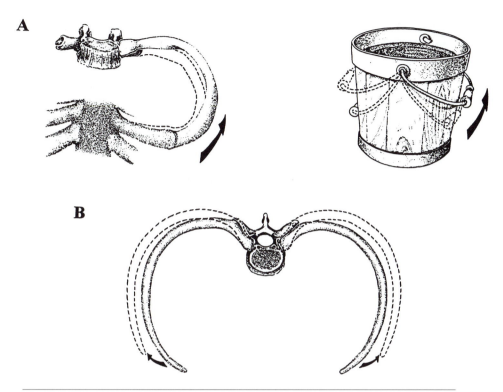

Figure 5.42   The bucket-handle movement of the typical vertebrosternal ribs and the caliper-like movement of the lowest ribs.
Adapted from MacConaill, M.A. & Basmajian, J.V. (1969). *Muscles and movements. A basis for human kinesiology*. Baltimore: Williams & Wilkins.

joint is analyzed with regard to the shank. In the shoulder complex, however, the scapula and clavicle are not fixed with the torso; they displace underneath the skin during the movement. Their orientation is hard to measure, their anatomic position is not defined, and their positions are different in different people. Also, the principal joint movements, such as flexion-extension and abduction-adduction, are not defined for these bones. All of these issues make studying the shoulder complex kinematics perplexing. To avoid these obstacles, many investigators disregard the motion of the scapula and limit the study to the movement of the humerus with regard to the trunk. This is comparable to studying foot motion and ignoring movements at the hip, knee, and ankle joints. Many important aspects have been lost in such studies.

Because the resting positions for the thorax, clavicle, and scapula have not been defined, the *virtual reference position* is proposed to characterize shoulder motion (van der Helm, Pronk, 1995). The virtual position is defined with regard to the global coordinate system with the origin at the suprasternal notch

••• *ANATOMY REFRESHER* •••

Shoulder Complex

The shoulder complex consists of three bones, the clavicle, scapula, and humerus, and four articulations, the glenohumeral, acromioclavicular, sternoclavicular, and scapulothoracic, with the thorax as a stable base (Figure 5.43). In addition, some authors consider the connection between the coracoacromial arch and the humeral head as the subacromial joint. The sternoclavicular joint is the only joint that connects the shoulder complex to the axial skeleton.

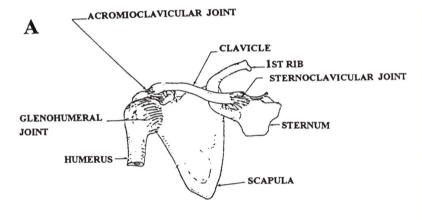

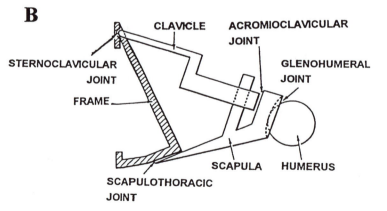

Figure 5.43   The shoulder complex.

The scapula and clavicle make up the *shoulder girdle*. To maintain its integrity, the shoulder girdle depends more on muscles than on joint structures. Such a construction provides *dynamic stability* to the complex, i.e. greater mobility and stability at the same time.

The *glenohumeral joint*, oftentimes called the *shoulder joint*, is a synovial ball-and-socket joint formed by the humeral head and the glenoid cavity of the scapula.

The *scapula* lies obliquely over the ribs between the frontal and coronal planes. The angle between the frontal plane and the plane of scapula, called *protraction* or *winging angle*, is 30° to 45°. The plane of the glenoid is approximately at right angles to the plane of scapula. The scapula has no bony and ligamentous connection with the chest.

The *scapulothoracic articulation* is formed by the female surface of the *scapula* and the male surface of the thorax. This is a bone-muscle-bone articulation that is not synovial and, in the true sense, is not a joint. However, considering it a joint is of some value when describing motion of the scapula over the thorax. The mating surfaces are separated by the subscapularis and serratus anterior muscles, which glide over each other during motion. The scapulothoracic articulation allows translation and rotatory movement of the scapula with respect to the rib cage.

The *sternoclavicular joint* is a compound joint in which the clavicle articulates with the manubrium of the sternum and the cartilage of the first rib. The joint cavity is divided into two compartments by the articular disk.

The *acromioclavicular joint*, also called the *claviculoscapular joint*, is formed by the lateral end of the clavicle and the acromion of the scapula. In the transverse plane, the angle between the scapular spine and the clavicle is approximately 50°. In the joint, the scapula moves on the clavicle. The joint has three rotational DOF.

The bones of the shoulder girdle, the scapula and clavicle, commonly move as a unit. The movement in the scapulothoracic, sternoclavicular, and claviculoscapular joints is conjunct.

and the axes pointing laterally, cranially, and dorsally. In the virtual reference position, the sternum is along the vertical axis, the clavicle is along the frontal axis, the scapular plane is along the frontal plane, the humerus is parallel to the vertical axis, and the epicondyles of the distal end of the humerus are in the frontal plane. In reality, the clavicle and scapula cannot assume the virtual reference position. The virtual reference position of body links with regard to the global coordinate system is defined by the following orientation matrices: for the thorax [Th$_v$], for the clavicle [C$_v$], for the scapula [S$_v$], and for the humerus [H$_v$].

$$[Th_V] = \begin{bmatrix} 1 & 0 & 1 \\ 0 & 1 & 0 \\ 0 & 0 & 1 \end{bmatrix}; [C_V]; = \begin{bmatrix} 1 & 0 & 0 \\ 0 & 1 & 0 \\ 0 & 0 & 1 \end{bmatrix}; [S_V] = \begin{bmatrix} -1 & 0 & 0 \\ 0 & -1 & 0 \\ 0 & 0 & 1 \end{bmatrix}; [H_V] = \begin{bmatrix} 1 & 0 & 0 \\ 0 & -1 & 0 \\ 0 & 0 & -1 \end{bmatrix}$$

$$(5.6)$$

5.7.1 Individual Joints

When discussing shoulder movement, oftentimes the orientation of a shoulder bone is defined with regard to the thorax rather than with regard to the proximal bone, as we did previously. The reader should be aware of whether the joint motion is considered relative to the second bone forming the joint, or the movement is taken with regard to another reference frame, anchored either to the trunk or to the axial skeleton.

5.7.1.1 Glenohumeral Joint

Representative paper: Poppen & Walker, 1976.

In what follows, the motion of the humerus is considered with regard to the glenoid fossa. The glenohumeral joint surfaces are very close to spherical, with only small deviations from sphericity of less than 1% of the radius (Soslowsky et al. 1992). The mating joint surfaces are quite congruent and have radii within a 3 mm difference. To model the glenohumeral joint surfaces, the following equation of a sphere has been used:

$$A\left(x^2 + y^2\ z^2\right) + Bx + Cy + Dz + E = 0 \qquad (5.7)$$

where x, y, and z are coordinates of points on the sphere and A, B, C, D, and E are numerical parameters (Van der Helm et al., 1991). Because the joint surfaces are spherical and in congruence, the joint motion is basically rotational and the surface-on-surface motion is mainly gliding. During arm elevation, the center of the humeral head is approximately at the same place and the surface of humeral head glides downward on the surface of the glenoid fossa. The range of the centrode of the glenohumeral joint, that is, the amount of the displacement of its ICR, is considered by many as negligibly small. Therefore, the glenohumeral articulation is customarily modeled as a ball-and-socket joint.

The glenohumeral joint is not an ideal ball-and-socket joint, however. (Do ideal joints ever exist?) The radii of curvature of the glenoid and the humeral head are not completely identical, especially if the bone curvature (not the cartilage curvature) is compared (Table 5.5). The ratio of the radii of curvature of the matching humeral head to glenoid surfaces ranges from 0.89 to 1.09. This ratio means that, at least in some people, and especially in those with degenerated cartilage, the articular surfaces are not entirely congruent.

The difference between the radii of the humeral head and the glenoid articular surface of up to 3 mm causes a rather complex pattern of glenohumeral movement. In abduction, the humeral head migrates upward (Poppen, Walker, 1976). During the first 30° of arm elevation in the scapular plane the humeral head moves up by about 3 mm because of the combined rolling and translation. Beyond the initial 30°, the humeral head displaces approximately 1 ± 0.5 mm up or down per each 30° of angular rotation. During external rotation, the humeral head displaces backward and during internal rotation forward (Howell et al., 1988). For instance, the humeral head glides posteriorly on the glenoid fossa during the cocking phase of pitching a ball and it glides anteriorly during the delivery phase. These displacements are mainly by reason of different curvature of the humeral head and glenoid cavity.

Arm elevation can be considered a combined motion comprising two elementary rotations in opposite directions: around the glenoidal center with the velocity of $\dot{\alpha}_g$ and around the humeral head center with the velocity of $\dot{\alpha}_h$ (see Section **4.2.1.2** and Figure 4.22). As a result, the humeral head rotates around the ICR. If the glenohumeral joint were a pure ball-and-socket joint, the three

Table 5.5   Radii of Curvature of the Mating Surfaces of the Glenohumeral Joint*

Specimen	Cartilage		Bone	
	Male	Female	Male	Female
Humeral head	26.85 ± 1.40	23.27 ± 1.69	26.10 ± 1.41	23.15 ± 2.09
	(n = 15)	(n = 16)	(n = 13)	(n = 13)
Glenoid	26.37 ± 2.42	23.62 ± 1.56	34.56 ± 1.74	30.28 ± 3.16
	(n = 12)	(n = 13)	(n = 7)	(n = 5)

*Means ± standard deviations in millimeters
From Soslowsky, L.J., Flatow, E.L., Bigliani, L.U., & Mow, V.C. (1992). Articular geometry of the glenohumeral joint. *Clinical Orthopaedics and Related Research, 285*, 181-190.

centers would coincide. Whether the diversion can be neglected and the joint considered a ball-and-socket joint depends on the required accuracy of the analysis.

Because the glenohumeral joint possesses three rotational DOF, all of the real as well as pseudo problems of describing three-dimensional rotation, such as *false torsion* described for the eye movement (recall Section **1.3.1**) or induced twist phenomena (recall Section **2.1.4**) appear in the shoulder literature. In particular, the reported magnitude of the axial rotation of the humerus, if any, depends on the angular convention selected, i.e., whether it is the globographic representation or the various Euler's-Cardan's angles such as the standard Fick's and Helmholtz's coordinates. The reader specifically interested in the three-dimensional movement of the humerus in the glenohumeral joint is referred to Section **1.3.1**, where eye movement is discussed. The issues are very similar.

Two types of arm abduction are discerned, *frontal plane abduction* and *scapular plane abduction*. Note that the scapular plane is defined as the plane of the scapula in the resting position. During arm motion the scapula changes its orientation and no fixed scapular plane exists. In frontal plane abduction, the range of motion depends on the orientation of the humerus. With the humerus internally rotated, the abduction range is limited by impingement of the greater tubercle of the humerus on the acromion process and is only between 60° and 90°. With the humerus externally rotated, the range of motion increases to 90° to 120°. In scapular plane abduction, the humeral rotation does not influence the range of motion. When the arm is abducted in the plane of the scapula, which is angled 30° to 45° anterior to the frontal plane, there is less restriction to motion. The range of active abduction in this plane is on average 107° to 112° (Doody et al., 1970; Freedman, Munro, 1966). The range of motion in frontal plane abduction increases from 90° to 120° when the abduction is performed passively, that is, caused by external forces (Norkin, Levangie, 1992). However, this motion is difficult to separate from an accompanying scapular movement. The range of motion for flexion is on average 120° for both active and passive movement.

The surface area of the humeral head is much larger than the surface of the glenoid fossa, and only a part of the humeral head touches the glenoid (Table 5.6). Thus, the joint is inherently unstable.

When the movement of the humerus relative to the axial skeleton, rather than to the scapula, is the object of interest, the humerus is an end link of the following kinematic chain: trunk (sternum)→sternoclavicular joint→clavicle→acromioclavicular joint→scapula→glenohumeral joint→humerus. With regard to the sternum, the glenohumeral joint is located at a sphere with the clavicle as a radius. Orientation of the joint is defined by the position of the scapula.

Table 5.6   Cartilage Surface Areas*

Specimen	Male	Female
Humeral head	17.34 ± 2.04	13.36 ± 2.20
	(n = 15)	(n = 16)
Glenoid	5.79 ± 1.69	4.68 ± 0.93
	(n = 11)	(n = 13)

*Means ± standard deviations, in square centimeters
From Soslowsky, L.J., Flatow, E.L., Bigliani, L.U., Mow, V.C. (1992). Articular geometry of the glenohumeral joint. *Clinical Orthopaedics and Related Research, 285*, 181-190.

5.7.1.2 Scapular Motion: Scapulothoracic and Acromioclavicular Articulation

Representative paper: Poppen and Walker, 1976.

The main function of scapular motion is to orient the glenoid cavity for the best humeral contact. Functionally, the scapula is a part of a closed kinematic chain formed by three links (the scapula, clavicle, and trunk [sternum]) and the three joints (sternoclavicular, acromioclavicular, and scapulothoracic). Any scapular movement occurs simultaneously in the scapulothoracic and claviculoscapular articulations.

Not including the humerus, the scapula has only one bony connection, with the clavicle through the acromioclavicular joint. Because the clavicle moves in the sternoclavicular joint, the acromioclavicular joint moves substantially relative to the thorax. Hence, the scapula can rotate around three axes in the acromioclavicular joint and translate together with this joint in two directions. These movements are, however, coupled, and the scapula has only four, or may be even three, DOF. The motion of the scapula over the thorax is not planar and because of that an isolated rotation or translation about an individual axis cannot be performed. For instance, when the scapula glides over the rib cage in the lateromedial direction it rotates at the same time around a vertical axis, and when it migrates vertically it rotates around a frontal axis (Figure 5.44). Because of this dependence, any classification of the scapula motion is not very strict.

In the acromioclavicular joint, there are three angular motions: scapular rotation, winging, and tipping. The primary movement of the scapula is its *upward rotation* (the glenoid tilts upward) and *downward rotation* (the glenoid

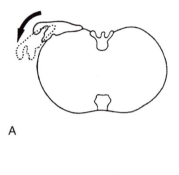

A

B

Figure 5.44 Coupled movements of the scapula. A. Movement of the scapula laterally is coupled with its rotation around a vertical axis (winging). B. Movement of the scapula upward and downward is coupled with its rotation around a frontal axis (tipping or tilting).

tilts downward) performed around an anteroposterior axis. The rotation occurs in both the acromioclavicular and scapulothoracic joints, and is often referred to as abduction-adduction.

Because of the complex character of the rotation, the ICR displaces substantially (Figure 5.45). Between 0° and 80° of arm elevation the ICR is near the root of the spine of the scapula, between 80° and 140° the center migrates toward the acromioclavicular joint, and beyond 140° the center is at the acromioclavicular joint.

Rotation of the scapula around a vertical axis is called *winging*. It is accompanied by the gliding of the scapula over the contour of the ribs in the scapulothoracic articulation. When this movement is performed jointly with anterior movement of the lateral end of the clavicle it is called *protraction* of the scapula. Protraction results in the translation of the scapula around the curved chest away from the vertebral column. The opposite movement is *retraction*. In a retracted position, the scapula approaches the vertebral spines; its vertebral border is parallel to the spinal column.

Rotation of the scapula around a frontal axis is called *tipping* or *tilting*. This is essentially an internal rotation of the scapula, a movement of the inferior tip of the scapula away from the thorax. An internal rotation of the scapula, during which the superior tip of the scapula moves posteriorly, takes place when arm elevation is accompanied by the external rotation of the humerus.

When people shrug their shoulders, the scapula is *elevated*. It is *depressed* when people drop their shoulders. Because all of the motions of the scapula are coupled, the elevation and depression of the scapula, which are considered predominantly translatory movements, are accompanied by scapular rotations.

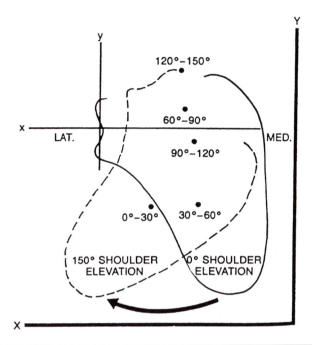

Figure 5.45 Instantaneous centers of rotation of the scapula during shoulder elevation in the scapular plane. The location of the ICR was determined for each 30° of arm lifting. During the abduction, the ICR migrates upward from the lower midportion of the scapula to its upper portion. The outlines of the scapula corresponding to the 0° and 150° abduction positions are also shown.
From Poppen, N.K. & Walker, P.S. (1976). Normal and abnormal motion of the shoulder. *Journal of Bone and Joint Surgery, 58A,* 195-201. Adapted by permission.

A quantitative description of scapular motion is far from simple. Scapular movement is hard to register and difficult to describe. Any movement is essentially three-dimensional and combines translation with rotation. Rotation of the scapula around an axis, such as during tipping, changes the orientation of other axes of rotation, i.e., the axis of winging is not vertical anymore. Hence, the classic anatomic angles are not determined flawlessly and Euler's angles should be used to define the scapular orientation. The scapular motion is not independent; it is an element of the shoulder girdle movement that allows a conscious control of the shoulder position. Accordingly, learning and training may influence movement pattern. This results in large differences among individual people. Anatomic peculiarities, such as a chest curvature or dissimilar winging angle (ranging 30° to 45°), contribute to this variability. The scapulothoracic gliding plane can modeled by an ellipsoid (van der Helm et al., 1991). The equation of such an ellipsoid with the axes of the ellipsoid placed in the coordinate planes and the x coordinate of the ellipsoid center at the x = 0 plane is

$$\left(\frac{x - m_x}{a_x}\right)^2 + \left(\frac{y - m_y}{a_y}\right)^2 + \left(\frac{z - m_z}{a_z}\right)^2 = 1 \qquad (5.8)$$

where x, y, z are coordinates of points on the ellipsoid; m_x, m_y, m_z are coordinates of the ellipsoid center ($m_x = 0$); and a_x, a_y, a_z are radii of the ellipsoid in the x, y, and z directions.

Because the scapula has no definite resting orientation, its attitude is defined with regard to the virtual reference position. As was mentioned earlier, this is a position in which the scapular plane is parallel to the frontal plane and the spine of the scapula is along the frontal axis. An example of such a representation is given in Figure 5.46. See also "Describing Orientation of the Scapula" on page 34.

Generally, scapular movement is described either relative to the sternum or relative to the trunk (a longitudinal axis of the sternum system is inclined to the vertical axis of the trunk approximately 15°). A product of two transformation matrices is calculated. The first matrix defines the orientation of the clavicle relative to the sternum and thorax, and the second the orientation of the scapula with regard to the clavicle. In total, 18 direction cosines or 6 Euler's angles are needed for such a calculation.

5.7.1.3 Sternoclavicular Joint

Kinematically, the sternoclavicular joint functions as a ball-and-socket joint with three DOF. The movements of the clavicle are: elevation-depression (the lateral end of the clavicle moves up or down), protraction-retraction (back-to-front and front-to-back movement), and axial rotation. The range of clavicular motion in elevation is on average 45°, in depression it is 15°, in protraction-retraction it is 30°, and in axial rotation it is 30° to 45°. During circumductory motion, the acromial end of the clavicle sweeps an ellipse with the major axis in the inferior-superior direction. The clavicle operates as a bony strut with two articulations at the ends. The clavicular motions are coupled with movements of the scapula (except during pure scapular abduction-adduction).

5.7.2 Movement of the Shoulder Complex: The Scapulohumeral Rhythm

Representative papers: Engin & Chen, 1986; van der Helm & Pronk, 1995.

The shoulder complex can be viewed as a complex mechanism comprising two individual mechanisms: (a) the shoulder girdle, consisting of the clavicle

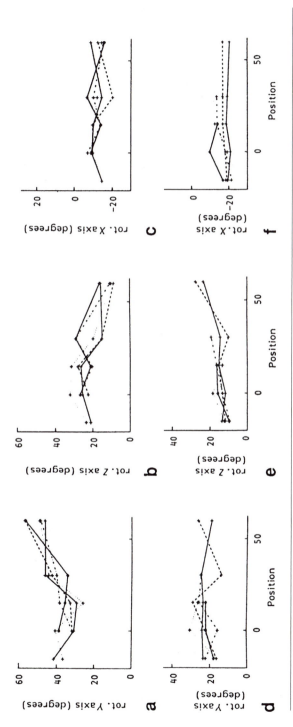

Figure 5.46   Orientation of the scapula with regard to the virtual position in a simulated wheelchair push. The orientation is expressed as consecutive Cardan's angles in the order $y \rightarrow z \rightarrow x$ (winging→upward-downward rotation→tipping). The scapula orientation was measured at different hand positions on the rim: −15°, 0°, 15°, 30°, and 60°, with 0° defined as the second metacarpal at the top center or 12 o'clock. The results are from two subjects, upper and bottom row; up to five attempts with different resistance were made per subject.

Adapted from *Clinical Biomechanics, 8,* Veeger, H.E.J., van der Helm, F.C.T., & Rozendal, R.H., Orientation of the scapula in a simulated wheelchair push, 81-90, 1993, with kind permission from Elsevier Science Ltd, The Boulevard, Langford Lane, Kidlington OX5 1GB, UK.

and scapula as the moving links and the sternum with the rib cage as a frame, and (b) the humerus as a moving link and the scapula and clavicle as a frame. Both mechanisms are spatial with the first mechanism being closed and the second open.

Three-dimensional description of motion of the shoulder complex can be done in various ways. For example, different coordinate systems or different sequences of Euler's angles can be used. From a pure mechanical standpoint these techniques are analogous, but some techniques are preferred because they permit an easier interpretation. In particular, the following techniques have been recommended (van der Helm, & Pronk, 1995). The humeral position is described by the Euler's angles in the following order: (a) rotation with regard to the vertical global axis defines the *plane of elevation,* (b) rotation around the horizontal axis that is normal to the plane of elevation defines the *elevation angle,* and (c) rotation around the longitudinal axis of the humerus defines axial rotation. Bone rotations should be described with regard to the virtual reference position as rotations in the global reference system. Joint rotations should be defined with regard to the starting position as rotations in the global reference system to which the proximal bone is fixed (see Table 5.7 for final bone orientation and the corresponding rotation matrices; see "Describing Arm Position with Respect to the Trunk" on page 121 for the symbols as well as for comparison).

Because each of the shoulder complex joints has three DOF, one would expect the shoulder complex as a whole to have 12 DOF. However, the clavicle and scapula, being parts of the shoulder girdle, move conjointly. As a result, the shoulder complex provides seven DOF for the arm movement, four at the shoulder girdle (the glenoid with regard to the trunk can translate up and down and forth and back and can rotate around the anteroposterior and vertical axes) and three at the glenohumeral joint. To describe the shoulder configuration with regard to the sternum system nine Euler's angles, three for each bone system, are necessary. In what follows, either the angles in the plane of motion or the projection angles on the reference planes are reported.

During upper extremity elevation, the scapula rotates at the sternoclavicular and the acromioclavicular joints. As a result, the glenoid fossa, on which the humerus head moves, tilts upward. This concerted movement of the humerus, scapula, and clavicle is called the *scapulohumeral rhythm,* or the *shoulder rhythm.* The rhythm becomes evident after about 30° of abduction or 60° of flexion. The initial phase of the humeral movement, during which the scapula either rotates a little or remains fixed, is called the *setting phase.* Beyond the setting phase, the humerus, scapula, and clavicle move conjointly.

In its entirety, the ratio of glenohumeral to scapular rotation is approximately 2:1 (Inman et al., 1944). In the Inman study, the projection angle of the scapular spine on the frontal plane was measured. Of 180° of full arm elevation, the

Table 5.7 Recommended Representation of the Shoulder Bone and Joint Rotations

Bone orientation as a result of rotation from the virtual to final position		Final bone orientation	Bone orientation as a result of rotations in the joints from the resting positions	
Final bone orientation	Rotation matrix		Final bone orientation	Rotation matrices
$[Th_i] = [Rt_{vi}][Th_v]$	$[Rt_{vi}] = [T_i][Th_v]^T$	$[Th_i] = [Rt_{0i}][Th_0]$	$[Rt_{0i}] = [T_i][Th_0]^T$	
$[C_i] = [Rc_{vi}][C_v]$	$[Rc_{vi}] = [C_i][C_v]^T$	$[C_i] = [Rt_{0i}][Rc_{0i}][C_0]$	$[Rc_{0i}] = [Rt_{0i}]^T[C_i][C_0]^T$	
$[S_i] = [Rs_{vi}][S_v]$	$[Rs_{vi}] = [S_i][S_v]^T$	$[S_i] = [Rt_{0i}][Rc_{0i}][Rs_{0i}][S_0]$	$[Rs_{0i}] = [Rc_{0i}]^T[Rt_{0i}]^T[S_i][S_0]^T$	
$[H_i] = [Rh_{vi}][H_v]$	$[Rh_{vi}] = [H_i][H_v]^T$	$[H_i] = [Rt_{0i}][Rc_{0i}][Rs_{0i}][Rh_{0i}][H_0]$	$[Rh_{0i}] = [Rs_{0i}]^T[Rc_{0i}]^T[Rt_{0i}]^T[H_i][H_0]^T$	

From van der Helm & Pronk (1995).

glenohumeral rotation contributes close to 120° and the scapular rotation around 60°. If the humerus were fixed to the glenoid, the scapular upward rotation alone would result in about 60° arm elevation. Other researchers have measured angles during arm abduction in the scapular plane and they found the overall glenohumeral and scapulothoracic contributions were 112.5° + 58.6° = 171.1° (Doody et al., 1970) and 104.3° + 63.8° = 168.1° (Bagg & Forrest, 1988). Because of the existence of the setting phase, during which the scapula does not rotate, the glenohumeral-to-scapular ratio for the period when the scapula does rotate is less than the overall ratio, on average about 1.25:1 (Figure 5.47).

The relationship between arm elevation angle and scapular orientation is not linear. The ratio of glenohumeral rotation to scapular rotation varies through-

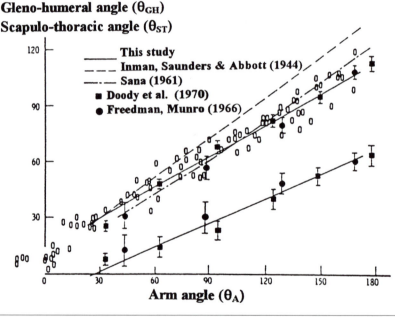

Figure 5.47   The glenohumeral (top) and scapular (bottom) attitude at various arm abduction angles in the scapular plane (abscissa). During arm elevation, the scapular plane changes, rotating upward around an anteroposterior axis. The scapula also wings (rotates around a vertical axis) and tilts (rotates around a horizontal axis). These rotations are not shown here. The relationship between the abduction-adduction angle in the scapular plane and the tipping-tilting angle, both in angular degrees, was estimated as $\alpha_{tipping} = 0.59\alpha_{rotation} + 5.4$ (r = 0.83). The tipping angle was 0.59 times the scapulothoracic angle.

From Poppen, N.K. & Walker, P.S. (1976). Normal and abnormal motion of the shoulder. *Journal of Bone and Joint Surgery, 58A*, 195-201. Adapted by permission.

out the range. Glenohumeral rotation predominates in the beginning and at the end of arm elevation. The interindividual differences in the glenohumeral-to-scapular rotation ratio are very large. The slope of the regression of the gleno-humeral component on total arm abduction—calculated over the whole range of motion—varies from 0.5 to 0.75, with a mean of 0.66 (Michiels & Grevenstein, 1995). The slope 0.66 is essentially the same as the previously mentioned ratio of glenohumeral to scapular rotation 2:1. The slope does not depend on the abduction velocity and external load. For any one person, the shoulder rhythm is essentially reproducible.

During arm elevation, the clavicle rotates about an anteroposterior axis at the sternoclavicular joint. The rotation results in the elevation of the acromion and lasts until the arm is elevated up to 90° to 100° (Figure 5.48). In rounded numbers, raising the arm to the horizontal involves 30° of scapular rotation, mainly due to the elevation in the sternoclavicular joint. Raising the arm from the horizontal to the vertical position involves 30° of scapular rotation produced by the acromioclavicular motion itself. Beyond the 90° to 100° region, the clavicle rotates around its long axis. The superior aspect of the clavicle

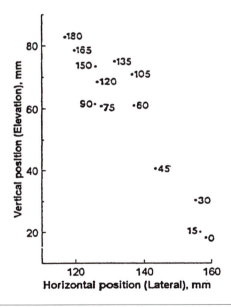

Figure 5.48 Position of the acromioclavicular joint with regard to the sterno-clavicular joint during the arm elevation. A projection on the frontal plane. The position was recorded at intervals of 15°.

Reprinted from *Journal of Biomechanics, 11,* Dvir, Z. & Berme, N., The shoulder complex in elevation of the arm: A mechanism approach, 219-225, 1978, with kind permission from Elsevier Science Ltd, The Boulevard, Langford Lane, Kidlington 0X5 1GB, UK.

spins posteriorly. During the setting phase, the angle formed by the clavicle and the spine of scapula in the transverse plane increases.

Because of the displacement of the shoulder girdle during arm movements, the ICR for the shoulder complex differs greatly from the ICR for the shoulder joint. The path of the centrode is very large (Figure 5.49).

5.8 THE ELBOW COMPLEX

The elbow complex includes the elbow joint and the radioulnar joints, superior and inferior. The elbow joint is a compound joint consisting of two joints: the humeroradial, between the capitulum and radial head, and the humeroulnar, between the trochlea and the trochlear notch of the ulna. The humeroradial joint is a ball-and-socket joint; however, its close association with the humeroulnar and superior radioulnar joint restricts the joint motion from three to two DOF. The humeroulnar joint is a hinge joint. In the elbow joint, the humerus is the male member. The two radioulnar joints function as one joint. The elbow joint, when considered as an entirety, is a hinge joint, and the radioulnar joints are pivot joints with 1 DOF.

As a whole, the elbow joint complex allows two DOF, flexion-extension and pronation-supination. Because of the close coupling of motions in the elbow, we consider cardinal joint motions rather than the individual joints.

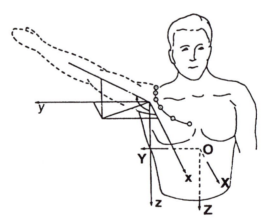

Figure 5.49 Centrode curve for the shoulder complex during arm elevation in the frontal plane.

Reprinted from *Journal of Biomechanics, 13,* Engin, A.E., On the biomechanics of the shoulder complex., 575-590, 1980, with kind permission from Elsevier Science Ltd, The Boulevard, Langford Lane, Kidlington 0X5 1GB, UK.

5.8.1 Flexion and Extension

The plane of motion for flexion-extension is normal to the interepicondylar line. In this plane, the surfaces of the trochlea and capitulum have approximately circular profiles. Because of that, elbow joint movement is nearly pure rotation. During elbow flexion and extension, the instant center of rotation displaces less than 3 mm (Figure 5.50).

The surface-on-surface motion is predominantly gliding. The concave surface of the ulna glides over the convex surface of trochlea and the concave radial head glides along the capitulum. The gliding changes to rolling in the final 5° to 10° of full flexion and extension.

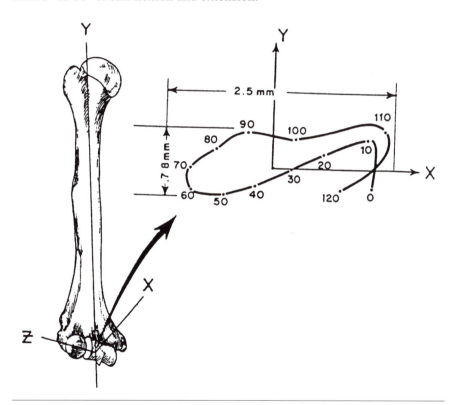

Figure 5.50 Migration of the instant center of rotation during elbow flexion. The numbers near the centrode line represent the joint angular position.
Reprinted from *Journal of Biomechanics, 11*, Chao, E.Y. & Morrey, B.F., Three-dimensional rotation of the elbow, 57-73, 1978, with kind permission from Elsevier Science Ltd, The Boulevard, Langford Lane, Kidlington 0X5 1GB, UK. Original data are due to Fisher, O. (1909). Zur Kinematik der Gelenke vom Typus des humeroradial Gelenkes. *Abhandlungen der Mathem-Phys. cl. sk. k. Sächs. Gesellschaft*, Bd. 32, S. 3-77.

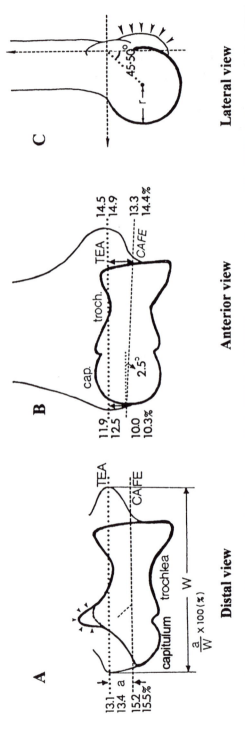

Distal view   **Anterior view**   **Lateral view**

Figure 5.51   Location of the axis of rotation in the elbow joint. TEA, transepicondylar axis; CAFE, centroidal axis of flexion-extension. A. In the distal view, the rotation axis runs parallel to the transepicondylar line, slightly anterior to it. The axis is at a distance of 13.1% to 13.4% of the interepicondylar distance (W) for women and 15.2% to 15.3% for men. B. In the anterior view, the axis slopes down toward the medial side at 2.5° relative to the transepicondylar line. The distances are 11.9% to 12.5% at the lateral end for women and 10.0% to 10.3% for men. At the medial side the distances are 14.5% to 14.9% for women and 13.3% to 14.4% for men. C. In the lateral view, the rotation axis is anterior and distal to the transepicondylar line; on a line drawn at approximately 45° from the shaft of the humerus for women and 50° for men.
Reprinted from Shiba, R. & Sorbie, C. (1985). Axis of flexion-extension of humeroulnar joint. In D. Kashiwagi (Ed.), *Elbow joint* (pp. 19-24). New York: Elsevier Science.

The location of the axis of rotation with regard to the transepicondylar line, which can be determined by palpating the bony landmarks, is shown in Figure 5.51.

When the elbow joint is fully extended and supinated, the long axis of the upper arm and that of the forearm form a *carrying angle* (Figure 5.52A). On average, this angle is 11° for adult men and 14° for adult women; it averages 6° for children (Amis & Miller, 1982). The axis of rotation approximately

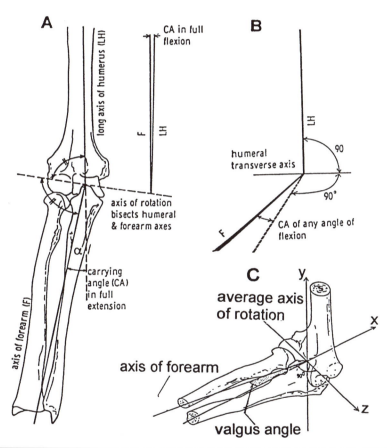

Figure 5.52   A. The carrying angle in full arm extension. B. The carrying angle at elbow flexion—the acute angle between the long axis of the forearm and the projection of the humeral axis on the plane of forearm. C. The valgus angle—the acute angle between the perpendicular to the average axis of rotation and the long axis of the forearm, viewed on the plane of the forearm. CA, carrying angle; F, long axis of forearm; LH, long axis of humerus.

From Deland, J.T., Garg, A., & Walker, P.S. (1987). Biomechanical basis for elbow hinge-distractor design. *Clinical Orthopaedics and Related Research, 215,* 303-312. Adapted by permission of the author.

bisects the angle $\beta = \pi - \alpha$, where α is the carrying angle (see Figure 5.52A). For arm positions other than complete extension, the carrying angle is defined as an Euler's angle. This is the angle in the *plane of the forearm*, the plane made by the axis of the forearm and the axis of rotation. In the plane of the forearm, two acute angles are then determined: the carrying angle between the long axis of the forearm and the projection of the long axis of the humerus onto the plane (Figure 5.52B), and the *varus angle* between the forearm and a normal to the average axis of rotation (Figure 5.52C). The normal is not coincident with the long axis of the humerus. The varus angle does not alter during elbow flexion, but the carrying angle changes from valgus to varus attitude (decreases) in a linear fashion (Figure 5.53).

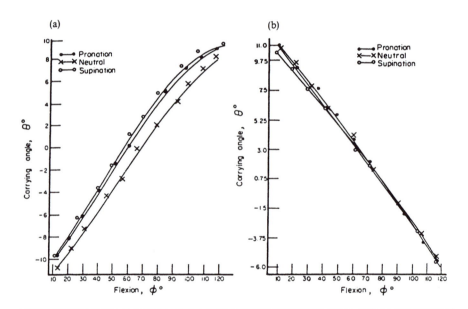

Figure 5.53 Change of carrying angle during elbow flexion with the forearm in pronated, neutral, and supinated positions: (a) left arm (specimen 1), (b) right arm (specimen 2). The carrying angle is a second Euler's angle, with the angle for flexion-extension being the first. The angle was measured between the long axis of the humeral shaft and the forearm axis, the axis through the center of the radial head proximally and the center of the styloid process of the ulna distally. See figure 2.10 for the definition of the axes and the angles involved, and "Three-dimensional Rotation of the Elbow Joint" on page 95 for the mathematical methods used.

Reprinted from *Journal of Biomechanics, 11,* Chao, E.Y. & Morrey, B.F., Three-dimensional rotation of the elbow, 57-73, 1978, with kind permission from Elsevier Science Ltd, The Boulevard, Langford Lane, Kidlington 0X5 1GB, UK.

The amount of motion available at the elbow joint is determined by the bony geometry as well as by the resistance provided by the soft tissues. The difference between the arc of the female and male surfaces in the joint is on average 140°. In the ulnohumeral joint, the angular value of the female surface is approximately 320° and that of the male surface is 180° (see Figure 4.32). In the radiohumeral joint, the angular value of the surface of the capitulum is approximately 180° and the arc of the proximal radial head, which is the male surface, is 40°. The difference is 140°. Because during active motion the bulk of the contracting muscles on the anterior surface of the upper arm prevents advancing the forearm to the humerus, the range of passive motion is usually larger than that of the active motion.

5.8.2 Supination and Pronation

Representative paper: Carret et al., 1976.

The axis of rotation extends from the center of the radial head in the elbow joint to the center of the distal ulna. The axis is oblique to the longitudinal axis of the forearm. In supination, the radius and ulna are parallel to each other. During pronation, the radius rotates in the two proximal joints, the humeroradial joint and proximal radioulnar, and crosses over the ulna. The proximal ulna practically does not rotate and the distal ulna slightly turns in the direction opposite to the rotation of the radius.

With the elbow in 90° of flexion, the average range of forearm pronation is approximately 70° and average supination is about 85°. When the arm is completely extended, the supination-pronation is conjoint with the internal-external rotation in the shoulder joint.

5.9 THE WRIST

Representative paper: Andrews & Youm, 1979.

The wrist, or carpus, is a deformable anatomic entity that connects the hand to the forearm. This is a collection of eight carpal bones and the surrounding soft tissue structures. The wrist contains several joints, including the radiocarpal joint, several intercarpal joints, and five carpometacarpal joints. The radiocarpal joint is a condyloid joint formed by the end of the radius and the three carpal bones: the scaphoid, the lunate, and the triquetrum. The intercarpal joints, as well as the carpometacarpal joints except for the thumb, are functionally similar to flat joints. The surface-on-surface movement is chiefly gliding. The movements of the carpal bones in the distal row approximate those of the hand.

On the contrary, the motion of the carpal bones of the proximal row is not proportional to the hand motions. Taken as an entity, the wrist joints are considered one joint, called the wrist joint. We will discuss the forearm-hand articulation as one joint and ignore the behavior of the individual carpal bones.

The wrist joint possesses two DOF, flexion-extension and abduction-adduction. Flexion is the bending of the wrist so the palm approaches the anterior surface of the forearm. Abduction, or *radial deviation*, is bending the wrist to the thumb side. The reverse movement is adduction, or *ulnar deviation*. Wrist motion is performed around an instantaneous center (Taylor & Schwarz, 1955). The path of the centrode is small, however. Customarily, the displacement of the ICR is ignored and the rotation axes for flexion-extension and abduction-adduction are considered fixed. The axes pass through the capitate, a carpal bone articulating with the third metacarpal (Figure 5.54). The rotation axes are offset. The abduction-adduction axis passes distally with regard to the flexion-extension axis by approximately 5 mm (Figure 5.55).

In everyday living, the functional range of wrist motion is 5° of flexion, 30° of extension, 10° of radial deviation, and 15° of ulnar deviation (Palmer et al., 1985). The maximal range of motion is much larger, especially in wrist flexion.

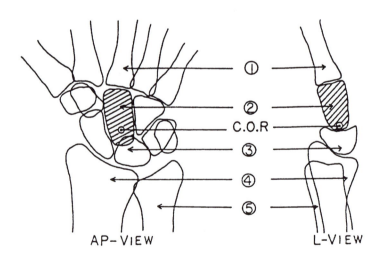

Figure 5.54   The centers of rotation (COR) of the wrist: (1) metacarpal, (2) capitate, (3) lunate, (4) radius, (5) ulna.

Reprinted from *Journal of Biomechanics, 12,* Youm, Y. & Yoon, Y.S., Analytical development in investigation of wrist kinematics, 613-621, 1979, with kind permission from Elsevier Science Ltd, The Boulevard, Langford Lane, Kidlington 0X5 1GB, UK.

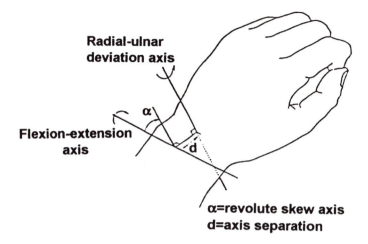

Figure 5.55   Wrist joint axes.

This material has been reproduced from *Proceedings of the Institution of Mechanical Engineers. Part H, Vol. 207,* Moore, J.A., et al., A kinematic technique for describing wrist joint motion: Analysis of configuration space plots, 211-218, 1993, by permission of the Council of the Institution of Mechanical Engineers.

5.10 THE JOINTS OF THE HAND

5.10.1 The Joints of the Thumb

Compared with the other fingers, the thumb has only two phalanges instead of three, does not lie in the same plane as the other digits, and rotates with regard to them. The thumb contains three joints: the carpometacarpal (CMC), metacarpophalangeal (MCP), and interphalangeal (IP).

5.10.1.1 The Carpometacarpal Joint

Representative paper: Hollister et al., 1992.

The CMC joint is an articulation of the first metacarpal and the trapezium (Figure 5.56). The neutral (zero) position of the CMC is defined by the orientation of the trapezium. With regard to the reference axes of the third metacarpal, the reference axes of the trapezium are flexed on average 48°, abducted 38°, and pronated (rotated medially) 80° (Cooney et al., 1981).

The CMC joint is a saddle joint with two axes of rotation. The axes are fixed, not perpendicular to each other or to the bones, and do not intersect. The axes do not coincide with the cardinal anatomic planes. Although the joint motion occurs in all the three anatomic planes, only two independent motions

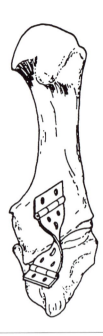

Figure 5.56   A schematic representation of the carpometacarpal joint. The joint has two axes of rotation. Neither axis lies in an anatomic plane. The articular motions are modeled as hinges.
From Hollister, A.M., Buford, W.L., Myers, L.M., Giurintano, D.J., & Novick, A. (1992). The axes of rotation of the thumb carpometacarpal joint. *Journal of Orthopaedic Research, 10,* 454-460. Reprinted by permission.

are possible. The first is abduction-adduction and the second is a complex articular motion combining thumb flexion with pronation and thumb extension with supination. Such a coupled movement is unique to humans in that it allows the best thumb position for opposition. The abduction-adduction axis passes through the base of the first metacarpal slightly distal to the dorsal and volar beaks. The average projection angles of the axis on the frontal plane are 78.3° in the transverse plane and 83.6° in the sagittal plane. The flexion-extension axis is located in the trapezium and runs parallel to its body.

When high accuracy is not required, the CMC joint center is usually estimated at the center of curvature of the circular arc of the trapezium.

5.10.1.2 The Metacarpophalangeal and Interphalangeal Joints

Representative papers: Barmakian, 1992; Imaeda et al., 1992.

The MCP joint arises from the articulation of the first metacarpal head and the distal phalange. The joint is considered a saddle joint. It provides up to 100° range of motion for flexion. Joint extension varies from 0° to 45° in different subjects. In some people, the MCP joint allows abduction and adduction up to 20°.

The IP joint is formed by the proximal and distal phalanges. It is a typical hinge joint with one DOF. The range of joint motion is 80° to 90° of flexion and 15° to 20° of extension.

5.10.2 The Joints of the Fingers

Representative paper: Buchholz et al., 1992.

The joints of the fingers are the metacarpophalangeal (MCP), proximal interphalangeal (PIP), and distal interphalangeal (DIP) joints. The MCP joints have two DOF, and the interphalangeal joints are one-DOF joints. The male surfaces of these joints are always on the bones that are proximal to the joint. The interphalangeal joints have fixed centers of rotation. The center of rotation for the MCP joint moves with joint rotation. The displacement is small, however, and in many applications is ignored. In the finger joints, a joint center of rotation coincides with the center of curvature of the head of the male bone. The centers can be determined anatomically from radiographs. They can also be estimated biomechanically using the Reuleaux method. The results from anatomic and biomechanical investigations are in good agreement. The most convenient technique of determining the finger joint centers, however, is from anthropometric measurements. The relative distance of the joint centers from the base of the proximal bone, the percentage of the bone length, is found to be nearly constant among individual persons. The coefficients of correlation squared between the joint center location and the bone length were in the range of 0.97 to 0.99. Hence, the bone length accounts for more than 97% of the variability in joint center position along the midline of the bone. The method is explained in Figure 5.57 and Table 5.8.

In most people, movement of the fingers is coupled. When someone flexes one finger with a large amplitude they also flex the neighboring fingers. The lack of finger independence is seen even in trained musicians, like pianists and string players. The fingers move conjointly because the tendons of the fingers are interconnected by fascia-like anatomic structures. Near the neutral hand position, the connections are slack and the fingers can move independently. However, when the range of motion increases, the connections become taut and flexing one finger causes the neighboring fingers to flex also.

5.11 THE TEMPOROMANDIBULAR JOINT

Representative paper: Baragar & Osborn, 1984.

The temporomandibular (TM) joint is involved in eating and articulating. During biting and chewing the joint sustains large forces; during speaking the joint should be finely controlled. The TM joint is used very often.

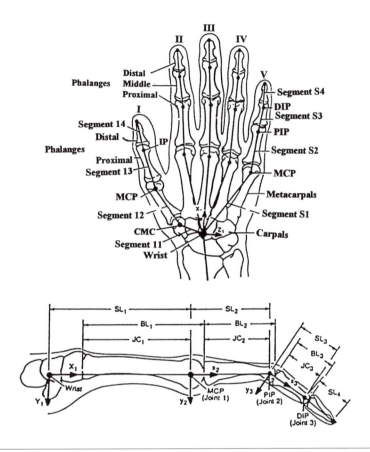

Figure 5.57 Kinematic skeleton of the hand (top panel) and the reference frame used for locating the finger joint centers (bottom panel). On the top panel, the skeleton is characterized by ideal joints. Segments are numbered distally from the wrist joint. Consequently, segment 1 is the carpometacarpal segment for each of the four fingers and the carpal segment for the thumb. The distal phalangeal segment is therefore segment 4. The origin of the root coordinate system of the hand is at the wrist joint center—at the intersection of the distal wrist crease and the midline of the third metacarpal. The x axis is pointing down the third metacarpal to the estimated position of the third MCP center. The z axis is projected along the distal wrist crease in the ulnar direction. On the bottom panel, the local coordinate systems are defined for each joint. The origin of a system is located at the next proximal joint center, with the x axis pointing toward the estimated position of the given joint center. The y axis is directed volarly so that flexion of the joint is a positive angle. Protrusions on the base of the phalanges are used as markers (open circles). Segment lengths (SL_i), bone lengths (BL_i), and joint center positions with regard to the origin of the local reference frames (JC_i) are also shown.

From Buchholz, B., Armstrong, T.J., & Goldstein, S.A. (1992). Anthropometric data for describing the kinematics of the human hand. *Ergonomics, 35,* 261-273. Adapted by permission of the author.

Table 5.8 Estimating Finger Joint Center Location From Bone Length[1]

Digit[2]	Joint 1[3]	Joint 2	Joint 3
I	-	0.832 ± 0.005	0.849 ± 0.004
II	0.9	0.909 ± 0.003	0.887 ± 0.005
III	0.9	0.912 ± 0.003	0.894 ± 0.004
IV	0.9	0.907 ± 0.002	0.895 ± 0.001
V	0.9	0.899 ± 0.004	0.870 ± 0.005
II-V	0.9	0.908 ± 0.002	0.889 ± 0.002

[1]Coefficients A_{ij} from equations $JC_{ij} = A_{ij} \times BL_{ij}$. Means ± standard deviations.

[2]For digit 1, joint 1 is CMC, joint 2 is MCP, and joint 3 is IP. For digits II-V, joint 1 is MCP, joint 2 is PIP, and joint 3 is DIP.

[3]The data for joint 1 are from Youm, Y., Gillespie, T.E., Flatt, A.E., & Sprague, B.L. (1978). Kinematic investigation of normal MCP joint. *Journal of Biomechanics, 11*, 109-118. The data for joints 2 and 3 are from Buchholz, B., Armstrong, T.J., & Goldstein, S.A. (1992). Anthropometric data for describing the kinematics of the human hand. *Ergonomics, 35*, 261-273.

The TM joint consists of two articulations, right and left. Each articulation includes the mandibular condyle, the disk, the temporal bone, the mandibular fossa, and the eminence. The disk subdivides the articulation into two compartments, superior and inferior, also called the upper and lower joints. The upper joint is between the temporal bone and the superior surface of the disk. The lower joint is between the mandibular condyle and the inferior surface of the disk. During jaw movement the disk slides within the joint. The upper joint is functionally a flat joint, permitting gliding of the articulating surfaces; the lower joint is usually classified as a hinge joint. During mouth opening, the condyle of the mandible rotates within the articular fossa and translates down the articular eminence. Any mandibular motion involves movement in the four joints.

The main mandibular motions are *depression* and *elevation* (mouth opening and closing), *protrusion* and *retrusion* (moving the chin forward and backward), and *lateral deviation* or *laterotrusion*. Those movements are produced by rotation and translation in the upper and lower joints. The depression-elevation and protrusion-retrusion are performed in the same plane (the sagittal).

The depression is a two-stage motion. First, the condyle rotates on the disk in the lower joint. This rotation accounts for 11 to 25 mm of the mouth opening. During the second stage the disk-condyle complex translates in the upper joint inferiorly and anteriorly along the articular eminence. This movement accounts for the remainder of the opening. The mouth closing is performed in a reverse order. The locus of the ICR for the jaw opening and closing is shown in Figure 5.58. In healthy people, the maximally wide opening of the mouth is approximately 23° (Rasmussen, 1978).

Protrusion and retrusion are jaw rotations about an axis located above and anterior to the line joining the centers of rotation of the two eyes (the base line, see Section **1.3.1**). Because the radius of rotation is large, the curvature is small and the movement is often identified as translation. It is translation when the movement in the TM joint is considered. During protrusion, no rotation occurs in the TM joint; the translation takes place in the upper joints.

During laterotrusion, one condyle rotates and the second moves forward. For instance, when the chin moves to the right, the right condyle is rotating and the left is gliding forward. Small rotation of the jaw around an anteroposterior axis is also possible and is used regularly during chewing. During this motion one of the condyles rotates and the second depresses.

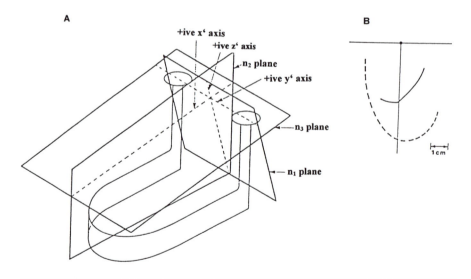

Figure 5.58 The centrode of the mouth opening (b) and the system of coordinates employed in the studies (a). The coordinate axis is along the line joining the "centers" of the left and right condyles. Solid line, data of Baragar and Osborn (1984); broken line, Grant's (1973) reconstruction of Hall's (1928) data. Adapted from *Journal of Biomechanics, 17,* Baragar, F.A. & Osborn, J.W., A model relating patterns of human jaw movement to biomechanical constraints, 757-767, 1986, with kind permission from Elsevier Science Ltd, The Boulevard, Langford Lane, Kidlington 0X5 1GB, UK.

5.12 SUMMARY

By definition, the nominal joint axes are (a) fixed, (b) orthogonal, (c) parallel to the main anatomic axes, and (d) intersect at one fixed center. Various techniques recommended in the literature for locating the nominal axes are presented. The kinematics of particular joints are also described in bottom-to-top order, from toes to fingers.

Typical movement in the MTP joints is performed about either a transverse axis through the joints of the first and second toe or through the metatarsal break, an oblique axis through the MTP joints of the second to the fifth toe. Joint movement is performed around the transverse axis when a high velocity is desired and it is performed around the oblique axis when a large force is needed.

The ankle joint complex is a combination of two one-DOF joints, the ankle joint and the subtalar joint. The axes of rotation in these two joints are offset in the anatomic planes, not perpendicular to the limb segments and not perpendicular to each other. As a result, the complex produces motion in all three anatomic planes. Rotation of the foot with regard to the shank, as observed from a fixed frame of reference, requires three Euler's angles for description. The angular coordinates are not independent, however, and if the axes of coordinates are chosen along the joint rotation axes, two angular coordinates are sufficient.

The knee joint axes are still a topic of discussion. The focus of controversy is the contribution of the muscle forces to the geometry of knee motion. According to one viewpoint, the knee motion is constrained by the articular surfaces and ligaments with the muscles working only as movers. Following this line of thought, researchers have determined the instantaneous axis of flexion-extension and the centrode, or axode, of the joint. Some researchers even conclude that the axis of rotation is fixed but offset relative to main anatomic planes and is therefore mistakenly perceived as a movable instantaneous axis. An opposite position is that knee motion depends on the acting muscle forces and may occur about an infinite number of axes. Hence, determining the axis of rotation is not an adequate way of understanding the knee motion. An envelope rather than the axis should be determined. The selection of the most suitable approach depends on the required accuracy. When requirements are not very strict, simplifying assumptions are useful.

In the patellofemoral joint, because of the shift of the contact point on the patella and because of its rocking, the lever arms of the quadriceps tendon and the patellar ligament vary during the knee motion.

The hip joint is usually modeled as a ball-and-socket joint with a fixed center and with restraints only at the terminal ranges. The HJC coincides with the geometric centers of the femoral head and acetabulum. Surface movement in a normal hip joint is gliding. Flexion in the hip joint is spin. The orientation of the hip joint movement with regard to the cardinal anatomic planes is influenced by the two femoral angles, the neck-to-shaft angle and the anteversion angle. The joint is hardly accessible; therefore, locating the joint center is not a simple task. Several methods for locating the joint center are described.

Spine kinematics deals either with the specific motion segments or with the entire region of the spine. Any spine movement combines displacement of several motion segments. Movement of the motion segments takes place by compressing and stretching the intervertebral disks. Because these disks can deform, the same resultant force and torque being applied to various vertebrae produces different movements depending on the disk stiffness and its dimensions, i.e., height and diameter. The amount of motion available at various motion segments depends mainly on the size of the disks and the orientation of the facet joint surfaces, which varies from region to region and defines the direction of the allowable movement. The vertebrae, when flexing and extending, move along circular arcs of different curvature. In a normal spine, the axes of rotation lie below the moving vertebra. The motion segments of the spine have six DOF. Because the intervertebral disks can be deformed, the vertebrae, in addition to rotation, can translate. The spine as a whole, however, can produce only three movements.

Trunk flexion involves spinal flexion and pelvic tilt. The coupled movement of the spine and pelvis is called the lumbar-pelvic rhythm. In lifting activities, the pelvis orientation depends on the trunk inclination and the knee flexion. Beyond approximately 30°, the pelvis rotates forward at the rate of 2° per each 3° of trunk flexion; the spine contributes the additional 1°. In contrast, when the thighs are flexed more than 10° to 15°, the pelvis rotates backward at a rate of 1° per each 3° of the thigh flexion. In complex movement involving both trunk bending and knee flexion, these two effects are subtracted from each other.

Head and neck movements occur around a distributed set of axes of rotation. However, customarily the movement is analyzed as a combined rotation around two fixed axes—at the C7-T1 and at the occipital-atlantal joint (the dual-joint model). In this technique, the trunk-neck-head system is considered a three-link planar kinematic chain. The instant center of rotation for the head in the reference frame anchored to the trunk is then determined using Kennedy's theorem. The multiple chain model includes several vertebrae, and the continuous model contains an infinite number of segments with vanishing length but totaling a fixed neck length.

The shoulder complex can be viewed as a complex mechanism comprising two individual mechanisms: (a) the shoulder girdle, consisting of the clavicle and scapula as the moving links and the sternum with the rib cage as a frame, and (b) the humerus as a moving link with the scapula and clavicle as a frame. Both mechanisms are spatial, with the first mechanism being closed and the second open. In the literature, movement of a shoulder bone is defined with regard to (1) the adjacent proximal segment; (2) the trunk; (3) the sternum, a longitudinal axis of the sternum is inclined to the vertical axis of the trunk; and (4) the virtual reference position. The humeral position is aptly described by (a) a plane of elevation, (b) an elevation angle, and (c) an axial rotation.

In the glenohumeral joint, mating joint surfaces have radii within a 3 mm difference. Depending on the required accuracy the joint can be either modeled as a ball-and-socket joint—in this case the difference in the radii is disregarded—or represented as a joint with a complex pattern of movement. In the latter case, the joint movement is considered a combined motion comprising two elementary rotations in opposite directions: around the glenoidal center and around the humeral head center. As a result, the humeral head rotates around the ICR. In the joint, two types of abduction are discerned, frontal plane abduction and scapular plane abduction. The scapular plane is defined as the plane of the scapula in the resting position.

The motion of the scapula over the thorax is not planar and because of that an isolated rotation or translation around or along an individual axis cannot be performed. For that reason, the classic anatomic angles cannot be determined and Euler's angles should be used to define the scapular orientation. Because the scapula has no definite resting orientation, its attitude is defined with regard to the virtual reference position. The sternoclavicular joint is a ball-and-socket joint with three DOF. The clavicle operates as a bony strut with two articulations at the ends.

The shoulder complex as a whole provides seven DOF for the arm movement, four at the shoulder girdle and three at the glenohumeral joint. When the movement of the humerus relative to the axial skeleton is the object of interest, the humerus is an end link of the following kinematic chain: trunk (sternum)→sternoclavicular joint→clavicle→acromioclavicular joint→scapula→glenohumeral joint→humerus. To describe the shoulder configuration with regard to the sternum system, nine Euler's angles, three for each bone system, are necessary. The collective movement of the humerus, scapula, and clavicle during arm elevation is called the scapulohumeral rhythm, or the shoulder rhythm. In its entirety, the ratio of glenohumeral to scapular rotation is approximately 2:1. Because of the displacement of the shoulder girdle during arm movements, the ICR for the shoulder complex differs greatly from the ICR for the shoulder joint.

The elbow joint complex allows two DOF, flexion-extension and pronation-supination. Flexion-extension is nearly pure rotation. When the elbow joint is fully extended and supinated, the long axis of the upper arm and that of the forearm form a carrying angle. The carrying angle for arm positions other than complete extension is defined as an Euler's angle. This is the angle between the long axis of the forearm and the projection of the long axis of the humerus onto the forearm plane. The varus angle is between the forearm and a normal to the average axis of rotation. The varus angle does not alter during elbow flexion, but the carrying angle changes from valgus to varus attitude (decreases) in a linear fashion. During pronation-supination the axis of rotation extends from the center of the radial head in the elbow joint to the center of the distal ulna. The axis is oblique to the longitudinal axis of the forearm.

The wrist contains several joints. Taken as an entity, the joints are considered one joint, the wrist joint. The wrist joint possesses two DOF, flexion-extension and abduction-adduction. The abduction-adduction axis passes distally with regard to the flexion-extension axis by approximately 5 mm.

The carpometacarpal joint is a saddle joint with two axes of rotation. Although the joint motion occurs in all three anatomic planes, only two independent motions are possible. The first is abduction-adduction and the second is a complex articular motion combining thumb flexion with pronation and thumb extension with supination. Such a coupled movement is unique to humans in that it allows the best thumb position for opposition. The metacarpophalangeal joints have two DOF. The interphalangeal joints are hinge joints with one DOF. The interphalangeal joints have fixed centers of rotation that coincide with the center of curvature of the head of the male bone.

The main mandibular motions are depression and elevation (mouth opening and closing), protrusion and retrusion (moving the chin forward and backward) and lateral deviation, or laterotrusion.

5.13 QUESTIONS FOR REVIEW

1. What are the nominal joint axes?
2. Locate the nominal joint axes of the following joints: ankle, knee, hip, shoulder, elbow, wrist.
3. Discuss the foot as a two-speed construction.
4. What is the metatarsal break?
5. What kind of foot deformity is hallux valgus?
6. Define Lisfranc's joint and Chopart's joint.
7. How are the axes of rotation of the ankle joint complex located?
8. Discuss the main kinematics models of the knee joint.
9. What is the screw-home mechanism?

10. Describe the movement of the patella during knee flexion.

11. How can location of the hip joint center of rotation be estimated?

12. Discuss the difference between the movement in synarthroses and diarthroses.

13. Describe the movement of the spinal motion segments. Compare different regions of the spine.

14. What is the lumbar-pelvic rhythm?

15. Estimate the pelvic inclination for a person standing with the knees flexed 90° and the trunk inclined 30° (use Figure 5.33).

16. Using the dual-joint model find the location of the instant center of rotation of the head with regard to the trunk. The angular velocity of the head with regard to the neck is 10 rad/sec; the angular velocity of the neck with regard to the trunk is 5 rad/sec; the length of the neck is 1.

17. Define the virtual reference positions for the bones of the shoulder complex. Why are the virtual positions introduced?

18. You are planning to describe a position of the humerus with regard to the trunk as a function of the shoulder bones' positions. How many transformation matrices should you use? Describe these matrices. In total, how many direction cosines should be known to solve the problem?

19. Discuss the kinematics of the glenohumeral joint. Assume that a high accuracy is required and the joint cannot be considered a ball-and-socket joint.

20. Define the scapular plane abduction.

21. Discuss scapular motion.

22. Name the sequence of the Euler's angles recommended for the defining of the humeral position with regard to the trunk.

23. Discuss the shoulder rhythm. When the arm is abducted 170°, what is (approximately) the magnitude of the scapular rotation?

24. Define the carrying angle and the varus angle for an arbitrary position of the elbow joint.

25. How are the axes of rotation of the wrist joint located?

26. Discuss kinematics of the carpometacarpal joint.

27. Using the data from Figure 5.57 and Table 5.8 find the location of the finger joint centers.

28. Discuss kinematics of the temporomandibular joint.

BIBLIOGRAPHY

For joints in general see the bibliography for Chapter **4**.

Nominal joint axes

Chandler, R.F., Clauser, C.E., McConville, J.T., Reynolds, H.M., & Young, J.W. (1975). *Investigation of inertial properties of the human body* (AMRL TR 74-137). Wright-Patterson Air Force Base, Ohio (NTIS No. AD-A016485).

Clauser, C.E., McConville, J.T., & Young, J.W. (1969). *Weight, volume and center of mass of segments of the human body* (AMRL TR 69-70). Wright-Patterson Air Force Base, Ohio (NTIS No. AD-710 622).

Dempster, W.T. (1955). *Space requirements of the seated operator* (Technical Report WADC-TR-55-159). Wright-Patterson Air Force Base, OH: Aerospace Medical Research Laboratory.

Leva, P. de. (1996). Joint center longitudinal positions computed from a selected subset of Chandler's data. *Journal of Biomechanics, 29,* 1231-1233.

Plagenhoef, S. (1971). *Patterns of human motion. A cinematographic analysis.* Englewood Cliffs, NJ: Prentice Hall.

The joints of the foot

Review papers

Gershman, S. (1988). A literature review of midtarsal joint function. *Clinics in Podiatric Medicine and Surgery, 5,* 385-391.

Original papers

Bojsen-Møller, F. (1978). The human foot—a two speed construction. In E. Asmussen & K. Jørgensen (Eds.). *Biomechanics VI-A* (pp. 261-266). Baltimore: University Park Press.

Elftman, H. (1960). The transverse tarsal joint and its control. *Clinical Orthopaedics and Related Research, 16,* 41-46.

Hicks, J.H. (1954). The mechanics of the foot. I. The joints. *Journal of Anatomy, 87,* 345-357.

Mann, R.A. & Inman, V.T. (1964). Phasic activity of intrinsic muscles of the foot. *Journal of Bone and Joint Surgery, 46A,* 469-476.

Manter, J.T. (1941). Movements of the subtalar and transverse tarsal joints. *Anatomical Record, 80,* 397-410.

Sammarco, G.J. (1989). Biomechanics of the foot. In M. Nordin & V.H. Frankel (Eds.), *Basic biomechanics of the musculoskeletal system* (pp. 163-181). Philadelphia: Lea & Febiger.

Shephard, E. (1951). Tarsal movements. *Journal of Bone and Joint Surgery, 33B,* 258-263.

Ankle

Books and review papers

Huson, A. (1987). Joints and movements of the foot: Terminology and concepts. *Acta Morphologica Neerlando-Scandinavica, 25,* 117-130.

Inman, V.T. (1976). *The joints of the ankle.* Baltimore: Williams & Wilkins.

Jones, F.W. (1944). *Structure and function as seen in the foot.* London: Bailliere, Tindal & Cox.

Original papers

Areblad, M., Nigg, B.M., Ekstrand, J., Olsson, K.O., & Ekström, H. (1990). Three-dimensional measurement of rearfoot motion during running. *Journal of Biomechanics, 23,* 933-940.

Barnett, C.H. & Napier, J.R. (1952). The axis of rotation at the ankle joint in man. Its influence upon the form of the talus and the mobility of the fibula. *Journal of Anatomy, 86,* 1-9.

Bogert, A.J. van den, Smith, G.D., & Nigg, B.M. (1994). In vivo determination of the anatomical axes of the ankle joint complex: An optimization approach. *Journal of Biomechanics, 27,* 1477-1488.

Chen, J., Siegler, S., & Schneck, C.D. (1988). The three-dimensional kinematics and flexibility characteristics of the human ankle and subtalar joint. Part II: Flexibility characteristics. *Journal of Biomechenical Engineering, 110,* 374-385.

Dempster, W.T. (1955). *Space requirements of the seated operator.* (Technical Report WADC-TR-55-159). Wright-Patterson Air Force Base, OH: Aerospace Medical Research Laboratory.

Dul, J. & Johnson, G.E. (1985). A kinematic model of the human ankle. *Journal of Biomechanical Engineering, 107,* 137-142.

Engsberg, J.R. (1987). A biomechanical analysis of the talocalcaneal joint in vitro. *Journal of Biomechanics, 20,* 429-442.

Engsberg, J.R., Krimston, S.K., & Wackwitz, J.H. (1988). Predicting talocalcaneal orientations from talocalcaneal/talocrural joint orientation. *Journal of Orthopaedic Research, 6,* 749-757.

Hicks, J.H. (1953). The mechanics of the foot. I. The joints. *Journal of Anatomy, 87,* 345-357.

Isman, R.I. & Inman, V.T. (1969). Anthropometric studies of the human foot and ankle. *Bulletin of Prosthetics Research, 10/11,* 97-129.

Jones, R.L. (1945). The functional significance of the declination of the axis of the subtalar joint. *Anatomical Record, 93,* 151-159.

Kirby, K.A. (1987). Methods for determination of positional variations in the subtalar joint axis. *Journal of the American Podiatric Medical Association, 77,* 228-234.

Langelaan, E.J. van (1983). A kinematic analysis of the tarsal joints. *Acta Orthopaedica Scandinavica, 54*(Suppl.), 1-267.

Lundberg, A., Goldie, I., Kalin, B., & Selvik, G. (1989a). Kinematics of the ankle/foot complex. 1. Plantarflexion and dorsiflexion. *Foot and Ankle, 9,* 194-200.

Lundberg, A., Goldie, I., Kalin, B., & Selvik, G. (1989b). Kinematics of the ankle/foot complex. 2. Pronation and supination. *Foot and Ankle, 9,* 248-253.

Lundberg, A., Goldie, I., Kalin, B., & Selvik, G. (1989c). Kinematics of the ankle/foot complex. 3. Influence of leg rotation. *Foot and Ankle, 9,* 304-309.

Lundberg, A. & Svensson, O.K. (1993). The axes of rotation of the talocalcaneal and talonavicular joints. *The Foot, 3,* 65-70.

Lundberg, A., Svensson, O.K., Nemeth, G., & Selvik, G. (1989). The axis of rotation of the ankle joint. *Journal of Bone and Joint Surgery, 71B,* 94-99.

Mann, R.A. (1976). Biomechanics of the foot. In *Atlas of orthotics: Biomechanical principles and applications* (pp. 257-266). American Academy of Orthopaedic Surgeons. St. Louis: C.V. Mosby, Co.

Manter, J.T. (1941). Movement of the subtalar and transverse tarsal joints. *Anatomical Record, 80,* 397-410.

Metz-Schimmerl, S.M. (1994). Visualization and quantitative analysis of talocrural joint kinematics. *Computerized Medical Imaging Graphics, 18,* 443-448.

Procter, P., Berme, N., & Paul, J.P. (1981). Ankle joint biomechanics. In A. Morecki, K. Fidelus, K. Kedsior, & A. Wit (Eds.), *Biomechanics VII-A* (pp. 52-56). Baltimore: University Park Press.

Procter, P. & Paul, J.P. (1981). Ankle joint biomechanics. *Journal of Biomechanics, 15,* 627-634.

Sammarco, G.J., Burstein, A.H., & Frankel, V.H. (1973). Biomechanics of the ankle: A kinematic study. *Orthopedic Clinics of North America, 4,* 75-96.

Scott, S.H. & Winter, D.A. (1991). Talocrural and talocalcaneal joint kinematics and kinetics during the stance phase of walking. *Journal of Biomechanics, 24,* 743-752.

Siegler, S., Chen, J., & Schneck, C.D. (1988a). The three-dimensional kinematics and flexibility characteristics of the human ankle and subtalar joint. Part 1: Kinematics. *Journal of Biomechanical Engineering, 110,* 364-373.

Siegler, S., Chen, J., & Schneck, C.D. (1988b). The three dimensional kinematics of the human ankle and subtalar joints. In G. de Groot, A.P. Hollander, P.A. Hujing, & G.J. van Ingen Schenau (Eds.), *Biomechanics XI-A.* (pp. 417-422) Amsterdam: Free University Press.

Siegler, S., Chen, J., Selikar, R., & Schneck, C.D. (1990). A system for investigating the kinematics of the human ankle and subtalar joint based on Rodrigues' formula and the screw axes parameters. In N. Berme & A. Cappozzo (Eds.) *Biomechanics of human movement: Applications in rehabilitation, sports and ergonomics* (pp. 278-283). Worthington: Bertec Corporation.

Singh, A.K., Starkweather, K., Hollister, A.M., Jatana, S., & Lupichuk, A. (1992). The kinematics of the talocural joint: A hinge axis model. *Foot and Ankle, 13,* 439-445.

Soutas-Little, R.W., Beavis, G.C., Verstraete, M.C., & Markus, T.L. (1987). Analysis of foot motion during running using a joint coordinate system. *Medicine and Science in Sports and Exercise, 19,* 285-293.

Stacoff, A., Steger, J., & Stussi, E. (1993). Lateral stability in sidewalk cutting movements. In *XIVth Congress of the International Society of Biomechanics, Abstracts* (pp. 1278-1279). Paris.

Wright, D.G., Desai, S.M., & Henderson, W.H. (1964). Action of subtalar and ankle-joint complex during the stance phase of walking. *Journal of Bone and Joint Surgery, 46A,* 361-381.

Knee

Books and review papers

Daniel, D.M., Akeson, W.H., & O'Connor, J.J. (Eds.). (1990). *Knee ligaments: Structure, function, injury, and repair.* New York: Raven Press.

Hefzy, M.S. & Grood, E.S. (1988). Review of knee models. *Applied Mechanics Review, 41,* 1-13.

Maquet, P.G. (1984). *Biomechanics of the knee* (2nd ed.). Berlin: Julius Springer.

Nissan, M. (1980). Review of some basic assumptions in knee biomechanics. *Journal of Biomechanics, 13,* 375-381.

Nordin, M. & Frankel, V.H. (1989). Biomechanics of the knee. In M. Nordin & V.H. Frankel (Eds.), *Basic biomechanics of the musculoskeletal system* (pp. 115-134). Philadelphia: Lea & Febiger.

Original papers

Blacharski, K. & Somerset, J.H. (1975). A three-dimensional study of the kinematics of the human knee. *Journal of Biomechanics, 8,* 375-384.

Blankevoort, L., Huiskes, R., & de Lange, A. (1988). The envelope of passive knee joint motion. *Journal of Biomechanics, 21,* 705-720.

Blankevoort, L., Huiskes, R., & de Lange, A. (1990). Helical axes of passive joint motions. *Journal of Biomechanics, 23,* 1219-1229.

Brunet, M.E., Kester, M.A., Cook, S.D., Haddad, R.J., & Skinner, H.B. (1986). Determination of the transverse centre of rotation of the knee using CAT scans. *Engineering in Medicine, 15,* 143-147.

Crowninshield, R., Pope, M.H., & Johnson, R.J. (1976). An analytical model of the knee. *Journal of Biomechanics, 9,* 397-405.

Eijeden, T.M.G.J. van, Kouwenhoven, E., Verburg, J., & Weijes, W.A. (1986). A mathematical model of the patello-femoral joint. *Journal of Biomechanics, 19,* 219-229.

Frankel, V.H., Burstein, A.H., & Brooks, D.B. (1971). Biomechanics of internal derangement of the knee. Pathomechanics as determined by analysis of the instant centers of motion. *Journal of Bone and Joint Surgery, 53A,* 945-962.

Freudenstein, F. & Woo, S.L. (1969). Kinematics of the human knee joint. *Bulletin of Mathematical Biophysics, 31,* 215-232.

Fukubayashi, T., Torzilli, P.A., Sherman, M.F., & Warren, R.R. (1982). An in vitro biomechanical evaluation of anterior-posterior motion of the knee. *Journal of Bone and Joint Surgery, 64A,* 258.

Goodfellow, J., Hungerford, D.S., & Zindel, M. (1976). Patello-femoral joint mechanics and pathology. 1. Functional anatomy of the patello-femoral joint. *Journal of Bone and Joint Surgery, 58B,* 287-290.

Hallen, L.G. & Lindahl, O. (1966). The screw-home movement in the knee joint. *Acta Orthopaedica Scandinavica, 37,* 97-106.

Harding, M.L. & Blakemore, M.E. (1980). The instant centre pathway as a parameter of joint motion: An experimental investigation of a method of assessment of knee ligament injury and repair. *Engineering in Medicine, 9,* 195-200.

Hart, R.A., Mote, C.D., & Skinner, H.B. (1991). A finite helical axis as a landmark for kinematic reference of the knee. *Journal of Biomechanical Engineering, 113,* 215-222.

Helfet, A.J. (1974). Anatomy and mechanics of movement in the knee joint. In A.J. Helfet (Ed.), *Disorders of the knee* (pp. 1-17). Philadelphia: J.B. Lippincott Co.

Hirokawa, S. (1991). Three-dimensional mathematical model analysis of the patellofemoral joint. *Journal of Biomechanics, 24,* 659-671.

Hollister, A.M., Jatana, S., Singh, A.K., Sullivan, W.S., & Lupichuk, A. (1993). The axes of rotation of the knee. *Clinical Orthopaedics and Related Research, 290,* 259-268.

Hungerford, D.S. & Barry, M. (1979). Biomechanics of the patellofemoral joint. *Clinical Orthopaedics and Related Research, 144,* 9-15.

Huson, A., Spoor, C.W., & Verbout, A.J. (1989). A model of the human knee, derived from kinematic principles and its relevance for endoprosthesis design. *Acta Morphologica Neerlando-Scandinavica, 27,* 45-62.

Kubein-Meesenburg, D., Nägerl, H., & Fanghänel, J. (1991). Elements of a general theory of joints. 5. Basic mechanics of the knee. *Anatomische Anzeiger, 173,* 131-142.

Lafortune, M.A., Cavanagh, P.R., Sommer, H.J., III, & Kalenak, A. (1992). Three-dimensional kinematics of the human knee during walking. *Journal of Biomechanics, 25,* 347-357.

Lange, A. de, Huiskes, R., & Kauer, J.M.G. (1990). Effects of data smoothing on the reconstruction of helical axis parameters in human joint kinematics. *Journal of Biomechanical Engineering, 112,* 107-113.

Laubenthal, K.N., Smidt, G.L., & Kettelkamp, D.B. (1972). A quantitative analysis of knee motion during activities of daily living. *Physical Therapy, 52,* 34-42.

Levens, A.S., Inman, V.T., & Blosser, J.A. (1948). Transverse rotation of the segments of the lower extremity in locomotion. *Journal of Bone and Joint Surgery, 30A,* 859-872.

Lindhal, O. & Movin, A. (1967). The mechanics of extension of the knee-joint. *Acta Orthopaedica Scandinavica, 38,* 226-234.

Markoff, K.L., Graff-Radford, A., & Amstutz, H.C. (1978). In vivo knee stability. *Journal of Bone and Joint Surgery, 60A,* 664-674.

Noyes, F.R., Grood, E.S., & Torzilli, P.A. (1989). The definition of terms for motion and position of the knee and injuries of the ligaments. [Current concept review]. *Journal of Bone and Joint Surgery, 71A,* 465-472.

Rajendran, K. (1985). Mechanism of locking at the knee joint. *Journal of Anatomy, 143,* 189.

Reuben, J.D., Rovick, J.S., Schrager, R.J., Walker, P.S., & Boland, A.L. (1989). Three-dimensional dynamic motion analysis of the anterior cruciate ligament deficient knee joint. *American Journal of Sports Medicine, 17,* 463-471.

Shiavi, R., Limbird, T., Frazer, M., Stivers, K., Strauss, A., & Abramovitz, J. (1987). Helical motion analysis of the knee. II. Kinematics of uninjured and injured knees during walking and pivoting. *Journal of Biomechanics, 20,* 653-665.

Smidt, G.L. (1973). Biomechanical analysis of knee flexion and extension. *Journal of Biomechanics, 6,* 79-92.

Soudan, K., Auderkercke, R.V., & Martens, M. (1979). Methods, difficulties and inaccuracies in the study of human joint mechanics and pathomechanics by the instant axis concept, example: The knee joint. *Journal of Biomechanics, 12,* 27-33.

Straiton, J.A., Todd, B., & Venner, R.M. (1987). Radiographic assessment of knee joint rotation. *Journal of Anatomy, 155,* 189-193.

Terajima, K., Terashima, S., Hara, H., Ishii, Y., & Koga, Y. (1994). Definition of three-dimensional motion parameters for the analysis of knee joint motion. In Y. Hirasawa, C.B. Sledge, &. S.L.-Y. Woo (Eds.), *Clinical biomechanics and related research* (pp. 383-391). Tokyo: Springer-Verlag.

Thoma, W., Jäger, A., & Schreiber, S. (1994). Kinematic analysis of the knee joint with regard to the load transfer on the cartilage. In Y. Hirasawa, C.B. Sledge, &. S.L.-Y. Woo (Eds.), *Clinical biomechanics and related research* (pp. 96-102). Tokyo: Springer-Verlag.

Townsend, M.A., Izak, M., & Jackson, R.W. (1977). Total motion knee goniometry. *Journal of Biomechanics, 10,* 183-193.

Walker, P.S., Shoji, H., & Erkman, M.J. (1972). The rotational axis of the knee and its significance to prosthesis design. *Clinical Orthopaedics and Related Research, 89,* 160-168.

Wang, C.-J. & Walker, P.S. (1974). Rotatory laxity of the human knee joint. *Journal of Bone and Joint Surgery, 56A,* 161-170.

Wismans, J., Veldpaus, F., Janssen, J., Huson, A., & Struben, P. (1980). A three-dimensional mathematical model of the knee joint. *Journal of Biomechanics, 13,* 677-685.

Yamaguchi, G.T. & Zajac, F.E. (1989). A planar model of the knee joint to characterize the knee extensor mechanism. *Journal of Biomechanics, 22,* 1-10.

Hip

Original papers

Bell, A.L., Brand, R.A., & Pedersen, D.R. (1989). Prediction of hip joint center location from external landmarks. *Human Movement Science, 8,* 3-16.

Bell, A.L., Pedersen, D.R., & Brand, R.A. (1990). A comparison of the accuracy of several hip center location prediction methods. *Journal of Biomechanics, 23,* 617-621.

Bullough, P.G., Goodfellow, J., Greenwald, A.S., & Connor, J.J. (1968). Incongruent surfaces in the human hip joint. *Nature, 217,* 1290.

Cappozzo, A. (1984). Gait analysis methodology. *Human Movement Science, 3,* 27-54.

Chao, E.Y.S., Rim, K., Smidt, G.L., & Johnston, R.C. (1970). The application of 4×4 matrix method to the correction of the measurements of hip joint rotations. *Journal of Biomechanics, 3,* 459-471.

Crowninshield, R.D., Johnston, R.C., Andrews, J.G., & Brand, R.A. (1978). A biomechanical investigation of the human hip. *Journal of Biomechanics, 11,* 75-85.

Davis, R.B. III, Ounpuu, S., Tyburski, D., & Gage, J.R. (1991). A gait analysis data collection and reduction technique. *Human Movement Science, 10,* 575-587.

Day, J.W., Smidt, G.L., & Lehmann, T. (1984). Effect of pelvic tilt on standing posture. *Physical Therapy, 64,* 510-516.

Delp, S.L. & Maloney, W. (1993). Effects of hip center location on the moment-generating capacity of the muscles. *Journal of Biomechanics, 26,* 485-499.

Drerup, B. & Hierholzer, E. (1987). Movement of the human pelvis and displacement of related anatomical landmarks on the body surface. *Journal of Biomechanics, 20,* 971-977.

Neptune, R.R. & Hull, M.L. (1995). Accuracy assessment of methods for determining hip movement in seated cycling. *Journal of Biomechanics, 28,* 423-437.

Nordin, M. & Frankel, V.H. (1989). Biomechanics of the hip. In M. Nordin & V.H. Frankel (Eds.), *Basic biomechanics of the musculoskeletal system* (pp. 135-151). Philadelphia: Lea & Febiger.

Reynolds, H.M. (1980). Three-dimensional kinematics in the pelvis girdle. *Journal of the American Osteopathic Association, 80,* 277-280.

Seidel, G.K., Marchinda, D.M., Dijkers, M., & Soutas-Little, R.W. (1995). Hip joint center location from palpable bony landmarks: A cadaver study. *Journal of Biomechanics, 28,* 995-998.

Spine

Books and review articles

Adams, M.A. & Dolan, P. (1995). Recent advances in lumbar spine mechanics and their clinical significance. *Clinical Biomechanics, 10,* 3-19.

Chaffin, D.B. & Andersson, G.B.J. (1984). *Occupational biomechanics.* New York: John Wiley & Sons.

Erdman, H. (1979). *Die körperlichen Untersuchung. Die Wirbelsäule in Forschung und Praxis* (vol. 83). Stuttgart: Hippokrates.

Junghanns, H. (1990).*Clinical implications of normal biomechanical stresses on spinal function.* Rockville, MD: Aspen Publishers, Inc.

Kapandji, J.A. (1972). *Physiologie articulaire* (vol. 3). Paris: Maloine.

MacConaill, M.A. & Basmajian, J.V. (1969). *Muscles and movements. A basis for human kinesiology.* Baltimore: Williams & Wilkins.

Norkin, C.C. & Levangie, P.K. (1992). *Joint structure and function. A comprehensive analysis* (2nd ed.) Philadelphia: F.A. Davis Company.

White, A.A. & Panjabi, M.M. (1975). Spinal kinematics. In M. Goldstein (Ed.), *The research status of spinal manipulative therapy* (pp. 93-102). Bethesda, Maryland: U.S. Department of Health, Education, and Welfare.

White, A.A. & Panjabi, M.M. (1978). The basic kinematics of the human spine: A review of past and current knowledge. *Spine, 3,* 12-20.

White, A.A. & Panjabi, M.M. (1990). *Clinical biomechanics of the spine* (2nd ed.). Philadelphia: J.B. Lippincott Co.

Lumbar and thoracic spine; sacroiliac joints

Original papers

Adams, M.A. & Dolan, P. (1991). A technique for quantifying bending moment acting on the lumbar spine in-vivo. *Journal of Biomechanics, 24,* 117-126.

Adams, M.A., Dolan, P., & Hutton, W.C. (1988). The lumbar spine in backward bending. *Spine, 13,* 1019-1026.

Adams, M.A., Dolan, P., Marx, C., & Hutton, W.C. (1986). An electronic inclinometer technique for measuring lumbar curvature. *Clinical Biomechanics, 1,* 130-134.

Adams, M.A. & Hutton, W.C. (1986). Has the lumbar spine a margin of safety in forward bending. *Clinical Biomechanics, 1,* 3-6.

Anderson, C.K., Chaffin, D.B., Herrin, G.D., & Matthews, L.S. (1985). A biomechanical model of the lumbosacral joint during lifting activities. *Journal of Biomechanics, 18,* 571-584.

Burton, A.K. & Tillotson, K.M. (1988). Reference values for "normal" regional lumbar sagittal mobility. *Clinical Biomechanics, 3,* 106-113.

Burton, A.K., Tillotson, K.M, & Troup, J.D.G. (1989). Variation in lumbar sagittal mobility with low back trouble. *Spine, 14,* 584-590.

Clayson, S.J., Newman, I.M., Debevec, D.F., Anger, R.W., Skowlund, H.V., & Kottke, F.J. (1962). Evaluation of mobility of hip and lumbar vertebrae of normal young women. *Archive of Physical Medicine and Rehabilitation, 43,* 1-8.

Cossette, J.W., Farfan, H.F., Robertson, G.H., & Wells, R.V. (1971). The instantaneous center of rotation of the third lumbar intervertebral joint. *Journal of Biomechanics, 4,* 149-153.

Davis, P.R., Troup, J.D.G., & Burnard, J.H. (1965). Movements of the thoracic and lumbar spine when lifting: A chrono-cyclophotographic study. *Journal of Anatomy (London), 99,* 13-26.

Evans, F.G. & Lissner, M.S. (1959). Biomechanical studies on the lumbar spine and pelvis. *Journal of Bone and Joint Surgery, 41A,* 379-385.

Frymoyer, J.W., Frymoyer, W.W., Wilder, D.G., & Pope, M.H. (1979). The mechanical and kinematic analysis of the lumbar spine in vivo. *Journal of Biomechanics, 12,* 165-172.

Gertzbein, S.D., Seligman, J., Holtby, R., Chan, K.H., Kapasouri, A., Tile, M., & Cruickshank, B. (1985). Centrode patterns and segmental instability in degenerative disc disease. *Spine, 10,* 257-261.

Gianturco, C. (1944). A roentgen analysis of the motion of the lumbar vertebrae in normal individuals and in patients with low back pain. *American Journal of Roentgenology, 52,* 261-268.

Gracovetsky, S., Farfan, H., & Lamy, C. (1981). The mechanism of the lumbar spine. *Spine, 6,* 249-262.

Gregersen, G.G. & Lucas, D.B. (1967). An in vivo study of the axial rotation of the human thoracolumbar spine. *Journal of Bone and Joint Surgery, 49A,* 247-262.

Gunzburg, R., Hutton, W., & Fraser, R. (1991). Axial rotation of the lumbar spine and the effect of flexion. *Spine, 16,* 22-29.

Haher, T.R., Bergman, M., O'Brien, M., Felmy, W.T., Choueka, J., Welin, D., Chow, G., & Vassiliou, A. (1991). The effect of the three columns of the spine on the instantaneous axis of rotation in flexion and extension. *Spine, 8,* 312-318.

Hindle, R.J., Pearcy, M.J., Cross, A.T., & Miller, D.H.T. (1990). Three-dimensional kinematics of the human back. *Clinical Biomechanics, 5,* 218-228.

Hirsch, C. & Lewin, T. (1968). Lumbosacral synovial joints in flexion-extension. *Acta Orthopaedica Scandinavica, 39,* 303-311.

Keessen, W., During, J., Beeker, T.W., Goudfroji, H., & Crowe, A. (1984). Recording of the movement at the intervertebral segment L5-S1: A technique for the determination of the movement in the L5-S1 spinal segment by using three specified postural positions. *Spine, 9,* 83-90.

Kippers, V. & Parker, A.W. (1989). Validation of single segment and three-segment spinal models used to represent lumbar flexion. *Journal of Biomechanics, 22,* 67-75.

Klein, J.A. & Hulins, D.W.L. (1983). Relocation of the bending axis during flexion-extension of lumbar intervertebral discs and its implication for prolapse. *Spine, 8,* 659-664.

LeVieu, B.F. (1974). Axes of joint rotation of the lumbar vertebrae during abdominal strengthening exercises. In R.C. Nelson & C.A. Morehouse (Eds.), *Biomechanics IV* (pp. 361-364). Baltimore: University Park Press.

Lumsden, R.M. & Morris, J.M. (1968). An in vivo study of axial rotation and immobilization at the lumbosacral joint. *Journal of Bone and Joint Surgery, 50A,* 1591-1602.

Lysell, E. (1969). Motion in the cervical spine. *Acta Orthopaedica Scandinavica* (Suppl. 123).

Macrae, I.F. & Wright, V. (1969). Measurement of back movement. *Annals of the Rheumatic Diseases, 28,* 584-589.

Mayer, T.G., Tencer, A.F., Kristoferson, S., & Money, V. (1984). Use of non-invasive techniques for quantification of spinal range-of-motion in normal subjects and chronic low-back dysfunction patients. *Spine, 9,* 588-595.

McNeill, T., Warwick, D., Anderson, G.B.J., & Schultz, A.B. (1980). Trunk strengths in attempted flexion, extension and lateral bending in healthy subjects with low back disorders. *Spine, 5,* 529-538.

Mellin, G.P. (1989). Comparison between tape measurements of forward and lateral flexion of the spine. *Clinical Biomechanics, 4,* 121-123.

Miles, M. & Sullivan, W.F. (1961). Lateral bending at the lumbar and the lumbosacral joints. *Anatomical Record, 139,* 387-398.

Miller, S.A., Mayer, T., Cox, R., & Catchel, R.J. (1992). Reliability problems associated with the modified Schober technique for true flexion measurement. *Spine, 17,* 345-348.

Moll, J.M.H. & Wright, V. (1971). Normal range of spinal mobility. *Annals of the Rheumatic Diseases, 30,* 381-386.

Nägerl, H., Kubein-Meesenburg, D., Becker, B., & Fanghänel, J. (1991). Elements of a general theory of joints. 3. Elements of theory of synarthrosis: Experimental examinations. *Anatomische Anzeiger, 172,* 187-195.

Nägerl, H., Kubein-Meesenburg, D., & Fanghänel, J. (1990). Elements of a general theory of joints. 2. Introduction into a theory of synarthrosis. *Anatomische Anzeiger, 171,* 323-333.

Ogston, N.G., King, G.J., Gerztbein, S.D., Tile, M., Karasouri, A., & Rubenstein, J.D. (1986). Centrode patterns in the lumbar spine: baseline studies in normal subjects. *Spine, 11,* 591-595.

Panjabi, M.M., Krag, M.H., Dimnet, J.C., Walter, S.D., & Brand, R.A. (1984). Thoracic spine centres of rotation in the sagittal plane. *Journal of Orthopaedic Research, 1,* 387-394.

Panjabi, M., Yamamoto, I., Oxland, T., & Crisco, J. (1989). How does posture affects coupling in the lumbar spine? *Spine, 14,* 1002-1011.

Pearcy, M.J. (1985). Stereoradiography of lumbar spine motion. *Acta Orthopaedica Scandinavica, 56*(Suppl. 212), 1-45.

Pearcy, M.J. (1993). Twisting mobility of the human back in flexed postures. *Spine, 18,* 114-119.

Pearcy, M. & Bogduk, N. (1988). Instantaneous axis of rotation in the lumbar intervertebral joints. *Spine, 13,* 1033-1041.

Pearcy, M.J., Gill, J.M., Whittle, M.W., & Johnson, G.R. (1987). Dynamic back movement measured using a three-dimensional television system. *Journal of Biomechanics, 20,* 943-949.

Pearcy, M.J. & Hindle, R.J. (1989). New method for the non-invasive three-dimensional measurement of human back movement. *Clinical Biomechanics, 4,* 73-79.

Pearcy, M., Portek, I., & Shepherd, J. (1984). Three-dimensional X-ray analysis of normal movement in the lumbar spine. *Spine, 9,* 294-297.

Pearcy, M.J. & Tibrewal, S.B. (1984). Axial rotation and lateral bending in the normal lumbar spine measured by three-dimensional radiography. *Spine, 9,* 582-587.

Plamondon, A., Gagnon, M., & Dejardins, P. (1996). Validation of two 3-D segment models to calculate the net reaction forces and moments at the L5/S1 joint in lifting. *Clinical Biomechanics, 11,* 101-110.

Portek, I., Pearcy, M.J., Reader, G.P., & Mowart, A.G. (1983). Correlation between radiographic and clinical measurement of lumbar spine movement. *British Journal of Rheumatology, 22,* 197-205.

Putto, E. & Tallroth, K. (1990). Extension-flexion radiograph for motion studies of the lumbar spine. A comparison of two methods. *Spine, 15,* 107-110.

Reichmann, S. (1969). Motion of the lumbar articular processes in flexion-extension and lateral flexion of the spine. *Acta Morphologica Neerlando-Scandinavica, 8,* 261-272.

Reichmann, S., Berglund, E., & Lundgren, K. (1972). Das Bewegungszentrum in der Lendenwirbelsäule bei Flexion und Extension. *Zeitschrift für Anatomie und Entwicklungsgeschichte, 136,* 283-287.

Reynolds, P.M.G. (1975). Measurement of spinal mobility: A comparison of three methods. *Rheumatology and Rehabilitation, 14,* 180-185.

Schultz, A.B., Warwick, D.N., Berkson, M.H., & Nachemson, A.L. (1979). Mechanical properties of human spine motion segments. Part 1: Responses in flexion, extension, lateral bending, and torsion. *Journal of Biomechanical Engineering, 101,* 46-52.

Stokes, I.A.F., Bevins, T.M., & Lunn, R.A. (1987). Back surface curvature and measurement of lumbar spine motion. *Spine, 12,* 355-361.

Sturesson, B., Selvick, G., & Uden, A. (1989). Movements of the sacroiliac joints: A roentgen stereophotogrammetric analysis. *Spine, 14,* 162-165.

Thurston, A.J. & Harris, J.D. (1983). Normal kinematics of the lumbar spine and pelvis. *Spine, 8,* 199-205.

Tilitson, K.M., & Burton, A.K. (1991). Non-invasive measurement of lumbar sagittal mobility. An assessment of the flexicurve technique. *Spine, 16,* 29-34.

Twomey, L.T. & Taylor, J.R. (1983). Sagittal movements of the human vertebral column: A qualitative study of the role of the posterior vertebral elements. *Archive of Physical Medicine and Rehabilitation, 64,* 322-324.

Cervical spine

Books and review articles

Dimnet, J. (1992). Biomechanical models of the head-neck system. In A. Berthoz, W. Graf, P.P. Vidal (Eds.), *The head-neck sensory motor system* (pp. 150-153). New York: Oxford University Press.

Original papers

Ålund, M. & Larsson, S.-E. (1990). Three-dimensional analysis of neck motion: A clinical method. *Spine, 15,* 87-91.

Amevo, B., Aprill, C., & Bogduk, N. (1992). Abnormal instantaneous axes of rotation in patients with neck pain. *Spine, 17,* 748-756.

Amevo, B., Macintosh, J.E., Worth, D., & Bogduk, N. (1991). Instantaneous axes of rotation of the typical cervical motion segments. 1. An empirical study of technical errors. *Clinical Biomechanics, 6,* 31-37.

Amevo, B., Worth, D., & Bogduk, N. (1991a). Instantaneous axes of rotation of the typical cervical motion segments. II. Optimization of technical errors. *Clinical Biomechanics, 6,* 38-46.

Amevo, B., Worth, D., & Bogduk, N. (1991b). Instantaneous axes of rotation of the typical cervical motion segments: A study in normal volunteers. *Clinical Biomechanics, 6,* 111-117.

Bogduk, N., Amevo, B., & Pearcy, M. (1995). A biological basis for instantaneous centres of rotation of the vertebral column. *Proceedings of the Institution of Mechanical Engineers. Part H. Journal of Engineering in Medicine, 209*(3), 177-183.

Dimnet, J., Pasquet, A., Krag, M.H., & Panjabi, M.M. (1982). Cervical spine motion in sagittal plane: Kinematic and geometric parameters. *Journal of Biomechanics, 12,* 959-969.

Fielding, J.W. (1957). Cineroentgenography of the normal cervical spine. *Journal of Bone and Joint Surgery, 39A,* 1280-1288.

Hinderaker, J., Lord, S.M., Barnsley, L., & Bogduk, N. (1995). Diagnostic value of C2-3 instantaneous axes of rotation in patients with headache of cervical origin. *Cephalgia, 15*(5), 391-395.

Hohl, M. & Baker, H.R. (1964). The atlanto-axial joint. Roentgenographic and anatomical study of normal and abnormal motion. *Journal of Bone and Joint Surgery, 46A,* 1739-1752.

Lysell, E. (1969). Motion in the cervical spine. *Acta Orthopaedica Scandinavica* (Suppl. 123).

Panjabi, M., Dvorak, J., Duranceau, J., Yamamoto, J., Gerber, M., Rausching, W., & Bueff, M. (1988). Three dimensional movements of the upper cervical spine. *Spine, 13,* 726-730.

Penning, L. (1978). Normal movements of the cervical spine. *American Journal of Roentgenology, 130,* 317-326.

Shapiro, I. & Frankel, V.H. (1989). Biomechanics of the cervical spine. In M. Nordin & V.H. Frankel (Eds.), *Biomechanics of the musculoskeletal system* (pp. 209-223). Philadelphia: Lea & Febiger.

Snijders, C.J., Hoek van Duke, G.A., & Roosch, E.R. (1991). A biomechanical model for the analysis of the cervical spine in static postures. *Journal of Biomechanics, 24,* 783-792.

Werne, S. (1957). Studies in spontaneous atlas dislocation. *Acta Orthopaedica Scandinavica,* (Suppl. 23).

Winters, J.M. & Peles, J.D. (1990). Neck muscle activity and 3-D head kinematics during quasi-static and dynamic tracking movements. In J.M. Winters & S.L.-Y. Woo (Eds.), *Multiple muscle systems: Biomechanics and movement organization* (pp. 461-480). New York: Springer-Verlag.

Woltring, H.J., Kong, K., Osterbauer, P.J., & Fuhr, A.W. (1994). Instantaneous helical axis estimation from 3-D video data in neck kinematics for whiplash diagnostics. *Journal of Biomechanics, 27,* 1415-1432.

Rib cage

Original papers

Minotti, P. & Lexcellent, C. (1991). Geometric and kinematic modeling of a human costal slice. *Journal of Biomechanics, 24,* 213-221.

Primiano, F.P. Jr. (1982). Theoretical analysis of chest wall mechanics. *Journal of Biomechanics, 15,* 919-931.

Roberts, S.B. (1977). A simple quantitative anatomical model for in-vivo human ribs. *Journal of Bioengineering, 1,* 443-454.

Saumerez, R.C. (1986). An analysis of possible movements of human upper rib cage. *Journal of Applied Physiology, 60,* 678-689.

Wade, O.L. (1954). Movements of the thoracic cage and diaphragm in respiration. *Journal of Physiology, 124,* 193-212.

Wilson, T.A., Rehder, K., Krayer, S., Hoffman, E.A., Whitney, C.G., & Rodarte, J.R. (1987). Geometry and respiratory displacement of human ribs. *Journal of Applied Physiology, 62,* 1872-1877.

Shoulder complex

Books and review articles

Codman, E.A. (1934). *The Shoulder.* Boston: Thomas Todd Co.

Culham, E. & Peat, M. (1993). Functional anatomy of the shoulder complex. *Journal of Orthopaedic and Sports Physical Therapy, 18,* 342-350.

Kent, B.E. (1971). Functional anatomy of the shoulder complex. *Physical Therapy, 51,* 867-887.

Matsen, F.A., Fu, F.H., & Hawkins, R.J. (Eds.). (1993). *The shoulder: A balance of mobility and stability.* Chicago: American Academy of Orthopaedic Surgeons.

Norkin, C.C. & Levangie, P.K. (1992). *Joint structure and function. A comprehensive analysis* (2nd ed.) Philadelphia: F.A. Davis Company.

Saha, A.K. (1961). *Theory of shoulder mechanism: Descriptive and applied.* Springfield, IL: Thomas.

Zuckerman, J.D. & Matsen F.A., III (1989). Biomechanics of the shoulder. In M. Nordin & V.H. Frankel (Eds.), *Basic biomechanics of the musculoskeletal system* (2nd ed., pp. 225-248). Philadelphia: Lea & Febiger.

Original papers

An, K.N., Browne, A.O., Tanaka, S., & Morrey, B.F. (1991). Three-dimensional kinematics of glenohumeral elevation. *Journal of Orthopaedic Research, 9,* 143-149.

Bagg, S.D. & Forrest, W.J. (1988). A biomechanical analysis of scapular rotation during arm abduction in the scapular plane. *American Journal of Physical Medicine and Rehabilitation, 67,* 238-245.

Blakely, R.L. & Palmer, M.L. (1984). Analysis of rotation accompanying shoulder flexion. *Physical Therapy, 64,* 1214-1216.

Browne, A.O., An, K.-N., Tanaka, S., & Morrey, B.F. (1990). Glenohumeral elevation studied in three dimensions. *Journal of Bone and Joint Surgery, 72B,* 843-845.

Casolo, F., Colnago, M., & Cosco, P.P. (1987). A technique for dynamic analysis of joint motion using bi-planar cineradiography. Preliminary investigations on the shoulder complex. In G. de Groot, A.P. Hollander, A.P. Huijing, & G.J. van Ingen Schenau (Eds.), *Biomechanics XI-B* (pp. 1062-1069). Amsterdam: Free University Press.

Conway, A.M. (1961). Movements at the sternoclavicular and acromioclavicular joints. *Physical Therapy Review, 11,* 421-432.

Dempster, W.T. (1965). Mechanisms of shoulder movement. *Archive of Physical Medicine and Rehabilitation, 46,* 49-70.

Dillman, C.J., Fleisig, G.S., & Andrews, J.R. (1993). Biomechanics of pitching with emphasis upon shoulder kinematics. *Journal of Orthopaedic and Sports Physical Therapy, 18,* 402-408.

Doody, S.G., Freedman, L., & Waterland, J.C. (1970). Shoulder movements during abduction in the scapular plane. *Archive of Physical Medicine and Rehabilitation, 51,* 595-604.

Dvir, Z. & Berme, N. (1978). The shoulder complex in elevation of the arm: A mechanism approach. *Journal of Biomechanics, 11,* 219-225.

Engen, T.J. & Spencer, W.A. (1968). Method of kinematic study of normal upper extremity movements. *Archive of Physical Medicine and Rehabilitation, 49,* 9-11.

Engin, A.E. (1980). On the biomechanics of the shoulder complex. *Journal of Biomechanics, 13,* 575-590.

Engin, A.E. & Chen, S.M. (1986). Statistical database for the biomechanical properties of the human shoulder complex. I. Kinematics of the shoulder complex. *Journal of Biomechanical Engineering, 108,* 215-221.

Engin, A.E. & Peindl, R.D. (1987). On the biomechanics of human shoulder complex. 1. Kinematics for determination of the shoulder complex sinus. *Journal of Biomechanics, 20,* 103-117.

Engin, A.E. & Tümer, S.T. (1989). Three dimensional kinematic modeling of the human shoulder complex. I: Physical model and determination of joint sinus cones. *Journal of Biomechanical Engineering, 111,* 107-112.

Freedman, L. & Munro, R.R. (1966). Abduction of the arm in scapular plane: Scapular and glenohumeral movements. *Journal of Bone and Joint Surgery, 48A,* 1503-1510.

Harryman, D.T. II, Sidles, J.A., Clark, J.M., McQuade, K.J., Tyler, D.G., & Matsen, F.A. III (1990). Translation of the humeral head on the glenoid with passive glenohumeral motion. *Journal of Bone and Joint Surgery, 72A,* 1334-1343.

Harryman, D.T. II, Sidles, J.A., Harris, S.L., & Matsen, F.A. III (1992). Laxity of the normal glenohumeral joint: A quantitative in vivo assessment. *Journal of Shoulder and Elbow Surgery, 1,* 66-76.

Hogfors, C., Peterson, B., Sigholm, G., & Herberts, P. (1991). Biomechanical model of the human shoulder. II. The shoulder rhythm. *Journal of Biomechanics, 24,* 699-709.

Hogfors, C., Sigholm, G., & Herberts, P. (1987). Biomechanical model of the human shoulder. 1. Elements. *Journal of Biomechanics, 20,* 157-166.

Howell, S.M., Galinat, B.J., & Renzi, A.J. (1988). Normal and abnormal mechanics of the glenohumeral joint in the horizontal plane. *Journal of Bone and Joint Surgery, 70A,* 227-232.

Inman, V.T. & Saunders, J.B. (1946). Observations on the functions of the clavicle. *California Medicine, 65,* 158-166.

Inman, V.T., Saunders, J.B., & Abbott, L.C. (1944). Observations on the function of the shoulder joint. *Journal of Bone and Joint Surgery, 26,* 1-30.

Johnston, G.R., Stuart, P.R., & Mitchell, S. (1993). A method for the measurement of three-dimensional scapular movement. *Clinical Biomechanics, 8,* 269-273.

Johnston, T.B. (1937). The movements of the shoulder-joint: A plea for the use of "the plane of scapula" as the plane of reference for movements occurring at the humero-scapular joint. *British Journal of Surgery, 25,* 252-260.

Kennedy, J.C. & Cameron, H. (1954). Complete dislocation of the acromioclavicular joint. *Journal of Bone and Joint Surgery, 36B,* 202-208

Kirsch, S., Nägerl, H., & Kubein-Meesenburg, D. (1993). Kinematics and statics of the human shoulder. *XIVth International Congress of Biomechanics, Abstracts* (pp. 692-693), Paris.

Kondo, M., Tazoe, S., & Yamada, M. (1984). Changes of the tilting angle of the scapula following elevation of the arm. In J.E. Bateman & R.P. Welsh (Eds.), *Surgery of the shoulder.* (pp. 12-16). St. Louis: C.V. Mosby Co.

Michiels, I. & Grevenstein, J. (1995). Kinematics of shoulder abduction in the scapular plane. *Clinical Biomechanics, 10,* 137-143.

Nägerl, H., Kubein-Meesenburg, D., Gotta, H., Fanghänel, J., & Kirsch, S. (1993). Biomechanische Prinzipen in Diarthrosen und Synarthrosen. Teil II: Die Articulation Humeri als dimeres Kugelgelenk. *Zeitschrift für Orthopädie und ihre Grengebiete, 131,* 293-301.

Pearl, M.L., Harris, S.L., Lippitt, S.B., & Sidles, J.A. (1992). A system for describing positions of the humerus relative to the thorax and its use by the presentation of several functionally important arm positions. *Journal of Shoulder and Elbow Surgery, 1,* 113-118.

Poppen, N.K. & Walker, P.S. (1976). Normal and abnormal motion of the shoulder. *Journal of Bone and Joint Surgery, 58A,* 195-201.

Pronk, G.M. (1988). Three-dimensional determination of the position of the shoulder girdle during humerus elevation. In G. de Groot, A.P. Hollander, P.A. Hujing, & G.J. van Ingen Schenau (Eds.), *Biomechanics XI-B* (pp. 1070-1076). Amsterdam: Free University Press.

Pronk, G.M. (1989). A kinematic model of the shoulder girdle: A resume. *Journal of Medical Engineering and Technology, 13,* 119-123.

Saha, A.K. (1983). Mechanism of shoulder movement and a plea for the recognition of "zero position" of the glenohumeral joints. *Clinical Orthopaedics and Related Research, 173,* 3-10.

Soslowsky, L.J., Flatow, E.L., Bigliani, L.U., & Mow, V.C. (1992). Articular geometry of the glenohumeral joint. *Clinical Orthopaedics and Related Research, 285,* 181-190.

Taylor, C.L. & Blaschke, A.C. (1951). A method for kinematic analysis of the shoulder, arm and hand complex. *Annals of the New York Academy of Sciences, 51,* 1251-1265.

Tümer, S.T. & Engin, A.E. (1989). Three dimensional kinematic modeling of the human shoulder complex. II. Mathematical modeling and solution via optimization. *Journal of Biomechanical Engineering, 111,* 113-121.

Van der Helm, F.C.T. & Pronk, G.M. (1995). Three-dimensional recording and description of motions of the shoulder mechanism. *Journal of Biomechanical Engineering, 117,* 27-40.

Van der Helm, F.C.T., Pronk, G.M., Veeger, H.E.J., & van der Woude, L.H.V. (1989). The rotation center of the glenohumeral joint. In R.J. Gregor, R.F. Zernicke, & W.C. Whiting (Eds.), *Proceedings of the 12th ISB Congress* (pp. 676-677). Los Angeles: University of California.

Van der Helm, F.C.T., Veeger, H.E.J., Pronk, G.M., van der Woude, L.H.V., & Rozendal, R.H. (1991). Geometry parameters for musculoskeletal modeling of the shoulder system. *Journal of Biomechanics, 25,* 129-144.

Veeger, H.E.J. van der Helm, F.C.T., & Rozendal, R.H. (1993). Orientation of the scapula in a simulated wheelchair push. *Clinical Biomechanics, 8,* 81-90.

Wallace, W.A. (1982). The dynamic study of shoulder movement. In L. Bailey & L. Kessel (Eds.), *Shoulder surgery* (pp. 139-143). Berlin: Springer-Verlag.

Wood, J.E., Meek, S.G., & Jacobsen, S.C. (1989). Quantification of human shoulder anatomy for prosthetic arm control. I. Surface modeling. *Journal of Biomechanics, 22,* 319-325.

Elbow; radioulnar joints

Original papers

Amis, A.A., Dowson, D.D., Unsworth, A., Miller, J.H., & Wright, V. (1977). An examination of the elbow articulation with particular reference to variation of the carrying angle. *Engineering in Medicine, 6,* 76-82.

Amis, A.A. & Miller, J.H. (1982). The elbow. *Clinics in Rheumatic Diseases, 8,* 571-593.

An, K.N. & Morrey, B.F. (1985). Biomechanics of the elbow. In B.F. Morrey (Ed.), *The elbow and its disorders* (pp. 43-61). Philadelphia: W.B. Saunders.

An, K.N., Morrey, B.F., & Chao E.Y.S. (1984). Carrying angle of the human elbow joint. *Journal of Orthopaedic Research, 1,* 369-378.

An, K.N., Morrey, B.F., Chao E.Y.S. (1985). Kinematics of the elbow. In D.A. Winter, R.W. Norman, R.P. Wells, K.C. Hayes, & A.E. Patla (Eds.), *Biomechanics IX-A* (pp. 154-159). Champaign, IL: Human Kinetics Publishers.

Carret, J.P., Fischer, L.P., Gonon, G.P., & Dimnet, J. (1976). Etude cinématique de la prosupination au niveau des articulations radiocupitalis (radio ulnaris). *Bulletin de l'Association Anatomie, 60,* 279-285.

Chao, E.Y. & Morrey, B.F. (1978). Three-dimensional rotation of the elbow. *Journal of Biomechanics, 11,* 57-73.

Deland, J.T., Garg, A., & Walker, P.S. (1987). Biomechanical basis for elbow hinge-distractor design. *Clinical Orthopaedics and Related Research, 215,* 303-312.

Hollister, A.M., Gellman, H., & Waters, R.L. (1994). The relationship of the interosseus membrane to the axis of rotation of the forearm. *Clinical Orthopaedics and Related Research, 298,* 272-276.

Ishizuki, M. (1979). Functional anatomy of the elbow joint and three-dimensional quantitative analysis of the elbow joint. *Journal of the Japanese Orthopaedic Association, 53,* 986-989.

King, G.K., McMurtry, R.Y., Rubinstein, J.D., & Gertzbein, S.D. (1986). Kinematics of the distal radioulnar joints. *Journal of Hand Surgery, 11A,* 798-804.

London, J.T. (1981). Kinematics of the elbow. *Journal of Bone and Joint Surgery, 63A,* 529-535.

Mori, K. (1985). Experimental study on rotation of the forearm: Functional anatomy of the interosseus membrane. *Journal of Japanese Orthopaedic Association, 59,* 611-622.

Morrey, B.F. Askew, L.J., An, K.-N., & Chao, E.Y.S. (1981). A biomechanical study of normal functional elbow motion. *Journal of Bone and Joint Surgery, 63A,* 872-876.

Morrey, B.F. & Chao, E.Y.S. (1976). Passive motion of the elbow joint. A biomechanical analysis. *Journal of Bone and Joint Surgery, 58A,* 501-508.

Palmer, A.K. & Werner, F.W. (1984). Biomechanics of the distal radioulnar joint. *Orthopedic Clinics of North America, 15,* 27-35.

Ray, R.D., Johnson, R.J., & Jameson, R.M. (1951). Rotation of the forearm: An experimental study of pronation and supination. *Journal of Bone and Joint Surgery, 33A,* 993-996.

Shiba, R. & Sorbie, C. (1985). Axis of flexion-extension of humeroulnar joint. In D. Kashiwagi (Ed.), *Elbow joint* (pp. 19-24). New York: Elsevier Science.

Wagner, C. (1977). Determination of the rotatory flexibility of the elbow joint. *European Journal of Applied Physiology, 37,* 47-59.

Youm, Y., Dryer, R.F., Thambyrajah, K., Flatt, A.E., & Sprague, B.L. (1979). Biomechanical analysis of forearm pronation-supination and elbow flexion-extension. *Journal of Biomechanics, 12,* 245-255.

Hand (wrist, thumb, fingers)

Books and review articles

An., K.-N., Berger, R.A., & Cooney, W.P. III (Eds.). (1991). *Biomechanics of the wrist joint.* New York: Springer-Verlag.

Chao, E.Y.S., An, K.-N., Cooney, W.P., & Linscheid, R.L. (1989). *Biomechanics of the hand. A basic research study.* Singapore: World Scientific.

Schuind, F., An, K.N., Cooney, W.P. III, & Garcia-Elias, M. (Eds.). (1994). *NATO ASI Series. Series A: Life Sciences: Vol. 256. Advances in the biomechanics of the hand and wrist.* New York: Plenum Press.

Wrist

Original papers

Andrews, J.G. & Youm, Y. (1979). A biomechanical investigation of wrist kinematics. *Journal of Biomechanics, 12,* 83-93.

Brumbaugh, R.B., Crowninshield, R.D, Blair, W.R., & Andrews, J.G. (1982). An in-vivo study of normal wrist kinematics. *Journal of Biomechanical Engineering, 104,* 176-181.

Cyriax, E.F. (1926). On the rotatory movement of the wrist. *Journal of Anatomy, 60,* 199-201.

Erdman, A.G., Mayfield, J.K., Dorman, F., Wallrich, M., & Dahlof, W. (1979). Kinematic and kinetic analysis of the human wrist by stereoscopic instrumentation. *Journal of Biomechanical Engineering, 101,* 124-133.

Evans, J.S., Blair, W.F., Andrews, J.G., & Crowninshield, J.S. (1986). The in-vivo kinematics of the rheumatoid wrist. *Journal of Orthopaedic Research, 4,* 142-151.

Jackson, W.T., Hefzy, M.S., & Guo, H. (1994). Determination of wrist kinematics using a magnetic tracking device. *Medical Engineering and Physics, 16,* 123-133.

Kauer, J.M.G. (1980). Functional anatomy of the wrist. *Clinical Orthopaedics and Related Research, 149,* 9-20.

Lange, A. de, Kauer, J.M.G., & Huiskes, R. (1985). Kinematic behaviour of the human wrist joint: A roentgen-stereophotogrammetric analysis. *Journal of Orthopaedic Research, 3,* 56-64.

Linscheid, R.L. (1986). Kinematics considerations of the wrist. *Clinical Orthopaedics and Related Research, 202,* 27-39.

Moore, J.A., Small, C.F., Bryant, J.T., Ellis, R.E., Pichora, D.R., & Hollister, A.M. (1993). A kinematic technique for describing wrist joint motion: Analysis of configuration space plots. *Proceedings of the Institution of Mechanical Engineers. Part H—Journal of Engineering in Medicine, 207,* 211-218.

Palmer, A.K. & Werner, F.W. (1984). Biomechanics of the distal radioulnar joint. *Clinical Orthopaedics and Related Research, 187,* 26-35.

Palmer, A.K., Werner, F.W., Murphy, D., & Glisson, R. (1985). Functional wrist motion: A biomechanical study. *Journal of Hand Surgery, 10A,* 39-46.

Ryu, J., Cooney, W.P., Askew, L.J., An, K.N., & Chao, E.Y.S. (1991). Functional ranges of motion of the wrist joint. *Journal of Hand Surgery, 16A,* 409-419.

Ryu, J., Palmer, A.K., & Cooney, W.P., III. (1991). Wrist joint motion. In K.-N. An, R.A. Berger, & W.P. Cooney, III. (Eds.), *Biomechanics of the wrist joint* (pp. 37-75). New York: Springer-Verlag.

Small, C.F., Bryant, J.T., & Pichora, D.R. (1992). Rationalization of kinematic descriptors for three-dimensional hand and finger motion. *Journal of Biomechanical Engineering, 14,* 133-141.

Taylor, C.L. & Schwarz, R.J. (1955). The anatomy and mechanics of the human hand. *Artificial Limbs, 2,* 22-35.

Yoon, Y.S. (1979). Analytical development in investigation of wrist kinematics. *Journal of Biomechanics, 12,* 613-621.

Youm, Y. & Flatt, A.E. (1980). Kinematics of the wrist. *Clinical Orthopaedics and Related Research, 117,* 24-32.

Youm, Y., McMurtry, R.Y., Flatt, A.E., & Gillespie, T.E. (1978). Kinematics of the wrist: An experimental study of radial-ulnar deviation and flexion-extension. *Journal of Bone and Joint Surgery, 60A,* 423-431.

Youm, Y. & Yoon, Y.S. (1979). Analytical development in investigation of wrist kinematics. *Journal of Biomechanics, 12,* 613-621.

Fingers, thumb

Original papers

Agee, J., Hollister, A., & King, F. (1986). The longitudinal axis of rotation of the metacarpophalangeal joint of the finger. *Journal of Hand Surgery, 11A,* 767-772.

An, K.N., Chao, E.Y., Cooney, W.P., & Linscheid, R.L. (1979). Normative model of human hand for biomechanical analysis. *Journal of Biomechanics, 12,* 775-788.

Barmakian, J.T. (1992). Anatomy of the joints of the thumb. *Hand Clinics, 8,* 683-691.

Berme, N., Paul, J.P., & Purves, W.K. (1977). A biomechanical analysis of the metacarpophalangeal joint. *Journal of Biomechanics, 10,* 409-412.

Buchholz, B. & Armstrong, T.J. (1992). A kinematic model of the human hand to evaluate its prehensile abilities. *Journal of Biomechanics, 25,* 149-162.

Buchholz, B., Armstrong, T.J., & Goldstein, S.A. (1992). Anthropometric data for describing the kinematics of the human hand. *Ergonomics, 35,* 261-273.

Buford, W.L., Hollister, A.M., & Myers, L.M. (1990). A modeling and simulation system for the human hand. *Journal of Clinical Engineering, 15,* 445-451.

Chao, E.Y., Opgrande, J.D., & Axmear, F.E. (1976). Three-dimensional force analysis of finger joints in selected isometric hand functions. *Journal of Biomechanics, 9,* 387-396.

Cooney, W.P. III, Lucca, M.J., Chao, E.Y.S., & Linscheid, R.L. (1981). The kinesiology of the thumb trapeziometacarpal joint. *Journal of Bone and Joint Surgery, 63A,* 1371-1381.

Flatt, A.E. & Fisher, G.W. (1968). Restraints of the metacarpophalangeal joints: A force analysis. *Surgical Forum, 19,* 459-460.

Haines, R.W. (1944). The mechanism of rotation at the first carpometacarpal joint. *Journal of Anatomy, 78,* 44-46.

Hollister, A.M., Buford, W.L., Myers, L.M., Giurintano, D.J., & Novick, A. (1992). The axes of rotation of the thumb carpometacarpal joint. *Journal of Orthopaedic Research, 10,* 454-460.

Imaeda, T. (1994). Kinematics of the normal trapeziometacarpal joint. *Journal of Orthopaedic Research, 12,* 197-204.

Imaeda, T., An, K., & Cooney, W.P. (1992). Functional anatomy and biomechanics of the thumb. *Hand Clinics, 8,* 9-15.

Kapandji, I.A. (1981). Biomechanics of the thumb. In R.Tubiana (Ed.), *The hand* (Vol. 1, pp. 410-422). Philadelphia: W.B. Saunders.

Kuczynski, K. (1974). Carpometacarpal joint of the human thumb. *Journal of Anatomy, 118,* 119-126.

Landsmeer, J.M.F. (1955). Anatomical and functional investigations on the articulation of the human fingers. *Acta Anatomica, 25,* 1-69.

Leijnse, J.N.A.L., Bonte, J.E., Landsmeer, J.M.F., Kalker, J.J., van der Meulen, J.C., & Snijders, C.J. (1992). Biomechanics of the finger with anatomical restrictions: The significance for the exercising hand of the musician. *Journal of Biomechanics, 25,* 1253-1264.

Leijnse, J.N.A.L. & Kalker, I.J. (1995). A two dimensional kinematic model of the lumbrical in the human finger. *Journal of Biomechanics, 28,* 237-249.

Leijnse, J.N.A.L., Snijders, C.J., Bonte, J.E., Landsmeer, J.M.F., Kalker, J.J., van der Meulen, J.C., Sonnefeld, G.J., & Hovius, S.E.R. (1993). The hand of the musician: Finger system with anatomical restrictions. *Journal of Biomechanics, 26,* 1169-1179.

Micks, J.E., Reswick, J.B., & Hager, D.L. (1978). The mechanisms of the intrinsic-minus finger: A biomechanical study. *Journal of Hand Surgery, 3,* 333-341.

Napier, J.R. (1955). The form and function of the carpo-metacarpal joint of the thumb. *Journal of Anatomy, 89,* 362-369.

Pagowski, S. & Piekarski, K. (1977). Biomechanics of metacarpophalangeal joint. *Journal of Biomechanics, 10,* 205-209.

Schultz, R.J., Storace, A., & Krishnamurthy, S. (1987). Metacarpophalangeal joint motion and the role of the collateral ligaments. *International Orthopaedics, 11,* 149-155.

Smith, S.A., Kuzinski, K. (1978). Observations on the joints of the hand, *Hand, 10,* 226-231.

Storace, A. & Wolf, B. (1982). Kinematic analysis of the role of the finger tendons. *Journal of Biomechanics, 5,* 381-391.

Tamai, K., Ryu, J., An, K.N., Linsscheid, R.L., Cooney, W.P., & Chao, E.Y.S. (1988). Three-dimensional geometric analysis of the metacarpophalangeal joint. *Journal of Hand Surgery, 13A,* 521-529.

Thompson, D.E. & Guirintano, D.J. (1989). A kinematic model of the flexor tendons of the hand. *Journal of Biomechanics, 22,* 327-334.

Toft, R. & Berme, N. (1980). A biomechanical analysis of the joints of the thumb. *Journal of Biomechanics, 13,* 353-360.

Youm, Y., Gillespie, T.E., Flatt, A.E., & Sprague, B.L. (1978). Kinematic investigation of normal MCP joint. *Journal of Biomechanics, 11,* 109-118.

Temporomandibular joint

Reviews

Hall, R.E. (1929). An analysis of work and ideas of investigators and authors of relations and movements of the mandible. *Journal of the American Dental Association, 16,* 1642-1693.

Sicher, H. (1964). Functional anatomy of the temporomandibular joint. In B.G. Sarnat (Ed.), *The Temporomandibular Joint* (2nd ed.). Springfield, IL: Charles C Thomas.

Winstanley, R.B. (1985). The hinge axis: A review of the literature. *Journal of Oral Rehabilitation, 12,* 135-159.

Original publications

Baragar, F.A. & Osborn, J.W. (1984). A model relating patterns of human jaw movement to biomechanical constraints. *Journal of Biomechanics, 17,* 757-767.

Barbenel, J.C. (1972). The biomechanics of the temporomandibular joint: A theoretical study. *Journal of Biomechanics, 5,* 251-256.

Bennet, N.G. (1908). A contribution to the study of movement of the mandible. *Proceedings of the Royal Society of Medicine, 1,* 79-98.

Bouchoucha, S. (1991). Axes of rotation of the temporomandibular joint. *Annals of Biomedical Engineering, 19,* 640-646.

Cappozzo, A., Della Croce, U., & Lucchetti, L. (1993). Instantaneous helical axis estimation in the human temporomandibular joint. *XIVth Congress of the International Society of Biomechanics, Abstracts* (pp. 236-237). Paris.

Chen, J. & Buckwalter, K. (1993). Displacement analysis of the temporomandibular condyle from magnetic resonance images. *Journal of Biomechanics, 26,* 1455-1462.

Chen, J. & Xu, L. (1994). A finite element analysis of the human temporomandibular joint. *Journal of Biomechanical Engineering, 116,* 401-407.

Conway, W.F., Hayes, C.W., Campbell, R.L., & Laskin, M.D. (1989). Temporomandibular joint motion: Efficacy of fast low-angle MR imaging. *Radiology, 172,* 821-826.

Faulkner, K.D.B. & Atkinson, H.F. (1984). Mandibular movement in lateral excursions. *Journal of Oral Rehabilitation, 11,* 103-109.

Grant, P. (1973). Biomechanical significance of the instantaneous center of rotation: The human temporomandibular joint. *Journal of Biomechanics, 6,* 109-113.

Hannam, A.G., DeCou, R.E., Scott, J.D., & Wood, W.W. (1980). The kinesiographic measurement of jaw displacement. *Journal of Prosthetic Dentistry, 44,* 88-93.

Hart, R.T., Hennebel, V.V., Thongpreda, N., Van Buskirk, W.C., & Anderson, R.C. (1992). Modeling the biomechanics of the mandible: A three-dimensional finite elements study. *Journal of Biomechanics, 25,* 261-286.

Hylander, W.L. (1975). The human mandible: Lever or link? *American Journal of Physical Anthropology, 43,* 433-456.

Kang, Q.S. (1993). Kinematic analysis of the human temporomandibular joint. *Annals of Biomedical Engineering, 21,* 699-707.

Koolstra, J.H. & Van Eijden, T.M.G.J. (1992). Application and validation of a three-dimensional mathematical model of the human masticatory system in vivo. *Journal of Biomechanics, 25,* 175-187.

Kubein-Meesenburg, D. & Nägerl, H. (1990). Basic principles of relation of anterior and posterior guidance in stomatognatic system. *Anatomische Anzeiger, 171,* 1-12.

Kubein-Meesenburg, D., Nägerl, H., & Klamt, B. (1988). The biomechanical relation between incisal and condylar guidance in man. *Journal of Biomechanics, 21,* 997-1009.

Lundberg, M. (1962). Free movements in the temporomandibular joint. A cineradiographic study. *Acta Radiologica*, (Suppl. 220).

McCollum, B.B. (1960). Mandibular hinge axis and method for locating it. *Journal of Prosthetic Dentistry, 10,* 428-435.

McMillan, A.S., McMillan, D.R., & Darwell, B.W. (1989). Centers of rotation during jaw movements. *Acta Odontologica Scandinavica, 47,* 323-327.

Nägerl, H., Kubein-Meesenburg, D., Fanghänel, J., Thieme, K.M., Klamt, B., & Schwestka-Polly, R. (1991). Elements of a general theory of joints. 6. General kinematical structure of mandibular movements. *Anatomische Anzeiger, 173,* 249-264.

Ostry, D.J. (1992). Human jaw movement kinematics and control. *Advances in Psychology, 87, Tutorials in Motor Behavior II*, p. 647.

Posselt, U. (1984). Studies in the mobility of the human mandible. *Acta Odontologica Scandinavica,* (Suppl. 10), 1-160.

Price, C. (1990a). A method of quantifying disk movement on magnetic resonance images of the temporomandibular joint. Part 1: The method. *Dento-Maxillo-Facial Radiology, 19,* 59-62.

Price, C. (1990b). A method of quantifying disk movement on magnetic resonance images of the temporomandibular. Part 2: Application of the method to normal and deranged joints. *Dento-Maxillo-Facial Radiology, 19,* 63-66.

Rasmussen, O.C. (1978). Size of variables in mandibular movements in autopsy material. *Journal of Oral Rehabilitation, 5,* 241-248.

Rees, L.A. (1954). The structure and function of the mandibular joint. *British Dental Journal, 96,* 125-133.

Siegler, S. (1991). Three-dimensional kinesiology of the human temporomandibular joint. *American Society of Mechanical Engineers, Bioengineering Division (Publication), BED, 20,* 395-398.

Van Rensburg, L.B., Lemmer, J., & Lewin, A. (1974). Co-ordinates for jaw movements. *Journal of Oral Rehabilitation, 1,* 285-291.

Wu, J., Xu, X., & Sheng, J. (1988). Analysis of the open-closing movement of the human temporomandibular joint. *Acta Anatomica (Basel) 133,* 213-216.

GLOSSARY

Acceleration—The rate of change of velocity. Acceleration is a vector quantity.

Acceleration difference—The vector difference in acceleration of two points of a rigid body measured in a fixed reference frame.

Centripetal acceleration—Acceleration pointing toward the center of curvature; the rate of change of the direction of the velocity.

Coriolis' acceleration—The acceleration that is acting when a body is moving with regard to the rotating reference frame.

Normal acceleration, see *Centripetal acceleration*

Tangential acceleration—Acceleration pointing along the path; the rate of change of the magnitude of velocity.

Affine transformation—A geometric transformation, such as shift, rotation, stretching, etc., retaining the original parallel straight lines.

Alias problem—A problem of kinematics. Description of a position given in one reference system, transformed into another system of coordinates ("in other words").

Alibi problem—A problem of kinematics. Description of a body's displacement.

Anatomic planes, see *Planes, anatomic*

Angle of actual torsion (of the eye)—The angle between the retinal horizon and the plane of fixation.

Angle of torsion (of the eye)—The angle between the retinal horizon and the real horizon of external space, or between the retinal vertical and the vertical (the line of gravity).

Angular conventions (in eye kinematics)

Fick's convention—The horizontal position of the line of sight (longitude) is specified first, and the vertical rotation (latitude) second.

Helmholtz's convention—The vertical rotation of the line of sight (elevation) is specified first. Then the eye is turned sideways in the plane of regard (azimuth).

Angular momentum (of a rigid body)—The product of the body's moment of inertia and angular velocity.

Conservation of angular momentum—The angular momentum of a system is constant, unless external torque is exerted on the system.

Angular velocity couple (pair of rotation)—Rotation of a distal link about the proximal end with the angular velocity equal in magnitude and opposite in direction to the angular velocity of the proximal link in its rotation relative to the environment. As a result of the two rotations, the distal link is translating along the circular trajectory and does not rotate—its attitude is kept constant.

Arc—A part of a curve. Also, a line between two points on an articular surface.

Articular surfaces

> *Anticlastic surfaces*—The surfaces that are convex in one direction and concave in the perpendicular direction.
>
> *Conarticular surfaces*—The articular surfaces of a joint.
>
> *Equivalent surface*—The surface calculated as the difference in the profiles between two original articular surfaces.
>
> *Female surface*—A concave articular surface.
>
> *Male surface*—A convex articular surface.
>
> *Ovoid*—Spheroidal, egg-shaped, articular surfaces.
>
> *Plane surface*—A flat surface.
>
> *Saddle surfaces*, see *Anticlastic surfaces*
>
> *Sellars*, see *Anticlastic surfaces*
>
> *Synclastic surfaces*—The surfaces that are either completely concave or completely convex.

Axes of the human body

> *Anteroposterior axis*—Interception of the sagittal and transverse planes.
>
> *Lateromedial (frontal) axis*—Interception of the frontal and transverse planes.
>
> *Longitudinal axis*—Interception of the sagittal and frontal planes.

Axial rotation

> *Pseudo axial rotation*—The difference in the initial and the final attitudes of a human body segment. The attitude is measured with regard to the external axis that initially aligns with the longitudinal axis of the segment.
>
> *Real axial rotation*—The amount of rotation of a segment about its own long axis occurring during the motion.

Axode—The path of the instantaneous screw axis.

Base line (in kinematics of eye movement*)*—The line joining the centers of rotation of the two eyes.

Basic theorem of kinematics of articular surfaces—Any two consecutive slides (from point m to point n and then from n to p) are equivalent to a single slide from m to p together with the spin.

Berkenblit's hypothesis—In reaching, the joints are controlled independently. The task of joint j is to turn the vector $\mathbf{P}_j$ that goes from the joint to the end effector in the direction to the target.

Bernstein's problem—Elimination of the redundant degree of freedom as a basis for the control of human and animal movements.

Biomechanism—A biokinematic chain having one degree of freedom, for instance the rib cage.

Body

    Rigid (solid) body—A body is rigid if the distance between any two points within the body is constant.

Body scheme—Internal representation of the body's kinematic geometry in the central nervous system.

Cardan's angle convention (gyroscopic system, three-axes system)—An angular convention that comprises three different axes, for instance Y, x', and z" in the starting position.

Center of rotation (finite center of rotation)—A point in which the coordinates, both in global and local reference frame, do not change during a planar movement.

Centrifugal—Directed away from a center.

Centripetal—Directed toward a center.

Centrode—The trajectory of the instant center of rotation.

    Fixed centrode—The trajectory of the instant center of rotation relative to the global frame (the fixed plane).

    Moving centrode—The trajectory of the instant center of rotation relative to the local frame (the moving plane).

Chasles' theorem—Any rigid-body motion can be obtained from the sum, or sequence, of translation and rotation.

Chopart's joint—The calcaneocuboid and talonavicular joints considered together.

Chord (in geometry)—A straight line segment joining two points on an arc.

Chord (in joint kinematics)—The shortest arc connecting two points on a joint surface, same as *Geodesic*.

Class of a joint—The number of constraints imposed at the joint.

Codman's paradox—Supination-pronation of the arm without axial rotation, as a result of consecutive shoulder movements.

Commutative—Independent of the order of terms ($A + B = B + A$; $AB = BA$).

Complex number—A number expressed as $c = a + ib$, where $I = \sqrt{-1}$. Complex numbers consist of a real part, a, and an imaginary part, ib. Both a and b are real numbers.

Components—Elements into which a vector quantity can be resolved.

    Radial component (of a vector)—Directed along the position vector or a radius.

    Transverse component (of a vector)—Directed perpendicularly to the position vector.

Condition number (of the skin markers' configuration)—The number characterizing the magnitude of the measurement error associated with positioning marks in the body segment.

Constraints—Any restriction to free movement.

Actual constraints—Tangible physical obstacles to movement.

Anatomic constraints—The constraints imposed by the structure of the musculoskeletal system.

Geometric constraint—The constraint in transferring joint angular velocity into linear velocity of the end effector in a given direction. For example, the more the knee is extended the smaller the contribution of the angular knee extension velocity to the linear velocity of the leg extension.

Holonomic (geometric) constraints—The constraints imposed on the position of a body.

Instructional constraints—The motor task constraints defined by an instructor, experimenter, competition rules, etc.

Intentional constraints—The motor task constraints imposed by the performer.

Mechanical constraints—The constraints necessitated by mechanical phenomena (e.g., balance requirements).

Motor task constraints—The constraints imposed to execute the planned motor task.

Coordinate system, see *Reference system*

Cartesian coordinate system—Three fixed planes that intersect one another at right angles. The position of any point P in space is uniquely determined by the three perpendiculars from P on these planes.

Oblique coordinate system—Three fixed planes that intersect each other at oblique angles. The position of any point P in space can be determined either by the three perpendiculars from P on these planes (projections, covariant vectors), or by the three lines from P running in parallel with the planes (components, contravariant vectors).

Coordinates—A set of numbers that locates the position of a body in a reference system.

Extrinsic coordinates—The coordinates of an object external to the subject represented in the central nervous system.

Generalized coordinates—A set of quantities that specify the state of a system.

Intrinsic coordinates—The positioning of body parts represented within the central nervous system (See *body scheme*).

Coriolis' theorem—The absolute linear acceleration of a point P that is moving with regard to a local reference system which is also in motion is equal to the vector sum of (1) the acceleration which the point would have if it were fixed to the moving system, (2) the acceleration of P with regard to the local moving system, and (3) the Coriolis' acceleration.

Coupling (of a joint motion)—Consistent association of a joint motion about an axis with either another motion around a different axis in the same joint or with a motion in another joint.

Cross product of vectors P and Q (vector product)—The vector of magnitude $P \times Q \sin \alpha$ (α is the angle formed by **P** and **Q**), with direction perpendicular to the plane containing **P** and **Q**, according to the right-hand thumb rule.

Curvature—The change in direction per unit of arc.

> *Gaussian curvature (at a point)*—A product of the maximal and minimal values of the curvature at a point.
>
> *Principal curvatures (at a point)*—The minimal and maximal values of the curvature at a point.
>
> *Relative principal curvatures*—The principal curvatures of the equivalent surface at the point of initial contact with the plane.

Cyclorotation, see *Angle of actual torsion*

Degenerate joint configuration—A *Joint configuration* at which freedom of motion is constrained, e.g., *Singular joint configuration.*

Degree of freedom (DOF)—An independent coordinate used to specify a body, system or, position.

> *Essential DOF*—The DOF controlled unconditionally throughout the performance.
>
> *Nonessential DOF*—The DOF that are under rather loose control throughout the performance.
>
> *Permitted DOF (mechanically permitted DOF)*—Mechanically unconstrained DOF.

Denavit-Hartenberg convention—The local system of coordinates defined in the following way: x_i axis is directed along the common normal (perpendicular) to the axes of rotation in joints i and (i + 1); z_i axis is taken along the axis of rotation in the *i*th joint; and y_i completes the right-handed coordinate system.

Denavit-Hartenberg parameters—Four parameters defining relative location of two links of a kinematic chain: (1) the segment length; (2) the joint offset; (3) the twist angle; and (4) the angle of rotation. The segment length, joint offset, and twist angle are anatomic parameters that are constant. *Joint configuration* varies according to the angle of rotation only.

Determinant (of a square matrix)—A certain number associated with the matrix. The determinant of the matrix is zero when some rows or columns of the matrix are a linear function of other rows or columns.

Diarthroses—Synovial joints.

Direction cosines—Cosines of the angle that local reference axes make with each of the coordinate axes of the global system

Displacement—The difference between the position coordinates of the body in its final and initial position.

Displacement vector—A vector connecting the origins of two coordinate frames or the location coordinates of the body in its final and initial position.

Distal—Farthest from the center; opposed to *Proximal.*

Distance—Magnitude of a traveled path.

Donders' law—The angle of torsion is constant for any given orientation of the line of sight.

Eigenvector and **eigenvalue**, of a square matrix [T]—A nonzero vector $\mathbf{V}$ and a scalar λ that satisfy the equation $[T] \mathbf{V} = \lambda\mathbf{V}.$

End effector—The last link of an open kinematic chain.

Envelope of passive joint motion—The boundary between the regions of the joint motion with the low and with the high resistance from the joint supporting structures.

Euler's angles—The angles defining orientation of one reference frame with respect to another in three-dimensional space. The angles can be associated with three consecutive rotations of the movable frame. The rotations are being made: (1) about a reference axis of the global system of coordinates; (2) about a newly rotated axis; and (3) about the axis of the local reference frame.

> *Nautical angles*—Yaw, pitch, and roll; a specific Euler's convention.
> *Nutation (tilt)*—A second Euler's angle; the rotation about the floating axis (nodes axis).
> *Pitch*—The second nautical angle; rotation around the transverse axis, z', of the intermediate frame.
> *Precession*—A first Euler's angle; the rotation relative to an axis defined in the global reference system.
> *Roll*—The third nautical angle. The rotation about the longitudinal axis, x", of the local frame.
> *Twist (spin)*—A third Euler's angle; the rotation about a local reference axis.
> *Yaw*—The first nautical angle. An angle in rotation about the vertical axis, Y, of the global frame.

Euler's angle convention (as opposed to **Cardan's angle convention**)— In the starting position, all three axes of rotation are located in one plane, for instance the axes Yx'y" are in the XY plane.

Euler's parameters—A set of four interdependent variables defining body orientation in space.

Euler's theorem—Any motion of a rigid body with one point fixed is a single rotation about an axis through that fixed point.

False torsion (of the eye), see *Angle of torsion*

Floating axis—The second axis of rotation in the Joint Rotation Convention usage; the floating axis coincides with the *Line of nodes*.

Four-bar linkage/mechanism—A closed four-link planar kinematic chain with two intercrossing links and revolute joints. A popular model of the knee joint.

Frame, also *Reference frame*, see *Reference system*

Frobenius norm—The matrix generalization of the Euclidean vector norm. The square root of the sum computed over all the elements squared of a matrix.

Generalized motor program (the hypothesis)—The central nervous system stores just a plain outline of the motor program. During the movement execution, the program is specified according to the required movement duration, amplitude, etc.

Geodesic—The shortest surface line between two points on a curved surface.

Gimbal lock—A position at the gimbal at which an axis at the outer gimbal and an axis at the inner gimbal are colinear. The Euler angles cannot be exactly defined for this position.

Globographic presentation—Description of an attitude of a body segment in spherical coordinates.

Goniometer—A device to measure joint angles.

Gruebler's formula—The formula used to calculate the mobility of a kinematic chain.

Gutman's hypothesis—Humans use subjective time rather than real time for movement planning.

Gyroscopic system, see *Cardan's angle convention*

Hallux valgus—Lateral angulation of the first toe.

Helical axis, see *Screw axis*

Helical axis surface—A surface produced by the migration of the instantaneous axis during motion.

Homogeneous transform—A matrix describing both translation and rotation, i.e., a Transformation matrix. See, also, Matrix.

Indirect method (of defining body attitude in space)—The method based on the law of the conservation of the angular momentum. First, an instantaneous angular velocity is calculated as a ratio of angular momentum to moment of inertia, and then, an angular distance is calculated as an integral of the angular velocity over time.

Induced twist—Visibly twisting a body segment without its rotation around the long axis; see also *Codman's paradox.*

Instantaneous axis/center of rotation (IAR, ICR)—A point which is momentarily at rest in both the global and local reference frames.

Instantaneous center of zero velocity, see *Instantaneous axis/center of rotation*

Invariant—Unchanged by a coordinate transformation and hence independent of the reference frame chosen, e.g., the rotation angle and the direction vector remain unchanged under different choices of coordinate axes.

Jacobian—A matrix of partial derivatives. In biomechanics of human motion, the matrix represents the transformation between the joint angular velocities and torques, on the one hand, and the end-effector velocity and generated force, on the other.

> *Orientation Jacobian*—The matrix whose elements are the partial derivatives of the end-effector orientation with regard to the joint angles.
>
> *Positional Jacobian*—The matrix whose elements are the partial derivatives of the end-effector position with regard to the joint angles.

Jerk—The time rate of change of acceleration.

Joint(s)

> *Ball-and-socket (spherical) joint*—A joint with three DOF.
>
> *Bicondylar joints*—The joints at which one bone articulates with another by two distinct surfaces, condyles.
>
> *Compound joint*—A joint that contains more than two articulating surfaces, e.g., the radiocarpal joint.
>
> *Ellipsoidal joint*—A synclastic joint with the curvature in one direction greatly exceeding the curvature in the perpendicular direction, e.g., the wrist (radiocarpal) joint.
>
> *Frozen joint*—A joint at which angular values do not change during a human movement.
>
> *Geometrically ideal joints,* see *Nominal joints*
>
> *Hinge (revolute) joint*—A joint with one DOF in which a cylinder-like surface fits into a concave surface, also any revolute joint with one DOF.
>
> *Mortise joints*—The joints that have a bar of bone on each side preventing movement in this direction.
>
> *Nominal joints*—The joints with the fixed orthogonal axes parallel to the main anatomic axes and intersecting at one center.
>
> *Pivot joints*—The joints with one DOF in which an arch-shaped surface rotates about a rounded pivot, such as between the radius and ulna.
>
> *Saddle joints*—The two-DOF joints with anticlastic articular surfaces.

Joint angle

Anatomic joint angle (external joint angle)—The angle between a segment's anatomic position and the position of interest.

Included joint angle (internal joint angle)—The angle located between the longitudinal axis of the two segments defining a joint ($\angle \leq 180°$).

Joint configuration—The set of joint angles representing the relative orientation of human body segments.

Degenerate joint configuration—A joint configuration in which freedom of motion is constrained.

Singular joint configuration—The degenerate joint configuration at which the determinant of the Jacobian matrix equals zero.

Joint congruency—Similarity of the profiles of two mating articular surfaces.

Joint motion

Coupled joint motion—Correlated, fixed motion of two or more joints or the correlated motion of a single joint with respect to the different DOF. For example, eversion of the foot includes pronation, abduction, and dorsiflexion.

Joint position

Close-packed position—The position of the full congruency. In this position, the male surface is in contact with the female surface in each point of the latter.

Loose-packed positions—Positions other than the close-packed position.

Kennedy's theorem—When three bodies, 1, 2, and 3, have plane motion, their instant centers, C_{12}, C_{13}, and C_{23}, lie on a straight line.

Kinematics—Description of motion disregarding mass and force involved in the motion.

Kinematic chain—A serial linkage of rigid bodies.

Branched (complex) kinematic chain—A chain containing at least one link entering more than two kinematic pairs.

Closed kinematic chain—The chain constrained on both its ends.

Open kinematic chain—The chain with one end free to move.

Redundant kinematic chain—An open kinematic chain with more than six DOF.

Serial kinematic chain—A chain in which each of the links enters no more than two kinematic pairs.

Kinematic pair—The chain consisting of two adjacent links connected by a joint.

Kinematic problems

Direct kinematic problem—The joint coordinates are known; the end-effector position is sought.

Inverse kinematic problem—The position of the end effector is known; the joint coordinates are sought.

Kinetics—The study of the relations between the forces and the resulting motion.

Line of centers—The straight line on which the instant centers of rotation of a three-link chain are located; see *Kennedy's theorem*.

Line of fixation, see *Line of sight*

Line of nodes, see *Nodes axis*

Line of sight—The line drawn from the point of fixation to the eye's center of rotation.

Link—A body in a kinematic chain.

Lisfranc's joint—The tarsometatarsal joints considered as a unit.

Listing's law—The eye assumes only those positions that may be reached by a single rotation from the primary position around the axis resting in the Listing's plane.

Listing's plane—The frontal vertical plane through the center of eye rotation.

Local proportional scaling—The method used for studying complex multiphasic movements in which consecutive phases are adjusted independently.

Locus of the point, see *Path of the point*

Lumbar-pelvic rhythm—The coupled movement of the spine and pelvis in the sagittal plane.

Maneuverability (of a kinematic chain)—The number of DOF of the chain minus six.

Manipulability ellipses—Ellipses used to describe mobility of a biokinematic chain in a given joint configuration.

Matrix—A rectangular array of numbers.

Angular velocity matrix—The skew-symmetrical matrix whose off-diagonal elements are components of the angular velocity vector and diagonal elements equal zero.

Canonical form of the rotation matrix—A rotation matrix in its simplest form, in the event that the rotation axis is parallel to the reference axis.

Displacement matrix—A matrix that transforms global coordinates given in position 1 into global coordinates in position 2.

Generalized inverse matrix—A Moore-Penrose pseudoinverse matrix with weighted coefficients.

Helical matrix—A 4×4 transformation matrix whose elements are a function of the helical parameters.

Inverse matrix—The matrix, $[A]^{-1}$, that satisfies the condition $[A]^{-1}[A] = [I]$, where $[A]$ is a square nonsingular matrix and $[I]$ is the unit matrix.

Moore-Penrose pseudoinverse matrix—The matrix that provides a minimal norm solution to the inverse kinematics problem.

Nonsingular matrix—A matrix with a nonzero determinant, $|A| \neq 0$.

Orientation matrix, see *Rotation matrix*

Orthogonal matrix—The orthogonal matrix satisfies the condition $[A]^T[A] = [A]^{-1}[A] = [I]$.

Orthonormal matrix—An orthogonal matrix in which all column vectors have unit length.

Position matrix, see *Transformation matrix*

Pseudoinverse matrix, see *Moore-Penrose pseudoinverse matrix*

Rotation matrix (orientation matrix, matrix of direction cosines)—A 2×2 or 3×3 matrix whose elements are direction cosines used to represent the attitude and/or rotation of one reference frame with respect to another. Note that in the literature the following terminology is often used: the orientation matrix defines an initial orientation of the local frame with regard to the global frame and the rotation matrix defines rotation from the initial to the final position.

Transformation matrix (position matrix)—A 4×4 matrix that incorporates both translation and rotation of a local reference system relative to the global one. Note that in the literature the following terminology is often used: the *position matrix* defines an initial position of the local frame with regard to the global frame and the *transformation matrix* defines transformation from the initial to the final position.

Transpose matrix—The transpose matrix $[A]^T$ of the matrix $[A]$ is the matrix with rows and columns interchanged.

Mechanism—The kinematic chain having one DOF.

Metatarsal break—The oblique axis through the five metatarsophalangeal joints.

Mobility (of a kinematic chain)—The total number of DOF in a kinematic chain.

Models (of the human body)

Anthropomorphic (skeletal) models—The models that visually resemble construction of the human body: the body segments are modeled as solid links and human joints as the joints of the model.

Functional models—In functional models, the body segments are modeled as nodes of a graph (of a tree) and the joints as arcs connecting the nodes.

Gross body models—The models in which small joints (e.g., interphalangeal) are not included.

Kinematic models—The models that represent motion only and neglect other aspects (for instance, the mass distribution).

Moment of inertia—A measure of the body's inertial resistance to changes in angular motion.

Motion

Angular motion, see *Rotation*

General motion—A combination of translation and rotation.

Linear motion, see *Translation*

Arm (leg) flexion/extension—A motion that decreases or increases the arm's (leg's) stretch.

Arm (leg) rotation—A change in orientation of the radius-vector drawn from the proximal joint of the kinematic chain to the distal terminal point of its end effector.

Multiangle (on a curved joint surface)—A closed n-sided figure, all the sides of which are chords.

Nodes axis (line of nodes)—An axis at the intersection of two planes, one from the global system and the second from the local.

Nominal (joint) axes—The imaginary joint axes that are (a) fixed, (b) orthogonal, (c) parallel to the main anatomical axes, and (d) intersect at one fixed center.

Ovoid of motion—Locus of all the attainable positions of a point fixed to the distal end of a bone; the ovoid of motion is a spheroidal convex surface.

Pair of rotation, see *Angular velocity couple*

Path of the point, or **Locus of the point**—The line representing successive positions of the point.

Pitch (of the screw)—The magnitude of translation per one revolution.

Reduced pitch of the screw—The magnitude of translation per one radian of the rotation.

Planes, anatomic

Cardinal plane—An anatomic plane passing through the center of mass of the human body.

Coronal plane, see *Frontal plane*

Frontal plane—A plane that divides the body into anterior and posterior sections.

Principal plane, see *Cardinal plane*

Midsagittal plane—The plane dividing the human body in two symmetrical halves, same as *cardinal sagittal plane,* or *principal sagittal plane.*

Sagittal plane—A plane dividing a human body into left and right sections.

Transverse plane—For a subject in an upright posture, the horizontal plane.

Plane of fixation—The plane containing both the object of regard, at which the axis of sight is directed, and the *Base line.*

Poisson's equation—The equation relating angular velocity with the rate of change of direction cosines and the body attitude.

Pole of rotation, see *Center of rotation*

Polode, see *Centrode*

Polygone (on a curved joint surface)—A closed figure, at least one side of which is not a chord.

Position of the human body—A human body's location, attitude, and posture.

Principal directions (of a surface at a point)—The directions of the principal curvatures.

Projection angles—The angles formed by the projections of a vector on the orthogonal planes of the global reference frame and the axes of this frame.

Proximal—Nearest the center; opposed to *Distal*.

Psychophysical law for time—A power function $t' = bt^{\alpha}$, where t' and t are the subjective and real time correspondingly, and b and α are parameters. See also *Gutman's hypothesis*.

Quaternions—Mathematical objects used for defining body orientation and consecutive rotations in space. Quaternions can be written as the sum of a scalar and a three-dimensional vector with the Euler's parameters as coefficients.

Radius of curvature—The reciprocal of the curvature at the point.

Range of motion—The difference between the two extremes of joint movement.

Redundancy, of a kinematic chain—The same end-point position can be assumed by various joint configurations.

Reference system (reference frame; space)—System of coordinates.

Clinical reference system—A joint configuration system in which a joint motion is described as "flexion-extension," "adduction-abduction," and "external-internal rotation" (or "pronation-supination").

Global reference system (world frame)—A system of coordinates with origin affixed to an unmovable point in the vicinity of the performer.

Joint rotation convention (JRC)—A joint reference system with the first axis fixed perpendicularly to the sagittal plane of the proximal segment and the third axis directed along the long axis of the distal segment. The second axis is the floating axis (the cross product of the first and third axes).

Local reference system—A system of coordinates attached to a moving body.

Principal somatic system—The somatic system with the reference axes directed along the principal axes of inertia.

Segment coordinate systems—A local reference system fixed within a body segment.

Somatic (moving) reference system—A local reference system located within the body and changing its orientation in space when body orientation is changed.

Split-body system—The reference system at which one (longitudinal) axis passes through the centers of mass of the upper and bottom parts of the body; the second axis is directed along the lateromedial axis of the pelvis; and the third axis is determined as the cross product of the first two.

Technical reference systems—The local systems fixed with technical devices, such as goniometers or skin markers, that are used to register human body movement.

Translatory reference system—A local reference system oriented similarly to the global reference system that moves translatory with the body while maintaining its orientation in space.

Relative acceleration method—The method of obtaining the acceleration difference.

Residual (in joint kinematics)—The difference between the sum of the internal angles of a triangle (S) and π, $r = S - \pi$. Residual is positive for any ovoid surface, negative for a saddle surface, and zero for a flat surface.

Retinal horizon—The horizontal plane through the eye center of rotation. This is the xy plane of the local system that moves with the eye.

Retinal vertical—The vertical plane through the center of rotation, which is the yz plane of the local system that moves with the eye.

Reuleaux method—The method of locating the ICR from experimental recordings.

Right-hand thumb rule—The convention used to determine the direction of vector **V,** which is a cross product of vectors **P** and **Q**: if the fingers of the right hand are curved in the direction from vector **P** to vector **Q**, the extended thumb represents the direction of vector **V**.

Rodriguez' parameters—A set of three independent variables defining body orientation in space.

Rolling (in a joint)—The rotation of a bone around an axis located on the articular surface.

Rotation—Movement of a body about an axis or center during which all particles of the body travel in the same direction through the same angle.

Bone rotation, see *Segment rotation*

Joint rotation—Movement of a body segment described with regard to a local reference system fixed within the adjacent body segment.

Segment rotation—A succession of body part orientations defined with regard to a global reference system.

Rotation surface—The surface containing the rotation vectors.

Scapular plane—The plane of the scapula in the resting position.

Scapulohumeral rhythm, see *Shoulder rhythm*

Screw axis (helical axis)—A line in space. At any given instant, the translation and rotation of a body occur along and around a screw axis.

Screw-home mechanism (in the knee joint)—Knee extension combined with the external rotation of the tibia.

Shoulder rhythm—The concerted movement of the humerus, scapula, and clavicle during arm adduction.

Singularity—A specific position for which a given mathematical operation is not defined or its result equals zero.

Singular joint angular position—The joint angular position for which the Euler's angles cannot be exactly defined. At this position, a global reference axis and a local reference axis are collinear.

Singular matrix—A matrix with the determinant that equals zero.

Singular position of a kinematic chain (singular joint configuration)—The chain configuration at which the inverse Jacobian does not exist (the Jacobian is singular). At the singular position, which is also a *Degenerate joint configuration,* the end effector of the chain cannot move freely in all directions.

Singular position of vectors—The collinear position of two vectors for which the cross product of the vectors cannot be determined (because $\sin \alpha = 0$ when α equals $0°$ or $180°$).

Skid bed (Skidding bed)—In human joints, when skidding occurs, the skid bed is a female surface.

Skidder—In human joints, when skidding occurs, the skidder is a male surface.

Skidding (in a joint)—A surface-on-surface movement in which the same region of an articular surface is in continuous contact with different regions of another surface.

Slide (in the algebra of articular surfaces)—Motion of an imaginable infinitesimal rod perpendicular to the surface. The rod can either spin, or slide, or both.

Sliding—One of two variants of skidding—the same point of the male surface is in constant contact with various areas of the female surface (compare with *Spinning*).

Slip ratio—The ratio $d_m:d_f$, where d_m and d_f are the distances between two pairs of contact points on a male and female surface, correspondingly.

The slip ratio is used to characterize the contribution of rolling and gliding to joint motion.

Snap—The fourth-time derivative of the displacement.

Space, see *Reference system*

> *Joint space*—Joint coordinate system. A reference frame, fixed to each link and used to describe the body posture in terms of joint configuration.
>
> *Segment space*, see *Task space*
>
> *Task space*—External coordinate system used to describe the body position (location, attitude, and posture) in terms of the position of the body segments with respect to the environment.

Speed—The time rate of covering distance. Speed is a scalar quantity (a number).

Spin—Rotation of joint surfaces around an axis perpendicular to the surfaces at the point of contact. Also, see *Euler's angles.*

Spinning—One of two variants of skidding: the same point of the female surface is in constant contact with various areas of the male surface (compare with *Sliding*).

Stretch (extremity stretch, arm stretch, leg stretch)—The distance between the proximal terminal point of the first link and the end effector of a kinematic chain.

> *Full stretch*—The distance between the proximal terminal point of the first link and the distal tip of the end effector of a kinematic chain.
>
> *Reduced stretch*—The distance between the proximal terminal point of the first link and the proximal tip of the end effector of a kinematic chain.

Structure equation (of a kinematic chain)—The product of corresponding transformation matrices defining the position of the end effector in terms of the relative positions of each link in the chain.

Swing—Any joint movement other than spin.

> *Pure swing*—A swing without an accompanying spin.

Synarthroses—Bony articulations in which the bones are joined by a deformable tissue.

Synergy (functional synergy)—Temporary coupling of motion in two or several joints during a human movement. Coupled joints are controlled as a unit.

Tangent method—The method of locating the instantaneous center of rotation from experimental recordings.

Tau hypothesis—The expected time to the approach (called "tau," or "tau margin") is estimated by the performer and the rate of the time change ("tau dot") is controlled.

Tau prime—The time it would take the subject to stop, if he or she were to continue at a constant deceleration.

Tensor—A mathematical object, such as a vector or matrix, whose components obey a certain law of transformation under changes of the coordinate system.

Three-axes system, see *Cardan's angle convention*

Trace (of a square matrix)—The sum of the elements of the main diagonal.

Transformation, see *Matrix (transformation matrix)*

> *Joint transformation*—The transformation, or rotation, matrix that transforms coordinates given in one segment reference frame into coordinates in the second frame affixed to the adjacent body segment. The transformation depends on the joint configuration.
>
> *Link transformation*—A matrix that transforms coordinates given in one segment reference frame into coordinates in the second frame affixed to the same body segment. The transformation does not depend on the joint configuration.

Translation—Movement of a body so that any line fixed with the body remains parallel to its original position.

Triangle (on a curved joint surface)—The three-sided figure all the sides of which are chords.

Trigone (on a curved joint surface)—The three-sided figure at least one side of which is not a chord.

Two-axes system, see *Euler's angle convention*

Vector—A quantity having magnitude and direction; a unidimensional array of numbers $(a_1, a_2, \ldots, a_n)$.

> *Contravariant vectors*—Component coordinates. Contravariant vectors can be summed vectorially; they obey the parallelogram rule. In the textbook, only contravariant vectors are used.
>
> *Covariant vectors*—Projection coordinates in an oblique coordinate system. Covariant vectors cannot be summed vectorially; they do not obey the parallelogram rule.
>
> *Direction vector*—The three-dimensional unit vector along the rotation vector.
>
> *Invariant vectors*—Vectors not changing their direction by a coordinate transformation.
>
> *Rotation vector*—The vector along the axis of rotation whose length is proportional to the magnitude of rotation.

Velocity—The time rate of change of position. Velocity is a vector quantity.

Angular velocity—The rate of movement in rotation. Angular velocity can be represented by a vector having a direction along the axis of rotation and a sense according to the right-hand thumb rule.

Joint velocity—Angular velocity at a joint.

Segment velocity—Velocity of a body segment as viewed by an external observer, in an absolute reference frame. Segment velocity can be both linear and angular.

Skidding velocity—The relative velocity of two points, P_m and P_f, defined at the point of contact. Point P_m is fixed in the male surface and P_f is fixed in the female surface.

Virtual reference position—A theoretical reference position of the bones of the shoulder girdle.

Working envelope—The set of boundary points that can be reached by the end effector.

Working range (volume)—The set of all points that can be reached by the end effector.

Zigzag arrangement—The three-link leg structure that necessitates rotation of the adjacent segments during support and takeoff in opposite directions.

INDEX

Page numbers in italics distinguish definitions.

ABOUT THE AUTHOR

Vladimir M. Zatsiorsky, PhD, is a world-renowned expert in the biomechanics of human motion. He has been a professor in the Department of Kinesiology at The Pennsylvania State University since 1991. He also is the director of the university's biomechanics laboratory.

Prior to coming to North America in 1990, Dr. Zatsiorsky served for 18 years as professor and department chair of the Department of Biomechanics at the Central Institute of Physical Culture in Moscow. For 26 years he served as consultant to the national Olympic teams of the USSR. He also was director of the USSR's All-Union Research Institute of Physical Culture for three years.

In addition to his academic pursuits in the classroom, laboratory, and field, Dr. Zatsiorsky is a prolific writer who has authored or coauthored more than 240 scientific papers and several books on various aspects of biomechanics. In recognition of his achievements, he has received several awards, including the Geoffrey Dyson Award from the International Society of Biomechanics in Sport (the society's highest honor) and the USSR's National Gold Medal for the Best Scientific Research in Sport in 1976 and 1982.

Dr. Zatsiorsky is a member of the American Society of Biomechanics and the International Society of Biomechanics.

He and his wife Rita live in State College, Pennsylvania. They have two children and two grandchildren.